MARKET BEHAVIOUR AND MACROECONOMIC MODELLING

Market Behaviour and Macroeconomic Modelling

Edited by

Steven Brakman
Associate Professor
University of Groningen
The Netherlands

Hans van Ees
Professor of Economic Theory and Economic Education
University of Groningen
The Netherlands

and

Simon K. Kuipers
Professor of Macroeconomic and Monetary Theory
University of Groningen
The Netherlands

First published in Great Britain 1998 by
MACMILLAN PRESS LTD
Houndmills, Basingstoke, Hampshire RG21 6XS and London
Companies and representatives throughout the world

A catalogue record for this book is available from the British Library.

ISBN 0–333–71836–4

First published in the United States of America 1998 by
ST. MARTIN'S PRESS, INC.,
Scholarly and Reference Division,
175 Fifth Avenue, New York, N.Y. 10010

ISBN 0–312–21493–6

Library of Congress Cataloging-in-Publication Data
Market behaviour and macroeconomic modelling / edited by Steven
Brakman, Hans van Ees, Simon K. Kuipers.
p. cm.
Includes bibliographical references and index.
ISBN 0–312–21493–6 (cloth)
1. Macroeconomics—Econometric models.   2. Free enterprise.
3. Labor market.   I. Brakman, Steven.   II. Van Ees, Hans, 1955–
III. Kuipers, S. K. (Simon Klaas)
HB172.5.M365   1998
339'.01'5118—dc21   98–13604
                                                                CIP

This book is printed on paper suitable for recycling and made from fully managed and
sustained forest sources.

10   9   8   7   6   5   4   3   2   1
07   06   05   04   03   02   01   00   99   98

Printed and bound in Great Britain by
Antony Rowe Ltd, Chippenham, Wiltshire

Contents

List of Tables

List of Figures

Preface

This book contains the papers of the workshop Market Behaviour and Macroeconomic Modelling that was held at the University of Groningen, the Netherlands on June 6–7, 1996. The workshop was jointly organized by the Centre for Economic Research (a research group of economists affiliated with the University of Groningen and the University of Twente), the Netherlands Bureau for Economic Policy Analysis (CPB) and the Ministry of Economic Affairs. The aim was to bring together economic scholars and economists engaged in policy formulation to discuss the implications of market imperfections for macroeconomic modelling from the perspective of both economic theory and policy. The papers in this volume deal with imperfections in commodity markets, capital markets and labour markets, as well as with their implications for economic policy design. In line with recent economic theory, all papers reveal the need for a sound theoretical framework. At the same time the various studies illustrate that a uniform microfoundation for market imperfections is still lacking in macro models, which consequently leads to a rather diverse range of models, sometimes with different implications for economic policy. This does not necessarily imply the existence of a weakness within this line of research; in a world as diverse as ours it may be a blessing in disguise that one can choose from many models with which to analyse the problems at hand. What to include in empirical models is highly dependent on the specific (policy) problem under consideration. Both the art of modelling and the skill of the researcher resolves this paradox, allowing the construction of sound theoretical models that produce interesting outcomes. There is no need to say that this is a difficult task. The papers in this volume illustrate both the progress that has been made in this area, as well the work that still has to be done. We would like to express our gratitude to Peter van Bergeijk, at that time at the Ministry of Economic Affairs and Ruud Okker of the Netherlands Bureau for Economic Policy Analysis for their help in the organization of the workshop. Similarly, Lies Baars and Nanne Brunia provided great organizational assistance and last but not least, we acknowledge the superb TEX-editing skills of Thijs Knaap. Finally, we thank the Royal Netherlands Academy of Sciences, the University of Groningen, and the Netherlands Central Bank for making this second CCSO workshop possible.

Groningen, August 1997   S. Brakman
                                                          H. van Ees
                                                          S.K. Kuipers

List of Contributors

M. Aloi	University of York
	York
A. van den Berg	Free University
	Amsterdam
P.A.G. van Bergeijk	Netherlands Central Bank
	Amsterdam
S. Brakman	University of Groningen
	Groningen
D.P. Broer	CPB - Netherlands Bureau for Economic
	Policy Analysis, The Hague
W.H. Buiter	Cambridge University
	Cambridge
D.W. Carlton	University of Chicago
	Chicago
K.B. Church	University of Warwick
	Coventry
H.D. Dixon	University of York
	York
D.A.G. Draper	CPB - Netherlands Bureau for Economic
	Policy Analysis, The Hague
H. van Ees	University of Groningen
	Groningen
T. Ellingsen	University of Oslo
	Oslo
H. Garretsen	University of Nijmegen
	Nijmegen
S.G. Gilchrist	Boston University
	Boston
L. de Haan	Netherlands Central Bank
	Amsterdam
J. de Haan	University of Groningen
	Groningen
R.C.G. Haffner	Ministry of Economic Affairs
	The Hague
S. Holden	University of Oslo
	Oslo

List of Contributors

A. Houweling	CPB - Netherlands Bureau for Economic Policy Analysis, The Hague
F.H. Huizinga	CPB - Netherlands Bureau for Economic Policy Analysis, The Hague
P.A. de Jongh	CPB - Netherlands Bureau for Economic Policy Analysis, The Hague
K. Kletzer	University of California Santa Cruz
S.K. Kuipers	University of Groningen Groningen
G.H. Kuper	University of Groningen Groningen
P. Lawler	University of York York
P.R. Mitchell	University of Warwick Coventry
A. Nieuwenhuis	CPB - Netherlands Bureau for Economic Policy Analysis, The Hague
G. Ridder	Free University Amsterdam
T.J.A. Roelandt	Ministry of Economic Affairs The Hague
J. van Sinderen	Ministry of Economic Affairs The Hague
E. Sterken	University of Groningen Groningen
J.-E. Sturm	University of Groningen Groningen
H. Uhlig	CentER, Tilburg University Tilburg
N. de Visser	Free University Amsterdam
K.F. Wallis	University of Warwick Coventry
E. Zakrajšek	Federal Reserve Board Washington

PART I
Imperfect
Competition
(A)

1 The Multiplier in an Economy with Monopolistic Output Markets and Competitive Labour Markets

Marta Aloi, Huw Dixon and Phillip Lawler

1.1 Introduction

The purpose of this paper is to derive some new results and to link together these results with an existing literature. The focus of the paper is a macro-economy characterized by monopolistic competition in output markets, but with a perfectly competitive labour market. We explore the implications of imperfect competition for the conduct and effectiveness of fiscal policy for output and employment in both the short run (with the number of firms fixed) and the long run (with free entry and exit of firms).[1]

This paper develops the analysis of Dixon and Lawler (1996): the extension is primarily in the introduction of a monetary sector into the model (Dixon and Lawler, 1996, has only output and leisure). The origins of this analysis lie in three papers: Startz (1989), Dixon (1987) and Mankiw (1988). In the papers by Dixon and Mankiw, the relationship between the fiscal multiplier and the degree of competition, was analyzed in the short run with a fixed number of firms. Startz (1989) examined the effect of free-entry on the model. In fact, all three papers share certain key assumptions which are crucial for the results which they obtain, as shown by Dixon and Lawler (1996). In particular they assume (a) constant marginal product of labour (CMPL) production technology and (b) household preferences which give rise to constant marginal budget shares.[2] We denote these two assumptions as defining the SMD (Startz - Mankiw - Dixon) approach. Two key conclusions emerged from the SMD analysis:

1. the short-run fiscal multiplier is larger the greater the degree of monopoly - Dixon (1987, p. 144, Proposition 3), Mankiw (1988, pp. 10-11, Equations (15), (16) and (17)), Startz (1989, p. 744, Equation (11)).

2. in the presence of monopoly power the short-run fiscal multiplier exceeds the corresponding long-run multiplier (Startz, pp. 749-50).

Dixon and Lawler (1996) examined the generality of these results using a model which encompassed the SMD framework as a special or limiting case. One of the main innovative features of that model lay in its treatment of firms' production technology. As in Startz a fixed production cost was assumed, but this was combined with the assumption of diminishing marginal productivity of labour. Together these assumptions imply the familiar U-shaped average cost curve and upward sloping marginal cost curve. Given this specification of technology, the long-run free entry condition serves to determine, independently of the rest of the model, the long-run equilibrium values of the real wage, and employment and output per firm. An appealing feature of these technological assumptions is that they are perfectly consistent with both monopolistic and perfect competition in contrast to Startz's framework, which is incompatible with a Walrasian equilibrium.

In this paper we make the same assumptions about technology and preferences as in Dixon and Lawler, but we employ our framework to examine the effects of fiscal policy in both the short and long runs. In so doing we show that neither of the conclusions (1) and (2) referred to above are general propositions about the impact of fiscal policy in an imperfectly competitive economy; rather both results reflect the particular assumptions with regards to household preferences and production technology which characterize the SMD framework. In particular we demonstrate, first, that there is no unambiguous relationship between the size of the fiscal multiplier (either short or long run) and the degree of monopoly power. Secondly, we show that Startz's ranking of the short and long-run output multipliers is reversed for 'sufficiently competitive' monopolistic economies. Thirdly, we find an unambiguous ranking of employment multipliers: in particular the long-run employment multiplier is always, that is regardless of the degree of monopoly power, larger than the corresponding short-run multiplier. Finally, we indicate the relationship between our own conclusions and the SMD results by demonstrating how each of the latter derives from the particular assumption of CMPL technology, constant marginal expenditure shares , or both. This confirms most of the results of Dixon and Lawler (1996) within the context of a monetary economy.

In Section 1.5 of the paper, we present a geometric and visual representation and derivation of the results of the paper. Whilst the main body of the paper uses a very general representation of preferences, in the graphical analysis we adopt the special case of homothetic preferences. We show how the relationship between the multiplier and the degree of imperfect competition depends crucially on division of full income between consumption and leisure by the household. In Section 1.6, we relate the static results to the continuous time intertemporal dynamic framework. This is based very much on Dixon (1997), and relates to the discrete-time framework of Rotemberg and Woodford (1995) and Devereux et al. (1996). We develop a formal and geometric analysis of the steady-state long-run multipliers, and show that the analysis is very similar to the static case, except for new factors introduced by capital accumulation and intertemporal preferences. However, we note that much of the literature in the intertemporal macroeconomic setup often makes strong assumptions about the functional form in order to solve or simulate the model (these often include Cobb-Douglas intra-temporal preferences and technology). The static analysis of Dixon and Lawler (1996) and this paper would indicate that the results of these papers might be very limited by the specific nature of these assumptions.

1.2 The model

In this section we outline the central assumptions which characterize our model of a monopolistically competitive economy: the Walrasian case of perfect competition can be viewed as a special limiting case within the framework, and will be treated as such. Briefly, three sets of agents, households, firms and the government interact in the markets for labour, goods and money. Whilst the labour market is taken to be perfectly competitive, the goods market is assumed to be populated by Dixit-Stiglitz monopolistic competitors, whose output is purchased by both households and government. Households, who act as price takers, consume goods, sell labour and hold money, the only asset in the model. Money is issued by the government and provides an alternative to lump-sum taxation as a means of financing government expenditure. We now turn to consider the individual components of the model in some detail.

1.2.1 Households

Our model of product-differentiation is essentially that of Judd (1985) which develops the basic Dixit-Stiglitz model (see Grossman and Helpman, 1991 for a useful exposition). This model is then embedded within a more-or-less standard treatment of the 'macroeconomic consumer.' There is a continuum of firms, indexed by j, uniformly distributed over the interval $[0, n]$, where n is the measure ('number') of firms. Each firm j sets price $p(j)$ and produces output $y(j)$. All households are taken to be identical and hence their behaviour can be encapsulated in the form of a single representative household, whose preferences over consumption goods are assumed to be separable from the other arguments of the utility function. We define the sub-utility function C :

$$C = n^{\frac{\mu}{\mu-1}} \left[\int_0^n c\,(j)^{1-\mu}\,\mathrm{d}j \right]^{\frac{1}{1-\mu}} \tag{1.1}$$

where $c\,(j)$ represents household consumption of firm j's output and μ, assumed to lie in the half-open interval $[0,1)$, is a preference parameter. For $\mu = 0$, firms outputs are viewed as perfect substitutes (hence C provides a direct measure of total consumption) and as μ increases the degree of substitutability between different goods declines.

Household utility is a function of the consumption index C, defined by (1.1), leisure (that is the household time-endowment, E, less labour supply L^s) and real money balances, that is nominal end-of-period money holdings, M, deflated by an appropriate price index P, to be defined below.

A1 : Household Preferences:

$$U \ : \ \mathbb{R}^3_+ \to \mathbb{R}_+ \text{ where } U = U\left(C, E - L^s, \frac{M}{P}\right)$$

with U (at least) twice continuously differentiable, strictly quasi-concave and increasing in each of its arguments.

In subsequent sections a special case of A1 is used on occasion as a reference point, namely that U is Cobb-Douglas

$$U = C^\alpha\,(E - L^s)^\beta \left(\frac{M}{P}\right)^\gamma, \quad \alpha,\beta,\gamma > 0, \alpha + \beta + \gamma = 1$$

As already noted, Cobb-Douglas preferences are assumed in Dixon and Mankiw, with Startz's Stone-Geary specification equivalent for all relevant purposes.[3] Other New Keynesian papers also use special cases of A1. Blanchard and Kiyotaki, for example, assume preferences to be additively separable, with utility Cobb-Douglas with respect to consumption and real money holdings but linear in leisure.[4]

The households budget constraint is

$$\int_0^n p\,(j) \cdot c\,(j)\,\mathrm{d}j + M \le W \cdot L^s + \Pi + M^0 - T \tag{1.2}$$

where W is the nominal wage rate, Π the nominal value of distributed profits, M^0 initial holdings of nominal money balances and T the nominal value of lump-sum taxation. The household maximizes utility as described by A1, subject to (1.1) and (1.2).

Since preferences are separable the households decision process can be represented as a two-stage budgeting problem.[5] In the first stage the household chooses, given its endowments, the optimal values for consumption expenditure, leisure and money balances. Then, in the second stage, it allocates its consumption expenditure between the outputs of different firms.

Considering the second stage, suppose that total nominal expenditure on consumption has been chosen to be C^N. The consumer then solves

$$\max_{c(j)} n^{\frac{\mu}{\mu-1}} \left[\int_0^n c(j)^{1-\mu}\, dj \right]^{\frac{1}{1-\mu}} \tag{1.3}$$

subject to

$$\int_0^n p(j) \cdot c(j)\, dj \le C^N \tag{1.4}$$

The solution to (1.3-1.4) is the demand for the output of each firm,

$$c(i) = \frac{p(i)^{\frac{-1}{\mu}}}{\int_0^n p(j)^{\frac{\mu-1}{\mu}}\, dj} C^N \tag{1.5}$$

The appropriate price index for subutility (1.1) can then be defined as

$$P = \left[\frac{1}{n} \int_0^n p(j)^{\frac{\mu-1}{\mu}}\, dj \right]^{\frac{\mu}{\mu-1}} \tag{1.6}$$

Note that, since (1.5) satisfies the budget constraint (1.4), $PC = C^N$. Hence using the quantity and price indices, (1.1) and (1.6) respectively, we are able to aggregate the outputs of the individual firms and treat them as a single (composite) commodity.

Returning to the first stage of the optimization process, the decision problem facing the household is

$$\max_{C,L^s,M} U\left(C, E - L^s, \frac{M}{P}\right)$$

subject to

$$P \cdot C + M \le W \cdot L^s + \Pi + M^0 - T$$

the solution to which yields the standard Marshallian consumption and labour supply functions

$$C = C\left(w, \frac{M^0 - T + \Pi}{P}\right); \quad C_1 > 0, C_2 > 0 \quad (1.7)$$

$$L^s = L^s\left(w, \frac{M^0 - T + \Pi}{P}\right); \quad L_1^s > 0, L_2^s > 0 \quad (1.8)$$

where $w = W/P$, the real wage, $C_1 = \partial C/\partial w$ and *etc.*

The indicated signs for the partial derivatives reflect the assumptions that each of the arguments of the utility function is a normal 'good' and that substitution effects outweigh income effects. The latter assumption ensures, of course, that labour supply is strictly increasing in the real wage, whilst the former implies a negative income or wealth effect on labour supply. In what follows we assume throughout that the functions C and L^s are twice - continuously differentiable.

1.2.2 The government

The government is assumed to formulate its expenditure plans in real terms. For any given price level this gives rise to a particular value of nominal expenditure, G, which is then allocated across firms according to government preferences, taken to be identical to those of the household sector. We assume the government to leave its potential monopsony power unexploited and hence its optimization process is entirely analogous to that of the household. In particular, the government chooses to purchase quantity $g(j)$ from firm j to solve:

$$\max_{g(j)} n^{\frac{\mu}{\mu-1}} \left[\int_0^n g(j)^{1-\mu} \, \mathrm{d}j\right]^{\frac{1}{1-\mu}} \quad (1.9)$$

subject to

$$\int_0^n p(j) \cdot g(j) \, \mathrm{d}j = G \quad (1.10)$$

As for household consumption the price index defined by (1.6) allows us to deflate nominal expenditure, to arrive at our representation of *real* government spending, that is $g = G/P$.

The finance of government expenditure may, in principle, be by means of lump-sum taxation, by money creation or by both used in combination, with the level of government spending and the means of finance linked by the government budget constraint:

$$M - M^0 = G - T$$

In our analysis of fiscal policy in Section 1.4 we shall, in fact, focus exclusively on the effects of a balanced budget rise in government spending. This apparent limitation to the scope of our analysis is not so restrictive as it might seem, however; as will be explained, the real effects of a change in government expenditure are invariant to the means of finance of the policy.

1.2.3 The firm

Each firm $j \in [0, n]$ employs $l(j)$ units of labour to produce $y(j)$ units of its own variety of output. All firms share a common production technology described by

> A2 : Technology: for all $j \in [0, n]$, we have
> f : $\mathbb{R}_+ \to \mathbb{R}_+$ such that $y(j) = f(l(j) - \lambda)$

with f (at least) twice continuously differentiable, derivatives such that $f'' < 0 < f'$, $\lim_{l(j) \to \lambda} f' = \infty$ and $\lim_{l(j) \to \infty} f' = 0$. Thus, as in Startz, there is a fixed set up level of employment, λ, at or below which the level of output is zero. As employment increases beyond λ output expands but at a diminishing rate with successive increments of labour input; this assumption of a diminishing marginal physical product of labour represents one of the central differences between our own model and the SMD framework. The essential feature of the latter is the assumption of a constant marginal product of labour (CMPL) with $f'' = 0$ for $l \geq \lambda$. The SMD specification of production technology can thus be viewed as a special case of A2.[6]

Nominal profits of the firm are given by

$$\pi(j) = p(j) \cdot f(l(j) - \lambda) - W \cdot l(j) \tag{1.11}$$

The firm chooses $(p(j), l(j))$ to maximize (1.11) subject to its demand curve:

$$f(l(j) - \lambda) = \frac{1}{n} \left(\frac{p(j)}{P} \right)^{\frac{-1}{\mu}} (C + g) \tag{1.12}$$

In the maximization process the firm takes the *general* price level, P, as given and independent of its own actions; this is the essence of monopolistic competition in the Dixit-Stiglitz model. It is particularly attractive and plausible in the macroeconomic context, where P is the price index not of an industry but of the whole economy. Under such circumstances nominal profit maximization seems the appropriate assumption to make.[7] Clearly, treating as given exogenously means that (1.12) is a constant elasticity demand function, with elasticity $\varepsilon = 1/\mu$. The profit maximizing prices yields the mark-up of price, $p(j)$, over marginal cost, W/f':

$$\frac{p(j) - \frac{W}{f'(l(j)-\lambda)}}{p(j)} = \frac{1}{\varepsilon} = \mu \qquad (1.13)$$

Hence the parameter is equivalent to Lerner's index of monopoly (the price-cost margin). An alternative way of expressing (1.13) is in terms of the relationship between the marginal physical product of labour and the real wage. Defining the firms own-product real wage as $w(j) = W/p(j)$, we have:

$$w(j) = (1 - \mu) \cdot f'(l(j) - \lambda) \qquad (1.14)$$

That is the firm chooses such that the real wage equals the marginal physical product of labour scaled down by $1 - \mu$; for the limiting case of perfect competition ($\mu \to 0$), the real wage and the marginal product are equated.

Since demand is symmetric across firms, each firm chooses the same price and employment level; hence $p(j) = P$ (implying $w(j) = w$) and $l(j) = l$. With employment per firm identical across firms aggregate employment, L, defined by:

$$L = \int_0^n l(j)\,\mathrm{d}j$$

is given by $L = n \cdot l$. Similarly, aggregate output, Y, is related to output per firm, $y(j) = y$, in an obvious fashion, that is $Y = n \cdot y$. Real profits per firm are simply

$$\frac{\pi}{P} = y - w \cdot l$$

and with all firms earning identical profits, aggregate real profits are given by

$$\frac{\Pi}{P} = n\frac{\pi}{P} = n(y - w \cdot l) \qquad (1.15)$$

1.2.4 Free entry and the firm in long-run equilibrium

With free entry (and exit) long-run equilibrium is characterized by the zero profit condition. This condition serves to tie down the long-run equilibrium values of the real wage and output and employment per firm. To see this note

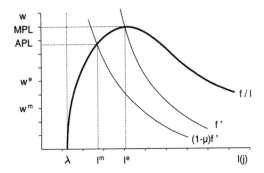

Figure 1.1: The firm in long-run equilibrium with free entry

that, since labour is the only input, profits are zero if and only if the real wage and the average product of labour are equal. Thus long-run equilibrium is characterized by

$$w^m = (1 - \mu) \cdot f'(l^m - \lambda) = \frac{f(l^m - \lambda)}{l^m} \qquad (1.16)$$

which clearly determines fully the long-run monopolistic equilibrium values of employment, l^m (hence output $y^m = f(l^m - \lambda)$) per firm and the real wage, w^m.

The zero profit condition is represented diagrammatically in Figure 1.1 below. The 'efficient' scale of production for the firm occurs at the intersection of the inverted U-shaped average product of labour schedule (f/l) and the upper downward sloping curve, representing the marginal product of labour (f'). This intersection point corresponds to the long-run equilibrium position of the firm in the Walrasian case $(\mu = 0)$, with employment per firm l^e, associated level of output y^e, and real wage w^e. For $\mu > 0$, however, the intersection point between the average product of labour schedule and the lower downward sloping curve, representing the marginal product of labour scaled down by $(1 - \mu)$ is the relevant one. Clearly the monopolistic long-run equilibrium occurs at a lower real wage and employment level, that is, $w^m < w^e$, $l^m < l^e$ for $\mu > 0$. In fact it is straightforward to see from the diagram that w^m, l^m and y^m are all decreasing in μ; as μ increases, the lower curve, $(1 - \mu) \cdot f'$, is displaced vertically downwards producing an intersection point with the average product of labour schedule associated with smaller values of l and w. The difference between l^m and l^e is, of course, the standard Chamberlinian excess capacity result; with monopolistic competition, the free entry long-run equilibrium is characterized by firms producing at below the optimal scale.

1.3 Equilibrium in a monopolistic economy

We now integrate the various components of the model outlined in the previous section in order to determine the characteristics of the short and long-run equilibria. The long-run equilibrium plays a central role in our analysis, due to the fact that it ties down the equilibrium real wage and hence output and employment per firm. Moreover, our policy analysis of Section 1.4 is conducted under the assumption that the economy begins from an initial position of long-run equilibrium. Accordingly we begin by exploring the properties of this equilibrium in some detail.

1.3.1 Long-run equilibrium

The economy comprises three markets; the markets for goods, for labour and for money. Walras' Law allows us to omit explicit consideration of the money market and focus our attention on the goods and labour markets. Long-run equilibrium is then characterized not only by the goods and labour market clearing conditions (Equations (1.17) and (1.18) below), but also by the free entry/zero profit condition (Equation (1.19))

$$C\left(w, \frac{M^0 - T + \Pi}{P}\right) + g = n \cdot f(1-\lambda) \qquad (1.17)$$

$$L^s\left(w, \frac{M^0 - T + \Pi}{P}\right) = n \cdot l \qquad (1.18)$$

$$w = (1-\mu) \cdot f'(1-\lambda) \qquad (1.19)$$

$$= \frac{f(1-\lambda)}{l}$$

The above system of equations implicitly defines n, l, w and P as functions of the exogenous parameters relating to household endowments, household preferences, production technology and government policy, that is, M^0, μ, λ, g and T.

The recursive structure of the system is readily apparent; Equation (1.19) fully determines the levels of employment, output and profits per firm, $l = l^m$, $y = y^m$, $\Pi(l^m) = 0$ as well as the real wage, $w = w^m$. A solution to (1.19) exists under A2 for $0 \leq \mu < 1$, λ bounded, allowing us to re-express

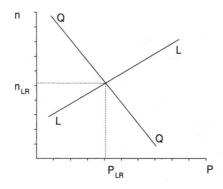

Figure 1.2: Long-run equilibrium

(1.17) and (1.18) as

$$C\left(w^m, \frac{M^0 - T}{P}\right) + g = n \cdot y^m \tag{1.20}$$

$$L^s\left(w^m, \frac{M^0 - T}{P}\right) = n \cdot l^m \tag{1.21}$$

This pair of equations now determines the long-run equilibrium values of the two endogenous variables, n and P. A diagrammatic representation of long-run equilibrium is provided in Figure 1.2. QQ represents goods market equilibrium and is downward sloping since, as the price level rises, real private sector wealth falls, thereby reducing consumption expenditure; consequently output must fall which, since output per firm is given in the long-run, requires a reduction in the number of firms, n.[8] The gradient of the goods market clearing locus is given by

$$\left.\frac{\mathrm{d}n}{\mathrm{d}P}\right|_{QQ} = -C_2 \frac{M^0}{P^2} / y^m < 0$$

The labour market clearing locus is positively sloped, since the fall in real money balances which results from a higher price level increases desired labour supply; thus, given the fixed long-run level of employment per firm, labour market equilibrium requires a larger number of firms. The gradient of LL is

$$\left.\frac{\mathrm{d}n}{\mathrm{d}P}\right|_{LL} = -L_2^s \frac{M^0}{P^2} / l^m > 0$$

It is readily seen that money is neutral in this economy. Assuming T, like G, is fully indexed, it is clear that both (1.20) and (1.21) are homogeneous of degree zero in (M^0, P). Hence an increase in the money supply will be reflected in a proportionate increase in all nominal magnitudes, leaving the values of all real variables undisturbed.[9]

In terms of Figure 1.2, a rise in M^0 leads to identical rightward horizontal shifts in the QQ and LL schedules, their intersection with respect to the vertical axis remaining unchanged.

An interesting and important issue is the nature of the dependence of the long-run equilibrium on the degree of monopoly, μ. Differentiation of (1.19) with respect to μ yields directly the influence of this parameter on the long-run values of the real wage, and output and employment per firm

$$\frac{dw^m}{d\mu} = \frac{\mu\left(f'\right)^2}{l^m \cdot \psi} < 0 \qquad (1.22)$$

$$\frac{dy^m}{d\mu} = \frac{\left(f'\right)^2}{\psi} < 0 \qquad (1.23)$$

$$\frac{dl^m}{d\mu} = \frac{f'}{\psi} < 0 \qquad (1.24)$$

where $\psi = (1 - \mu) \cdot f'' - \mu \cdot f'/l^m < 0$.

Given our discussion of Figure 1.1 in the preceding section, the above results carry no surprises. A higher value of μ implies an increased wedge between the marginal product of labour and the real wage. Given the larger mark-up, the zero profit condition then implies a lower real wage and reduced levels of employment and output per firm.

Whilst it is clear from Equations (1.22)-(1.24) above that the influence of μ on w^m, y^m and l^m depends purely on technological factors, its impact on the number of firms, on *aggregate* output, Y, and employment, L, and on the price level, is determined by the system's general equilibrium characteristics. Differentiating Equations (1.20) and (1.21), using where appropriate (1.22)-(1.24), we find, after some straightforward manipulation[10]

$$\frac{dn}{d\mu} = \frac{\left[\mu\left(L_1^s C_2 - C_1 L_2^s\right) f' + L\left(L_2^s f' - C_2\right)\right] f'}{\left(l^m\right)^2 \psi \left(C_2 - w^m L_2^s\right)} \gtrless 0 \quad (1.25)$$

$$\frac{dY}{d\mu} = \frac{\mu\left[\left(L_1^s C_2 - C_1 L_2^s\right) w^m + LC_2\right]\left(f'\right)^2}{l^m \psi \left(C_2 - w^m L_2^s\right)} \leq 0 \qquad (1.26)$$

$$\frac{dL}{d\mu} = \frac{\mu\left[L_1^s C_2 - C_1 L_2^s + LL_2^s\right]\left(f'\right)^2}{\psi \left(C_2 - w^m L_2^s\right)} \gtrless 0 \qquad (1.27)$$

$$\frac{\mathrm{d}P}{\mathrm{d}\mu} = \frac{\mu\left[C_1 - w^m L_1^s - L\right](f')^2 P^2}{l^m \psi \left(C_2 - w^m L_2^s\right) M^0} \gtrless 0 \tag{1.28}$$

Although, as indicated by Equation (1.25) the effect of μ on the total number of firms n is indeterminate, from (1.26) aggregate output is strictly decreasing in μ. This result is in accordance with that of Dixon (1987) and Mankiw (1988), though these papers are concerned with the properties of short-run equilibrium, that is, with the number of firms fixed. However, whilst the SMD assumption of a constant marginal product of labour implies directly that the inverse relationship between μ and aggregate output is reflected in a similarly negative relationship between μ and aggregate employment, in the present context the effect of μ on total employment can be seen from (1.27) to be ambiguous. Although, as noted above, employment per firm is negatively related to μ, the potential for a positive relationship between n and μ gives rise to the possibility that *aggregate* employment is increasing in μ. Of course, given the assumed production technology, this result is perfectly consistent with the finding of a negative relationship between total output and μ. Finally, it is apparent from (1.28) that the nature of the influence of μ on the price level is indeterminate. However two points are worth noting with regards to this result. First, the condition $C_1 - L < 0$ which is sufficient for $\mathrm{d}P/\mathrm{d}\mu > 0$ can be seen from (1.27) to be necessary for $\mathrm{d}L/\mathrm{d}\mu > 0$; given the real wage is negatively related to μ, a necessary condition for labour supply, and hence equilibrium employment to be increasing in μ, is a positive relationship between P and μ. Secondly, for the special case of Cobb-Douglas preferences, $C_1 - w^m L_1^s - L = (1 - \alpha - \beta) E < 0$, implying $\mathrm{d}P/\mathrm{d}\mu$ is unambiguously positive.

Before moving on to a discussion of short-run equilibrium within the model, we point to an interesting property of the relationship between the characteristics of long-run equilibrium and μ. That is, a small displacement of μ from its Walrasian value of zero has no effect on aggregate output, employment or the price level (1.26-1.28). The explanation for this result lies in the fact that in the Walrasian equilibrium each firm operates at the maximum point on its average product of labour curve. Consequently a small increase in μ from an initial value of zero and the implied fall in employment and output per firm have no first-order effect on average productivity and the long-run equilibrium real wage. Given the unchanged real wage, the requirement of simultaneous goods and labour market equilibrium then dictates that the decline in employment and output within each individual firm be precisely offset at the aggregate level by a compensating increase in the number of firms,[11] leaving the features of the long-run equilibrium otherwise undisturbed.

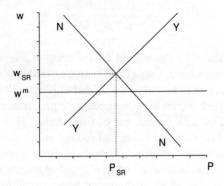

Figure 1.3: Short-run equilibrium

1.3.2 Short-run equilibrium

In the short run the number of firms is fixed. Equilibrium in the markets
for goods and labour is described by (1.17) and (1.18) respectively, where
of course, the real wage is given by $w = (1 - \mu) f'(l - \lambda)$. Thus with n
constant, but employment per firm variable, the market-clearing conditions,
together with the mark-up equation, can be viewed as determining the short-
run equilibrium values of L, P and w.[12]

Figure 1.3 provides a diagrammatic representation of the short-run equilibrium and its relationship to the long-run steady state, depicting combina-
tions of the price level and the real wage consistent with labour and goods
market equilibrium, for given n. The horizontal schedule, LR, intersects
the vertical axis at the value of the real wage, w^m, which must hold in long-
run given the free entry condition. The negative slope of NN, representing
equilibrium in the labour market, reflects the operation of the real balance
effect on labour supply; as the price level rises labour supply increases, with
the rise in employment necessary to equilibrate the labour market being reflected in a decline in the real wage.[13] With regards to the goods market,
however, an ambiguity is present. As the price level increases consumption
falls via the real balance effect, but the change in the real wage necessary to
maintain goods market equilibrium is indeterminate in direction.

In fact a sufficient condition for the goods market clearing locus YY to
be upward sloping as depicted in Figure 1.3 is that the positive direct effect
on consumption of a rise in the real wage outweighs the negative indirect
effect which operates via the induced fall in profits, a condition which is

certainly fulfilled for Cobb-Douglas preferences. The gradient of the labour
and goods market clearing loci are given by:

$$\left.\frac{\mathrm{d}w}{\mathrm{d}P}\right|_{NN} = \frac{-(1-\mu)\,f''L_2^s\frac{M^0}{P}}{n\,(1-\mu f'L_2^s)-(1-\mu)\,f''\,(L_1^s-LL_2^s)} < 0$$

$$\left.\frac{\mathrm{d}w}{\mathrm{d}P}\right|_{YY} = \frac{-(1-\mu)\,f''C_2\frac{M^0}{P^2}}{nf'\,(1-\mu C_2)-(1-\mu)\,f''\,(C_1-LC_2)} \gtrless 0^{14}$$

In the diagram NN and YY intersect at a value of the real wage above w^m.
Accordingly, the depicted short-run equilibrium is associated with each existing firm earning negative profits and subsequently firms respond to these
losses by ceasing production. Diagrammatically this decline in the number
of operative enterprises is reflected in downward shifts of both NN and
YY, leading to a fall in the short-run equilibrium real wage (but an ambiguous change in the price level) until the long-run steady state is achieved. In
fact an adjustment process of this nature underlies the distinction between
the short-run and long-run effects of fiscal policy to be discussed in the next
section.[15]

1.4 Fiscal policy

In this section we examine the effects on the economy of an increase in government spending, focusing in particular on the short and long-run output and
employment fiscal multipliers. Because our model encompasses the SMD
framework as a special case we are able to provide straightforward comparisons between our own results and those of previous work on fiscal policy
within an imperfectly competitive economy.

As already indicated we restrict our attention to balanced-budget fiscal
policy; that is a rise in government spending financed entirely by an increase
in lump-sum taxation. This has convenience value in that it allows us to
avoid the need to take into account the monetary dynamics associated with
the government budget constraint. At the same time it restricts the applicability of our analysis very little due to the fact that the real effects of a rise in
government spending are independent of whether the means of finance is via
taxation or by money creation. This feature is most easily seen by noting,
from Equations (1.20) and (1.21),[16] that following an increase in lump-sum
taxation, T, equilibrium is restored simply by a fall in the price level sufficient in magnitude to return the household sector's non-labour income to its
previous value. If follows that the only difference between the two alternative sources of finance of an expansionary fiscal policy derives from their

effects on the price level; thus the induced changes in output and employment, for example, are invariant to the means of finance of the policy.[17]

1.4.1   The short-run effects of fiscal policy

The effects of fiscal policy in the short run are found from (1.17) and (1.18), together with the mark-up relationship. Beginning from an initial position of long-run equilibrium (wherein $\Pi = 0$, $w = w^m$), differentiating these equations, holding n constant, and using Y and L to represent aggregate output and employment respectively, yields

$$\left.\frac{dY}{dg}\right|_{SR} = \Delta^{-1} > 0 \tag{1.29}$$

$$\left.\frac{dL}{dg}\right|_{SR} = \frac{1-\mu}{w^m}\Delta^{-1} > 0 \tag{1.30}$$

$$\left.\frac{dw}{dg}\right|_{SR} = \frac{(1-\mu)^2 f''}{n \cdot w^m}\Delta^{-1} < 0 \tag{1.31}$$

$$\left.\frac{dP}{dg}\right|_{SR} = -\frac{P^2}{M^0}\frac{1-\mu}{n \cdot w^m L_2^s} \cdot \Upsilon \cdot \Delta^{-1} \gtrless 0 \tag{1.32}$$

where

$$\Upsilon = n\left(1 - C_2 + w^m L_2^s\right) - (1-\mu) f'' \left(L_1^s \left(1 - C_2\right) + L_2^s \left(C_1 - L\right)\right)$$

$$\Delta = 1 - \frac{(1-\mu)}{w^m}\frac{C_2}{L_2^s} + \frac{(1-\mu)^2}{n \cdot w^m} f'' \left(\frac{C_2 L_1^s}{L_2^s} - C_1\right) > 1 \tag{1.33}$$

The rise in government spending creates excess demand in the goods market. To restore equilibrium, an expansion of output (1.29) and, hence, employment (1.30) is necessary. But, with the number of firms fixed, the diminishing marginal product of labour implies, given the mark-up, a decline in the real wage (1.31) as employment increases. The requirement for goods market equilibrium is made compatible with equilibrium in the labour market via the rise in labour supply which results from the increase in lump-sum taxation, and by an appropriate change in the price level. The direction of adjustment of the price level is determined by whether an excess demand or excess supply of labour results from the change in employment dictated by considerations of goods market equilibrium together with the change in desired labour supply prompted by the increase in lump-sum taxation and the fall in the real wage.[18]

Given $\Delta > 1$ (1.33) it follows that the output multiplier is less than unity, which is in accordance with the results of previous analyses of balanced budget fiscal policy within the SMD framework of an imperfectly competitive economy. It is straightforward to see precisely how a diminishing marginal product of labour affects the value of the multiplier compared with the SMD assumption of CMPL technology. In the latter case $f'' = 0$ of course, implying Δ is smaller in magnitude; hence the value of the output multiplier is larger. The reason lies in the fact that, with $f'' < 0$, as employment expands the real wage falls, reducing consumption expenditure directly and, at the same time, causing a fall in desired labour supply. This latter effect implies, *ceteris paribus*, a larger rise in the price level is required to equilibrate the labour market than is the case with a constant marginal product of labour resulting, via the real balance effect, in a larger fall in consumption spending. Hence, with $f'' < 0$ additional channels of crowding-out are present.

The precise value of the short-run output multiplier with a constant marginal product of labour is seen to be

$$\frac{dY}{dg}\bigg|_{SR} = \frac{1}{1 - \frac{1-\mu}{w^m}\frac{C_2}{L_2^s}} \tag{1.34}$$

Further, adopting the assumption of Cobb-Douglas preferences allows us to obtain an expression for the output multiplier which is directly comparable with the DSM results:

Proposition 1.1 *Assume Cobb-Douglas preferences, with utility function* $U = C^\alpha (E - L^s)^\beta (M/P)^\gamma$, *and CMPL technology,* $f'' = 0$. *Then:*

$$\frac{dY}{dg}\bigg|_{SR} = \frac{\beta}{\beta + (1-\mu)\alpha}$$

The value of the multiplier given by the above expression is identical to that of Dixon and equivalent to those contained in Mankiw and Startz.[19] For this special case it is readily apparent that the magnitude of the multiplier is increasing in μ. This particular result, which rests upon the dependence on μ of the relationship between aggregate profits (thereby disposable income) and aggregate output, is the feature which lends the Keynesian flavour to analyses of fiscal policy in models of imperfectly competitive economies. However, inspection of (1.33) indicates that Δ is related to μ in a highly complex fashion; in general, the nature of this relationship will depend on the various second-order derivatives, C_{ij}, L_{ij}^s, the precise form of the implicit functional dependence of both w^m and P on μ, and on the sign and magnitude of the third derivative of the production function. Thus, there is no general presumption that the size of the short-run output multiplier is increasing in μ.

1.4.2   The impact of fiscal policy in the long run

In the short run, with the number of firms fixed, the expansion in output and employment associated with a rise in government spending leads to an increase in profits from their long-run equilibrium value of zero. The prospect of positive profits prompts the entry of new firms until the inducement to enter has itself been eliminated, that is, profits have returned to zero. This expansion in the number of firms, of course, changes the characteristics of the equilibrium and, hence, modifies the impact of fiscal policy compared to the short run. We find the effects of fiscal policy in the long run by differentiation of (1.20) and (1.21) and solving for the changes in the endogenous variables n and P; the adjustments in aggregate output and employment are then found using $Y = ny^m$ and $L = nl^m$.

$$\left.\frac{dY}{dg}\right|_{LR} = \frac{1}{1 - \frac{1}{w^m}\frac{C_2}{L_2^s}} > 0 \tag{1.35}$$

$$\left.\frac{dL}{dg}\right|_{LR} = \frac{1}{w^m - \frac{C_2}{L_2^s}} > 0 \tag{1.36}$$

$$\left.\frac{dn}{dg}\right|_{LR} = \frac{1}{y^m\left(1 - \frac{1}{w^m}\frac{C_2}{L_2^s}\right)} > 0 \tag{1.37}$$

$$\left.\frac{dP}{dg}\right|_{LR} = \frac{P^2}{M^0}\frac{(1 - C_2 + w^m L_2^s)}{(C_2 - w^m L_2^s)} > 0 \tag{1.38}$$

With the real wage and output and employment per firm tied down by the zero profit condition, the increase in aggregate output (1.35), and associated rise in employment (1.36), necessary to maintain goods market equilibrium following the fiscal expansion are achieved purely by way of an increase in the number of firms (1.37). To induce the required rise in labour supply, the price level must increase (1.38) which, in turn, crowds out some private sector consumption. Hence the long-run output multiplier, like the short-run multiplier, is less than one in value.

Just as the relationship between μ and the magnitude of the short-run multiplier is indeterminate in direction, so is that between μ and the value of the long-run multiplier. However, an interesting result emerges for the special case of Cobb-Douglas preferences, where the long-run multiplier (1.35) becomes[20]

$$\left.\frac{dY}{dg}\right|_{LR} = \frac{\beta}{\beta + \alpha} \tag{1.39}$$

Thus for Cobb-Douglas preferences it is clear that the long-run multiplier is *independent* of μ. This result generalizes that of Startz (pp. 748-749), obtained under the assumption of CMPL technology:

Proposition 1.2 *If household preferences are Cobb-Douglas in form then, for any technology satisfying A2, the long-run output multiplier, given by (1.39), is independent of the degree of monopoly, μ.*

Although the result summarized in Proposition 1.2 is clearly a special one, the independence from μ of the value of the long-run multiplier extends to more general preferences in the neighbourhood of the Walrasian equilibrium. That is:

Proposition 1.3 *For any preferences satisfying A1 and any technology satisfying A2, then a small increase in μ from its Walrasian value of zero has no first order effect on the long-run output and employment multipliers.*

This proposition is straightforward to prove by differentiation of equation (1.35) and (1.36) with respect to μ[21] and follows directly from the fact that in the neighbourhood of Walrasian equilibrium the long-run levels of output and employment remain invariant to a small displacement of μ from zero. Note also that the local result of Proposition 1.3 relates to the employment as well as the output multiplier, whilst the global Proposition 1.2 applies only to the latter. In fact, with Cobb-Douglas preferences, the long-run employment multiplier is given by

$$\frac{\mathrm{d}L}{\mathrm{d}g} = \frac{1}{w^m} \frac{\beta}{\beta + \alpha}$$

which is increasing in μ for $\mu > 0$.[22]

1.4.3 Comparing the short and long-run output and employment multipliers

Given the distinction between equilibrium in the short and long runs and our discussion of the corresponding multipliers, a natural issue to examine is the question of whether fiscal policy is more powerful in the short run or in the long run. Within the SMD framework only Startz addresses this question; his findings are reflected in the results summarized in Propositions 1 and 2

$$\left.\frac{\mathrm{d}Y}{\mathrm{d}g}\right|_{LR} = \frac{\beta}{\beta + \alpha} < \frac{\beta}{\beta + (1 - \mu)\alpha} = \left.\frac{\mathrm{d}Y}{\mathrm{d}g}\right|_{SR}$$

for $0 < \mu < 1$. That is, with CMPL technology and Cobb-Douglas preferences, then for $\mu \in (0,1)$ the output multiplier is greater in the short run than in the long run. In fact, a comparison of (1.34) with (1.35) provides a rather more general result, summarized in Proposition 1.4:

Proposition 1.4 *Given CMPL technology and with $0 < \mu < 1$, then for any preferences satisfying A1*

$$\left.\frac{dY}{dg}\right|_{LR} = \frac{1}{1 - \frac{1}{w^m}\frac{C_2}{L_2^s}} < \frac{1}{1 - \frac{1-\mu}{w^m}\frac{C_2}{L_2^s}} = \left.\frac{dY}{dg}\right|_{SR}$$

Thus the assumption which underpins Startz's ranking of short and long-run output multipliers is seen to be that of CMPL technology; the structure of household preferences is clearly irrelevant for this result. Nonetheless the technological specification adopted by Startz is rather a special case. Comparing (1.29) and (1.35) we find:

$$\left.\frac{dY}{dg}\right|_{SR} \gtrless \left.\frac{dY}{dg}\right|_{LR} \text{ as } \frac{\mu}{(1-\mu)^2} \gtrless \frac{f''}{n}\left(\frac{C_1 L_2^s}{C_2} - L_1^s\right) \qquad (1.40)$$

and, in general, the direction of this inequality is indeterminate. An important result emerges however for the Walrasian case of $\mu = 0$; in this case it is apparent from (1.40) that the long-run multiplier is greater in magnitude than the short-run multiplier. Comparing (1.29) for $\mu = 0$ with (1.35) we have:

$$\left.\frac{dY}{dg}\right|_{SR(\mu=0)} = \frac{1}{1 - \frac{1}{w^m}\frac{C_2}{L_2^s} + \frac{f''}{n \cdot w^m}\left(\frac{C_2 L_1^s}{L_2^s} - C_1\right)}$$

$$< \frac{1}{1 - \frac{1}{w^m}\frac{C_2}{L_2^s}} = \left.\frac{dY}{dg}\right|_{LR(\mu=0)}$$

The explanation for this ranking of the Walrasian short and long-run output multipliers lies in the fact that, with $f'' < 0$, the short-run response to the rise in government spending involves an expansion of firms beyond the efficient scale, where production is located in the Walrasian long-run equilibrium. Subsequently, as the prospect of positive profits induces the entry of new firms, the associated increase in the average product of labour allows an expansion of aggregate output even in the absence of any increase in labour supply. In fact, the rise in the real wage, which accompanies the rise in the *marginal* product of labour as new firms enter, leads to an increase in both labour supply and consumption expenditure. Thus in the Walrasian case there is less crowding out in the long-run than in the short-run. Our Walrasian results are summarized in Proposition 1.5:

Proposition 1.5   *In the Walrasian case of $\mu = 0$:*

$$0 < \left.\frac{dY}{dg}\right|_{SR(\mu=0)} < \left.\frac{dY}{dg}\right|_{LR(\mu=0)} < 1$$

The significance of Proposition 1.5 is that it reverses the ranking of the short and long-run multipliers found by Startz for $\mu > 0$ and CMPL technology. In fact Proposition 1.5 can be generalized somewhat. Assuming the conditions of the Implicit Function Theorem are satisfied, both the short and long-run multipliers will be continuous in μ in the neighbourhood of $\mu = 0$. Hence for values of μ sufficiently close to zero the direction of the inequality in Proposition 1.5 will be maintained and we have:

Proposition 1.6   *A1, A2. If the derivatives f'', C_1, L_1^s, C_2, L_2^s are continuous in the neighbourhood of the Walrasian equilibrium, then there exists $\bar{\mu} > 0$, such that for $0 \le \mu < \bar{\mu}$,*

$$\left.\frac{dY}{dg}\right|_{SR} < \left.\frac{dY}{dg}\right|_{LR}$$

Thus Proposition 1.6 emphasizes that Startz's ranking of the short and long-run output multipliers is not at all a general result, but instead a reflection of the rather special assumptions which characterize the SMD framework and, in particular, that of CMPL technology.

Although an unambiguous comparison of the short and long-run *output* multipliers can be made only in the neighbourhood of the Walrasian equilibrium, a general ranking of the corresponding employment multipliers is possible. From Equations (1.30) and (1.36) we find:

$$\left.\frac{dL}{dg}\right|_{LR} - \left.\frac{dL}{dg}\right|_{SR}$$
$$= \left(\frac{\mu w^m}{1-\mu} + \frac{1-\mu}{n}f''\left(\frac{C_2 L_1^s}{L_2^s} - C_1\right)\right)\Delta^{-1}\left(w^m - \frac{C_2}{L_2^s}\right)^{-1} > 0$$

Therefore we have:

Proposition 1.7   *For all $\mu \in [0, 1)$,*

$$\left.\frac{dL}{dg}\right|_{SR} < \left.\frac{dL}{dg}\right|_{LR}$$

Thus despite the possibility of a larger output multiplier in the short run than in the long run, the long-run employment multiplier is *necessarily* greater in magnitude than the short-run employment multiplier. This ranking of the employment multipliers follows from the rise in the marginal product of labour which accompanies the entry of new firms. The associated increase in the real wage induces an expansion of labour supply and hence employment. The fixed overhead labour requirement, which implies, for $\mu > 0$, the average product falls with new entry, means that this increase in employment remains compatible with the potential fall in aggregate output.

1.5 A diagrammatic exposition of the fiscal multiplier with imperfect competition

1.5.1 The short-run relationship: results

Whilst the formal results have been shown for a monetary economy with a general production function, we can illustrate the results in a non-monetary economy (as in Dixon and Lawler, 1996). The only restriction on preferences we make for this diagrammatic analysis is that preferences U are homothetic. With homothetic preferences, the marginal rate of substitution between leisure and consumption depends only on the ratio $C/(E-L)$, where E is the endowment of leisure: utility maximization implies that the ratio of these two is therefore a function of the relative price w:

$$\frac{C}{E-L} = \gamma(w) \tag{1.41}$$

where γ is strictly decreasing in w, $\gamma' < 0$, since consumers have a strictly decreasing marginal rate of substitution. The Income Expansion Path (IEP) of consumers are therefore linear, and depend only on the relative price (real wage) w.

If we assume that there are CRTS, then we can normalize the MPL to unity, so that $w = 1 - \mu$. Hence, as μ increases, the income expansion in $(C, E-L)$ space becomes flatter. This reflects the fact that more imperfect competition leads to the price of consumption rising relative to that of leisure (w falls), and hence households substitute away from consumption to leisure, as depicted in Figure 1.4: the Walrasian IEP corresponds to $w = 1$ ($\mu = 0$), and more imperfectly competitive to higher value of μ.

Now we can turn to the production side. Leaving aside fixed costs, we can represent the production possibility frontier in $(C, E-L)$ space. The

PPF for the economy is represented by:

$$L - C - g = 0 \qquad (1.42)$$

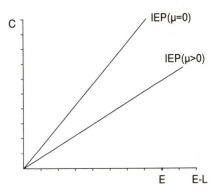

Figure 1.4: Income expansion paths

Hence, in $(C, E - L)$ space, we have a series of negatively sloped $45°$ lines: those closer to the origin corresponding to higher levels of g, as in Figure 1.5. The need to supply higher levels of g reduces the possibilities for private use. We can now put together Figures 1.4 and 1.5, and see the impact of imperfect competition the multiplier. In Figure 1.6, we depict an initial equilibrium for two values of μ: the Walrasian and an imperfectly competitive value. Turning first to the Walrasian : when $\mu = 0$, the equilibrium occurs at point W, and the marginal rate of substitution equals the Marginal Rate of Transformation (MRT) of leisure into consumption (MRT=1). The household budget constraint corresponds to the PPF. This is apparent, because when $\mu = 0$, constant returns to scale implies zero profits; hence the household budget constraint is:

$$C = wL - T \qquad (1.43)$$

with a balanced budget $T = g$, and when $\mu = 0$, $w = 1$, so that (1.43) is equivalent to (1.42).

Now let us turn to the imperfectly competitive case. Here, the MRS equals $-(1 - \mu)$, which is less than the MRT in absolute terms: giving up one unit of leisure appears to increase consumption by only $(1 - \mu)$. The equilibrium occurs at point M in Figure 1.6, where the PPF intersects the IEP(μ). The perceived budget constraint passes through M, with corresponding con-

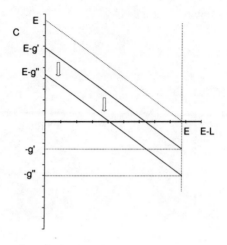

Figure 1.5: The PPF for consumption and leisure as g varies

sumption C_M. The vertical distance OC_0 represents the consumption level
when the household supplies no labour ($L = 0$, leisure equals E). This represents the profits of firms which are redistributed to the household less tax
T. In general, when $\mu > 0$ we have the budget constraint (for any C):

$$C = wL + \Pi - g \qquad (1.44)$$

where $\Pi = \mu \cdot (C + g)$ and $w = 1 - \mu$ (note that the household treats Π
as fixed, and the relation between Π and C comes from the firm's budget
constraint). Clearly, at point M, $C = C_M$, and the budget constraint can be
written as:

$$C = \mu \cdot (C_M + g) - g = \Pi - g \qquad (1.45)$$

At point M, the budget constraint passes through the PPF (since $L = C_M +
g$). However, at all other points this is not true. At $L = 0$, we have the
corresponding consumption C_0, where:

$$C_0 = \mu \cdot (C_M + g) - g = \Pi - g \qquad (1.46)$$

Now we consider the effect of an increase in g by Δg, which shifts the PPF in
$(C, E - L)$ space down by the vertical distance Δg. We can see this in Figure
1.7, where the new equilibria are W' and M' respectively. If we define total

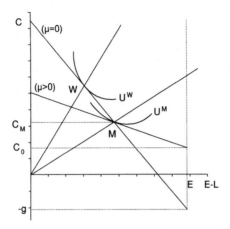

Figure 1.6: The equilibria compared

output as $Y = C + g$, the value of the multiplier is:

$$\frac{\Delta Y}{\Delta g} = 1 + \frac{\Delta C}{\Delta g} < 1 \qquad (1.47)$$

This is less than unity, since consumption is reduced by the increase in Δg. Now, from Figure 1.7, we can see that the ratio $\Delta C/\Delta g$ is more negative as we move down the Walrasian IEP from W to W' than it is in the imperfectly competitive case as we move from M to M'. This follows from the fact that the optimal consumption/leisure ratio is *larger*. In fact we can expand (1.47) to allow for the fact that $C = (E - L) \cdot \gamma (1 - \mu)$ and $L = C + g$:

$$\frac{\Delta Y}{\Delta g} = 1 + \frac{\Delta C}{\Delta L}\frac{\Delta L}{\Delta Y}\frac{\Delta Y}{\Delta g} = 1 - \gamma (1 - \mu) \frac{\Delta Y}{\Delta g} \qquad (1.48)$$

Hence:

$$\frac{\Delta Y}{\Delta g} = \frac{1}{1 + \gamma (1 - \mu)} \qquad (1.49)$$

Hence the ratio $\Delta Y/\Delta g$ is increasing in μ (since $\gamma' > 0$).

1.5.2 The long-run multiplier

With free entry the analysis is a little more complex. In this case, let us consider the more general technology. From (1.16), we can see that in the

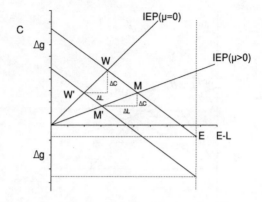

Figure 1.7: The multipliers compared

long run under free entry the average product of labour is fixed at the value:

$$a\left(\mu\right) = \frac{f\left(l^m - \lambda\right)}{l^m} \tag{1.50}$$

where from (1.16) and Figure 1.1, $\partial a/\partial \mu < 0$. Suppose that we normalize $a(\mu)$ so that in the Walrasian case $a(0) = 1$ (here we assume that $f'' < 0$). Then the long-run PPF (LPPF) will depend on μ. In Figure 1.8 we depict two LPPFs: both originate in $-g$ when $L = 0$: however, the monopolistic LPPF lies inside the Walrasian LPPF, since $a' < 0$ and the slope of the LPPF is a. In the long run, free entry eliminates profits, and all income is labour income: therefore the budget constraint is as in (1.43). In Figure 1.9, we depict the multiplier as before: an increase in g shifts the LPPF vertically downwards. However, whereas in the short run case considered above, we were looking at the same PPF and two different IEPs, we now have two different LPPFs as well.

On the issue of the short-run multiplier with a fixed number of firms versus the long-run multiplier, we need to consider the short-run PPF (SPPF) as well as the LPPF. The shape of the LPPF and SPPF depends a little on whether the technology has a CMPL (in which case we have a natural monopoly), or the case of $f'' < 0$. In this case, we can consider two LPPFs, one Walrasian and one with $\mu > 0$, as in Figure 1.10. As we move Northwest along both, as output rises more firms are producing the output: in the long run with free entry, output per firm is fixed (given μ), and total output is varied by varying the number of firms. Now, take a point (B) on the imperfectly competitive LPPF: there is a SPPF passing through B:

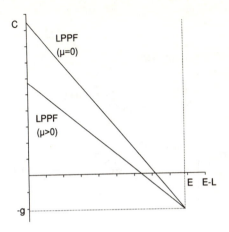

Figure 1.8: The long-run production possibility frontiers

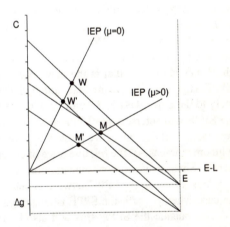

Figure 1.9: The long-run multipliers compared

if more output is produced, it can be produced more efficiently than along the LPPF (since the APL of labour is increasing), up to point B^W on the Walrasian LPPF (at that point APL is maximized). There after the SPPF passing through and B' will lie within the Walrasian LPPF, and may (or may not) intersect the LPPF($\mu > 0$). Starting from B, if less output is produced, the SPPF lies inside the LPPF($\mu > 0$).

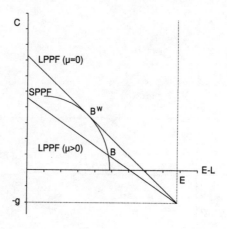

Figure 1.10: The short-run and long-run multipliers compared when $\mu > 0$

Now, consider an increase in g: this will shift the SPPF and the LPPF passing through B down to B', as in Figure 1.11. The long-run effect is a move down the same IEP to point C. However, in the short run, the move will be along the new SPPF which passes through B'. As drawn, the SPPF passing through B' lies above C: that is the new long-run equilibrium is within the old SPPF. This is not always the case: although the segment of the LPPF immediately to the north-west of B' will lie with the SPPF, it is quite possible that the SPPF will intersect the LPPF at a point between B' and C. When the real wage will decline as we move along the SPPF (although profit inclusive income increases): the relative price of consumption/leisure is $w = (1 - \mu)f'$. Since consumption appears more expensive, we will switch to an IEP which intersects the LPPF between B' and C. For example, in the Walrasian case when $\mu = 0$, the SPPF is tangential to the LPPF at B', and hence C lies outside the SPPF: this will also be true for μ small enough. If we look at Figure 1.11, we can draw a horizontal line through C. As drawn, it intersects the SPPF at D. Clearly, if the new short run lies below D, then the degree of crowding out is larger in the short run than in the long run: conversely, if the new short-run equilibrium lies above D, then the degree of crowding out in the short run must be less. Thus, it is possible for the short-run multiplier to be larger or smaller than the long-run. Since Startz looked at a case where the short-run multiplier exceeded the long-run, it is easy to see how this might be reversed: if μ is small enough, then the SPPF lies everywhere below the horizontal line passing through C, so that the short-run multiplier must be less than the long-run (as was demonstrated in Proposition 1.6).

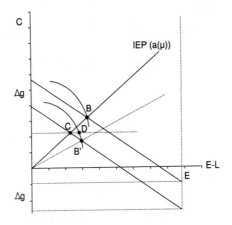

Figure 1.11: The short-run and long-run multipliers compared when $\mu > 0$

1.6 Dynamic equilibrium models

In this section, we will show how the results of the previous section are carried over into a dynamic context. We will adopt the continuous time Ramsey case, with an infinitely lived consumer. A full analysis is given in Dixon (1997), but here we will deal with steady-state effects of fiscal policy, under the assumption of instantaneous free entry. The flow utility function of the consumer depends on aggregate consumption $C(t)$ and leisure $1 - L(t)$. The production of the representative firm is given by the production function of the form (we suppress the firm subscript):

$$y(t) + \varphi = F(K(t), L(t)) \tag{1.51}$$

where $y(t)$ is output and φ is the overhead fixed cost per firm. With constant elasticity of demand for output, monopolistic firms will mark up price over marginal cost. We now have two factor demand equations implied by this:

$$w = (1 - \mu) F_L; \quad r = (1 - \mu) F_K \tag{1.52}$$

where r is the real interest rate. If we assume that F is homogeneous to degree 1, then we can write F in intensive form: $y(t) + \varphi = L(t) \cdot f(k(t))$

where $k(t) = K(t)/L(t)$, and $F_K = f'$ and $F_L = f' - k \cdot f'$. By Euler's equation, since F is homogeneous, $F = L \cdot F_L + K \cdot F_K$, so that the profits of the firm are, using (1.51):

$$\Pi(t) = \mu \cdot F - \varphi = \mu \cdot y(t) - (1 - \mu)\varphi \qquad (1.53)$$

Now, if we assume that there is free entry, then profits will be zero. As in the case where there was one factor of production, this implies that the output per firm is tied down:

$$y(t) = \left(\frac{1-\mu}{\mu}\right)\varphi \qquad (1.54)$$

Aggregating over firms, the total number of firms $n(t)$ is given by the relationship:

$$n(t) = \frac{Y(t)}{y(t)} = Y(t) \cdot \left(\frac{\mu}{1-\mu}\right)\frac{1}{\varphi} \qquad (1.55)$$

Hence, the free entry condition implies that number of firms varies as the aggregate output varies. We can combine (1.51) and (1.55), to obtain the aggregate production function given that there is imperfect competition and free entry:

$$Y(t) = (1 - \mu) \cdot F(K(t), L(t)) \qquad (1.56)$$

This equation is very important: it means that although the technology per firm displays increasing returns to scale ($\varphi > 0$), the aggregate technology displays constant returns due to free entry. This would apply so long as F is homothetic, and does not require F to be homogeneous to degree 1. Note that in the case of F being homogeneous to degree one or more, we have a natural monopoly, and that entry of additional firms is inefficient. Imperfect competition increases the inefficiency due to excess entry, since when μ is larger it raises profitability per firm for a given number of firms.

Given this, we can write down the constrained social planner s problem which yields the imperfectly competitive economy:

$$\max \int_0^\infty e^{-\beta \cdot t} \cdot U(C(t), 1 - L(t)) \, \mathrm{d}t \qquad (1.57)$$

$$\text{s.t. } \frac{\mathrm{d}K}{\mathrm{d}t} = (1 - \mu) F(K(t), L(t)) - C(t) - \delta K(t) - g(t)$$

In formulating the optimization, we are taking the free entry condition and imperfect competition as given, and thus solving the second-best problem. The first best would involve freely choosing the number of firms and output per firm.[23] The current value Hamiltonian for (1.57) is:

$$H(t) = U(C(t), 1 - L(t)) +$$

$$\lambda(t)((1 - \mu) F(K(t), L(t)) - C(t) - \delta K(t) - g(t)) \qquad (1.58)$$

The first order conditions for this are:

$$H_C = U_1 - \lambda = 0 \qquad (1.59)$$

$$H_L = -U_2 + \lambda(1 - \mu) F_L = 0 \qquad (1.60)$$

$$H_K = \lambda((1 - \mu) F_K - \delta) = -\frac{d\lambda}{dt} + \beta\lambda \qquad (1.61)$$

with the transversality condition and $H_\lambda = 0$. Equations (1.59-1.61) define the dynamic equilibrium of the imperfectly competitive economy. If we are concerned with the steady state only, then we can write these in the familiar form (using the intensive form production function):

$$(1 - \mu) f' = \delta + \beta \qquad (1.62)$$

$$(1 - \mu)(f(k) - k \cdot f') = \frac{U_L}{U_C} \qquad (1.63)$$

Equation (1.62) is the intertemporal optimality condition, sometimes called the modified golden rule (MGR); (1.63) is the intratemporal optimality condition equating the MRS between consumption and leisure with the real wage.

Clearly, the solution to (1.59-1.61) can define the entire dynamics of the system given $g(t)$: for a full analysis see Dixon (1997). Since in this paper we are only interested in the steady state effects, we will consider the steady state fiscal multiplier: the multiplier that occurs when there is an unanticipated permanent change in g from one level to another. This is easily represented diagrammatically. The MGR (1.62) can be solved for the optimal capital-labour ratio, $k * (\mu)$, with $k^{*\prime} < 0$. This then defines the optimal net output per unit labour $(1 - \mu) \cdot f(k^*(\mu)) - \delta k^*(\mu)$. Hence, we can represent the MGR in $(C, 1 - L)$ space, given g. It is the line which we call the intertemporal PPF (IPPF) defined by:

$$C = L((1 - \mu) \cdot f(k^*(\mu)) - \delta k^*(\mu)) - g \qquad (1.64)$$

The IPPF defines the trade off between leisure and consumption in the steady state given that accumulation satisfies the MGR.[24] Note that as μ increases, the curve rotates anticlockwise from the point $(-g, 1)$. The effect of imperfect competition is twofold: first, it reduces the returns to investment,

34   *Marta Aloi et al.*

yielding less saving/investment, and hence the MGR yields a lower steady state $k^*(\mu)$; secondly the inefficiency due to excess entry is increased. Furthermore, with free entry the equilibrium level of productivity is lower, so that for any given k, output per unit labour is lower. The intratemporal optimality condition is exactly the same as in (1.41), and can be represented by the IEP $\gamma(w)$, where now:

$$w(\mu) = (1 - \mu) \cdot (f(k^*(\mu)) - k^* \cdot f'(k^*(\mu)))$$ (1.65)

Note that μ reduces the steady state real wage (1.65) in two ways: first $k^*(\mu)$ falls, and second $(1 - \mu)$ falls. The Walrasian equilibrium only exists when $\mu = \varphi = 0$, since when $\varphi > 0$ there are increasing returns to scale.

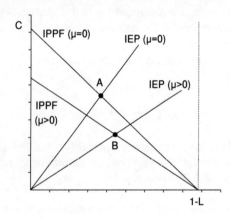

Figure 1.12: The Walrasian (A) and imperfectly competitive (B) equilibria

However, even when $\varphi > 0$, an equilibrium exists for all $\mu > 0$, and in this case we can rather loosely take the Walrasian economy as the limiting equilibrium as $\mu \to 0$. In Figure 1.12 we compare this limiting Walrasian economy ($\mu = 0$) with the imperfectly competitive economy : the Walrasian equilibrium is at point A, and the imperfectly competitive at point B. If we compare Figure 1.12 to Figure 1.6, the two look very similar. This is not surprising: in both cases there is a negative linear relationship between consumption and leisure. In the static case this was due to free entry; in the dynamic case, it is due to free entry combined with the intertemporal optimality condition. However, the visual similarities hide some important differences. First, the slope in the static case is equal to the real wage: this is not so in the dynamic case - the slope of the IPPF includes income from wages $(1 - \mu) \cdot [f(k^*(\mu)) - k^*(\mu) \cdot f(k^*(\mu))]$ and from the rental on capital $(1 - \mu) \cdot k^*(\mu) \cdot f(k^*(\mu)) - \delta k^*(\mu)$ which combine to make the total steady-state income per unit labour of $(1 - \mu) \cdot f(k^*(\mu)) - \delta k^*(\mu)$.

Let us first consider the steady -state effects of a permanent increase in g, as depicted in Figure 1.13. This will lead to the IPPF to shift downwards by the distance dg. As a result, both consumption and leisure will fall: this yields a multiplier which is less than one (since consumption falls), but greater than zero (since total output rises as leisure falls). Clearly, the household is made worse off because it has to pay the taxes to fund dg; it responds to this reduction in utility by reducing both leisure and consumption. Output rises, but by less than the increase in g. The story is almost exactly the same as in the static case, as is reflected in the similarity of Figures 1.13 and 1.7.

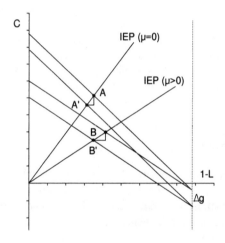

Figure 1.13: The Walrasian multiplier $(A - A')$ and the imperfectly competitive multiplier $(B - B')$

How is the multiplier affected by an increase in μ? Here, the IEP will become flatter, as will the IPPF. In Figure 1.13 we compare the multiplier for two different values of μ. There are two counteracting tendencies here. First, as μ increases the IEP becomes flatter: this means that the reduction in leisure is greater and the reduction in consumption smaller. This tends to make the employment multiplier larger, and the output multiplier smaller. Second, the IPPF also becomes flatter. This second effect tends to increase the reduction in consumption and leisure. Thus the combined effect is unambiguous for the employment multiplier: it is larger when μ is larger (confirming Proposition 1.7 in the static case). However, the effects on consumption, the crowding-out effects, are ambiguous: the fall in consumption could be larger or smaller, depending on the specific assumptions about functional forms.

In this section we have developed a simple analysis of the dynamic intertemporal model without the need to assume any explicit functional forms. As we have seen, if we restrict ourselves to steady state reactions to permanent changes in government expenditure, the analysis is very similar to the static case analyzed in Dixon and Lawler (1996) and the previous sections of this paper. For a continuous time intertemporal model similar to the one presented in this section, which extends the analysis to the short-run dynamics, see Heijdra (1997). However, the analysis of intertemporal models with imperfect competition has tended to be done in a discrete time RBC framework (see for example Rotemberg and Woodford, 1995, Devereux *et al.*, 1996). This approach often ends up using explicit functional forms, and it becomes unclear what results are general, and which are driven by the explicit functional forms. Whilst adopting explicit functional forms allows the calibration of the model and its confrontation with the data, it also can dramatically restrict the range of possible responses.

1.7 Conclusion

In this paper we have generalized existing models which seek to analyze fiscal policy within the context of an economy with monopolistic output markets and a Walrasian labour market. Our approach highlights the dangers of attempting to draw too general conclusions from models which make specific assumptions about functional forms. The main implications of the analysis of this paper are two-fold. First, the relationship between imperfect competition and the behaviour of the macroeconomy is not as simple as suggested by the SMD framework. Rather the results found within the latter rest on one or both of the crucial assumptions of constant marginal expenditure shares and a constant marginal product of labour. Secondly, whilst matters are not surprisingly more ambiguous in our own framework, we can still extract some general results:

1. if the economy is sufficiently competitive then the long-run output multiplier exceeds the corresponding short-run multiplier, with both lying between zero and unity.
2. whatever the degree of monopoly, including the Walrasian limit, the

long-run employment multiplier is greater than that obtaining in the short-run.

3. starting from the Walrasian limit, a small increase in the degree of monopoly has no first order effect on either the long run levels of output and employment or the corresponding long-run multipliers.

Thirdly, dynamic intertemporal models often make very restrictive assumptions about preferences and technology. These assumptions will tend to determine the results of these models as in the static framework, and perhaps we should not treat the results of these dynamic models as anything other than rather special cases. Just as in Dixon and Lawler (1996) and the present paper have generalized the static model of SMD, there is an urgent need to allow for much more general preferences and technology in dynamic models. Fourthly, we would strongly argue for the need to develop a visual and geometric understanding of macroeconomic models.

The existence of some general results in this paper is encouraging. It appears to indicate the possibility of moving beyond the highly specific models which typify the macroeconomics of imperfect competition literature without loosing the capability to obtain useful results. Let us hope that researchers in the coming years are able to develop models in this way.

Appendix

1. Long-run equilibrium and μ

Differentiating (1.20) and (1.21) with respect to μ, noting that, with T fully indexed, the real value of taxation is independent of μ:

$$C_1 \frac{\mathrm{d}w^m}{\mathrm{d}\mu} - \frac{M^0}{P} C_2 \frac{\mathrm{d}P}{\mathrm{d}\mu} = n \frac{\mathrm{d}y^m}{\mathrm{d}\mu} + y^m \frac{\mathrm{d}n}{\mathrm{d}\mu} \tag{1.66}$$

$$L_1^s \frac{\mathrm{d}w^m}{\mathrm{d}\mu} - \frac{M^0}{P} L_2^s \frac{\mathrm{d}P}{\mathrm{d}\mu} = n \frac{\mathrm{d}l^m}{\mathrm{d}\mu} + l^m \frac{\mathrm{d}n}{\mathrm{d}\mu} \tag{1.67}$$

Substituting the values for $\frac{\mathrm{d}w^m}{\mathrm{d}\mu}$, $\frac{\mathrm{d}y^m}{\mathrm{d}\mu}$, and $\frac{\mathrm{d}l^m}{\mathrm{d}\mu}$ given by equations (1.22)-(1.24) of the main text yields, after some rearrangement:

$$\begin{bmatrix} y^m & \frac{M^0}{P} C_2 \\ l^m & \frac{M^0}{P} L_2^s \end{bmatrix} \begin{bmatrix} \frac{\mathrm{d}n}{\mathrm{d}\mu} \\ \frac{\mathrm{d}P}{\mathrm{d}\mu} \end{bmatrix} = \begin{bmatrix} \frac{(\mu C_1 - L)(f')^2}{l^m \psi} \\ \frac{(\mu L_1^s f' - L)f'}{l^m \psi} \end{bmatrix} \tag{1.68}$$

The solution to (1.68) provides expressions for (1.25) and (1.28). The effect of μ on aggregate output (1.26) and employment (1.27) can then be found in an obvious way using $Y = ny^m$, $L = nl^m$.

2. Fiscal policy and long-run equilibrium

We differentiate (1.20) and (1.21) with respect to g, noting that w^m, y^m and l^m are all invariant to the stance of fiscal policy, whilst, given the balanced budget nature of the expansion, $d(T/P)/dg = 1$:

$$
\begin{bmatrix} y^m & \frac{M^0}{P}C_2 \\ l^m & \frac{M^0}{P}L_2^s \end{bmatrix} \begin{bmatrix} \frac{dn}{dg} \\ \frac{dP}{dg} \end{bmatrix} = \begin{bmatrix} 1 - C_2 \\ -L_2^s \end{bmatrix}
\tag{1.69}
$$

Solving (1.69) yields $dn/dg|_{LR}$ (1.37) and $dP/dg|_{LR}$ (1.38) directly, allowing the impact of the policy on Y (1.35) and L (1.36) to be found in a straightforward fashion.

3. Fiscal policy the short run

To find the short-run impact of fiscal policy we use the goods and labour market equilibrium conditions (1.17) and (1.18), noting that in the short-run n is fixed, whilst l, y and π may all diverge from their long-run equilibrium values. Using the facts that:

$$
\frac{dw}{dg} = (1 - \mu) f'' \frac{dl}{dg}; \qquad \frac{d(\Pi/P)}{dg} = n\left(\mu f' - l(1 - \mu) f''\right) \frac{dl}{dg}
$$

we have, upon differentiation of (1.17) and (1.18) with respect to g:

$$
\begin{bmatrix} nf'(1 - \mu C_2) - (1 - \mu) f''(C_1 - LC_2) & \frac{M^0}{P}C_2 \\ n\left(1 - \mu L_2^s f'\right) - (1 - \mu) f''\left(L_1^s - LL_2^s\right) & \frac{M^0}{P}L_2^s \end{bmatrix} \begin{bmatrix} \frac{dl}{dg} \\ \frac{dP}{dg} \end{bmatrix}
$$
$$
= \begin{bmatrix} 1 - C_2 \\ -L_2^S \end{bmatrix}
\tag{1.70}
$$

(1.70) can then be solved for $dP/dg|_{SR}$ (1.32) and dl/dg. The latter then allows the short-run effects of the policy on Y (1.29), L (1.30) and w (1.31) to be found.

Notes

1. The implications of imperfect competition, often combined with institutional or behavioural rigidities (for example menu costs) for the conduct of government policy have been explored in a number of papers which fall within the New Keynesian School. The main emphasis of this literature has been on the question of monetary non-neutrality (for example, Ball and Romer (1990), Blanchard and Kiyotaki (1987), Caplin and Spulber (1987). For views on the relationship between New Keynesian and both New Classical and traditional Keynesian macroeconomics see Greenwald and Stiglitz (1987), Mankiw and Romer (1991) and Mankiw (1992)

2. In Dixon and Mankiw the utility function is Cobb-Douglas, whilst Startz adopts the slightly more general assumption of Stone-Geary preferences.

3. The crucial assumption in the DMS framework is constancy of marginal budget shares. This requires only that the Engel curves are linear, irrespective of whether they pass through the origin.

4. Thus income effects on labour supply are precluded.

5. See for example, Deaton and Muellbauer (1980).

6. Note further that in Dixon and Mankiw $\lambda = 0$.

7. See Blanchard and Kiyotaki (1987).

8. Note that the vertical axis in Figure 1.3 can be viewed as representing total output and employment, since both are proportional to n in long-run equilibrium.

9. Note that in the absence of full indexation of T then an increase in the money supply would result in a *more* than proportionate increase in the price level. The new long-run equilibrium would then be characterized by lower real money balances and a lower real value of taxation such that $(M_0 - T)/P$ remained unchanged.

10. See Appendix.

11. Note, from (1.25) that for $\mu = 0$, $dn/d\mu$ is strictly positive.

12. It is straightforward to confirm that monetary neutrality, a feature of the long-run as already discussed, extends to the short-run.

13. This decline, of course, also acts to offset directly the initial increase in labour supply.

14. Given the signs of the other terms in the expression, the condition $C_1 - LC_2 > 0$, referred to above, is sufficient to ensure $dw/dP|_{YY}$ is positive. For Cobb-Douglas preferences $C_1 - LC_2 = \alpha\beta(E + \frac{1}{w}\frac{M^0-T+\Pi}{P})$.

15. We note that the short-run equilibrium values of aggregate employment and output, and the real wage are all decreasing in μ, whilst the relationship between æ and the price level is in general ambiguous (though positive for Cobb-Douglas preferences).

16. Which describe long-run equilibrium; the argument extends in an obvious fashion to short-run equilibrium.

17. For a detailed analysis of taxation in the SMD framework, see Molana and Moutos (1991).

18. Referring to (1.32) note that $1-C_2+wL_2^s$ is unambiguously positive. Hence a sufficient condition for $dP/dg|_{SR} > 0$ is $L_1^s(1-C_2) + L_2^s(C_1 - L) > 0$. With Cobb-Douglas preferences this latter expression becomes $(1 - \alpha - \beta)\frac{\beta}{w}\left[E + \frac{1}{w}\frac{M^0-T+\Pi}{P}\right] > 0$. Hence for the Cobb-Douglas case the price level must rise.

19. For both Mankiw and Startz, money is absent from the utility function and, hence, $\beta = 1 - \alpha$.

20. With Cobb-Douglas preferences, $C_2 = \alpha$, $L_2^s = -\beta/w$..

21. Noting that for $\mu = 0$, $dw^m/d\mu = 0$, $dP/d\mu = 0$. (See (1.22) and (1.28)).

22. Since $dw^m/d\mu < 0$ for $\mu > 0$.

23. In the literature, the term social planner is used to refer to the first best given only the technology constraint. Even in the Walrasian literature, however, the constraint of a fixed number of firms is often imposed. To make matters clear, we use the term constrained social planner and specify explicitly the constraints we are imposing.

24. The term intertemporal PPF is a bit of a hybrid. The PPF is usually used to refer to a frontier that is purely defined by the technology (as in the case of the corresponding golden rule lines. However, the intertemporal PPF includes the intertemporal optimality condition, which incorporates the subjective discount rate.

References

Ball, L. and D. Romer (1990), "Real Rigidities and the Non-neutrality of Money," *Review of Economics Studies*, 57:183-203.

Blanchard, O. and N. Kiyotaki (1987), "Monopolistic Competition and the Effects of Aggregate Demand," *American Economic Review*, 77:647-66.

Caplin, A. and D. Spulber (1987), "Menu Costs and the Neutrality of Money," *Quarterly Journal of Economics*, 102:703-25.

Chamberlain, E. (1933), *The Theory of Monopolistic Competition*, Harvard University Press, Cambridge, Massachussets.

Deaton, A. and J. Mulbauer (1980), *Economics and Consumer Behaviour,* Cambridge University Press, Cambridge.

Devereux, M. B., A.C. Head and B. J. Lapham, (1996), "Monopolistic Competition, Increasing Returns, and the Effects of Government Spending," *Journal of Money Credit and Banking*, 28:235-54.

Dixon, H. D. (1987), "A Simple Model of Imperfect Competition with Walrasian Features," *Oxford Economic Papers*, 39:134-60.

Dixon, H. D. and N. Rankin (1994), "Imperfect Competition and Macroeconomics: A Survey," *Oxford Economic Papers*, 46:171-99.

Dixon, H. D. and P. Lawler (1996), "Imperfect Competition and the Fiscal Multiplier," *Scandinavian Journal of Economics*, 98:219-31.

Dixon, H. D. (1997), "Lectures on Imperfect Competition and Macroeconomics," *mimeo*, York University. Lectures given at Paris-1 MAD (April 1996), and at Helsinki as part of the Finnish Doctoral programme (October 1996), and University College Cork (November 1996).

Greenwald, B. and J. Stiglitz (1987), "Keynesian, New Keynesian and New Classical Economics," *Oxford Economic Papers*, 39:119-32.

Grossman, G. and E. Helpman (1991), *Innovation and Growth*, MIT Press, Cambridge, Massachussets.

Heijdra, B. J. (1997), "Fiscal Policy Multipliers: The Role of Market Imperfection and Finite Lives," forthcoming in *Journal of Economic Dynamics and Control*.

Judd, K. L. (1985), "On Performance of Patents," *Econometrica*, 53:567-86.

Mankiw, N. G. (1988), "Imperfect Competition and the Keynesian Cross," *Economics Letters*, 26:7-14.

Mankiw, N. G. (1992), "The Reincarnation of Keynesian Economics," *European Economic Review*, 36:559-65.

Mankiw, N. G. and D. Romer (1991), "Introduction," in: N. G. Mankiw and D. Romer (eds.), *New Keynesian Macroeconomics,* MIT Press, Cambridge, Massachussets.

Molana, H. and T. Moutos (1991), "A Note on Taxation, Imperfect Competition and the Balanced Budget Multiplier," *Oxford Economic Papers*:68-74.

Rotemberg, J. and M. Woodford (1995), "Dynamic General Equilibrium Models with Imperfectly Competitive Product Markets," in: T. F. Cooley (ed.), *Frontiers of Business Cycle Research*, Princeton University Press, Princeton.

Startz, R. (1989), "Monopolistic Competition as a Foundation for Keynesian Macroeconomic Models," *Quarterly Journal of Economics*, 104:737-52.

2 Imperfect Competition and Aggregate Price Equations

Ate Nieuwenhuis

1.1 Introduction

A lot of traditional economic theory treats prices as parametric: No agent sets prices, except for the fictitious Walrasian auctioneer. As a result, macroeconomic model builders in the sixties and early seventies faced a dearth of theoretical results to base their price equations on. In fact, they had to make do with two sketchy theories (see Eckstein and Fromm, 1968).

One theory of price formation originates in general competitive equilibrium theory. It is the *Walrasian tâtonnement* mechanism, according to which prices respond to discrepancies between supply and demand:

$$\dot{p} = f\left(\frac{d-s}{s}\right), \qquad f(0) = 0, f'(\cdot) > 0 \qquad (1.1)$$

where $\dot{p}$ is the time rate of change of price p and d and s are the levels of demand and supply, respectively. Prices rise if demand exceeds supply, they fall if supply exceeds demand, and they adjust until equilibrium is reached. This process describes the effects of anonymous market forces, that is, the working of the *Invisible Hand*.

The other theory originates in business interviews (Hall and Hitch, 1939). In contrast to the first theory, it would apply to oligopolistic industries, where firms are price makers. Their market power allows them to base their pricing decisions on long-run considerations and to ignore the vagarious, temporary fluctuations in demand. This behaviour leads to prices that are stable for a relatively long period of time (Means' concept of 'administered prices'). Two versions of the theory exist. First, *target- rate-of-return pricing* or *normal-cost pricing* states that the firm sets its price so as to earn a target rate of return on capital at a standard volume of output:

$$p = (\rho \cdot k^* + p_l \cdot l^* + p_m \cdot m^*)/x^* \qquad (1.2)$$

41

where k, l and m are the inputs of capital, labour and materials, respectively, x is output, ρ is the target rate of return, a subscripted p is a factor price, and where the superscript * denotes the 'standard' or 'normal' value of a variable. Second, *full-cost pricing* states that the firm sets its price by applying a mark-up to standard unit variable cost:

$$p = (1 + \mu) \cdot (p_l \cdot l^* + p_m \cdot m^*) / x^* \qquad (2.3)$$

where μ is the mark-up.

Empirical researchers have freely drawn from both theories, combining their elements into one equation. They often specified it in rate-of-change form, explaining price changes from cost changes and demand factors. The inclusion of demand factors was commonly justified by the *ad hoc* assumption that the target rate of return or the mark-up would depend on a variable like capacity utilisation or the ratio of unfilled orders to sales. A major theme in the early empirical research was the relationship between market form and price formation. In particular, both theories discussed above suggested that demand factors would be more important in competitive industries, while cost factors would be more important in oligopolistic industries. Moreover, prices would be more sticky in oligopolistic industries.

Concern over the international transmission of inflation in the high-inflation era of the seventies added a variable to the price equation. Deardorff and Stern (1978) suggested that the prices of foreign competitors might have a direct impact on domestic prices, because domestic producers would raise their profit margins if competitors' prices rose. Calmfors and Herin (1979) incorporated this effect into their model by assuming in an *ad hoc* way that the mark-up is a homogeneous-of-degree-zero function of all prices.

Microeconomic models of imperfect competition have yielded theoretical advances in the specification of price equations. The results generally supported the early formulations, although some exceptions may be noted. For example, Nordhaus (1972) showed that target-rate-of-return pricing, which was believed to apply to oligopolistic industries, was theoretically justified only under the conditions of constant returns to scale in production and perfect competition. Over time, the theories became more sophisticated. Nordhaus (1972), de Menil (1974) and Maccini (1978) still opted for a simple static model without explicit uncertainty, but both Iwai (1974) and Bruno (1979) included uncertainty and investigated the consequences of firms' forecasting errors for short-run price dynamics. Maccini (1981, 1984), abstracting from explicit uncertainty and focusing on the effects of permanent shocks, specified an intertemporal optimisation problem to study the links between the firm's output, price and investment decisions. Others used intertemporal models to derive price-smoothing policies resembling normal-cost pricing for firms facing transitory shocks (for an overview, see Encaoua and Geroski (1986, pp. 48–52)). All these studies assume Bertrand competition, where firms are price-setting oligopolists or monopolistic competitors. The central result is that the optimal price follows from marking-up marginal cost; the mark-up depends on the price elasticity of demand, which is commonly treated as a *constant*.

The present paper reconsiders the implications of imperfect competition for aggregate price equations. It differs from most previous studies in several respects. First, it takes the non-constancy of the elasticity of demand as the main difference with the case of perfect competition, like Nieuwenhuis (1986) and Zeelenberg (1986) have done. Second, it studies not only Bertrand competition but also Cournot competition. Third, it deals explicitly with aggregation. The paper adopts a simple comparative-statics approach but retains some quasi-dynamic elements in the treatment of marginal cost by distinguishing the long run, where all factor inputs are variable, from the short run, where capital is fixed. Its fourth contribution is an analysis of the short-run marginal-cost function, using the theory of rationing to relate it to the long-run marginal-cost function.

The paper does not present new empirical research, but it does discuss existing results with an emphasis on two issues, the role of foreign-competitors' prices and the specification of demand factors. The reasons are, respectively, that recent model comparisons have revealed that the elasticities of foreign prices in the aggregate price equations have a strong impact on the reduced-form elasticities of a macro model, and that concern over hysteresis on product markets has revived an old debate, that on the level effect versus the rate-of-change effect of demand factors.

Section 2.2 briefly reviews some results from the microeconomic theory of consumer and producer behavior, and Section 2.3 deals with aggregation. Section 2.4 presents the theoretical results on aggregate price equations under Bertrand and Cournot competition. Section 2.5 confronts the theory with existing empirical results. Section 2.6 concludes.

2.2 Micro theory

2.2.1 Consumer behaviour

Let $\mathbf{p} = \{p_i\}$ denote the vector of market prices and $\mathbf{q} = \{q_i\}$ the vector of quantities consumed by some person with income Y; the subscript i

identifies consumer items. Every consumer maximises her utility U, $U = u(\mathbf{q})$, subject to the budget constraint $\mathbf{p}'\mathbf{q} = Y$. The optimal quantities are given by the consumer's *demand functions*, $\mathbf{q}_i = g_i(\mathbf{p}, Y)$, $i = 1, \ldots, I$. Their local properties are summarised by the matrix E of price elasticities, $E \equiv \{e_{ij}\} \equiv \{\partial \ln g_i / \partial \ln p_j\}$, and the vector $\mathbf{e}$ of income elasticities, $e \equiv \{e_i\} \equiv \{\partial \ln g_i / \partial \ln Y\}$. The theory entails a number of restrictions on these elasticities. For the present purpose it suffices to mention that

$$\sum_j e_{ij} + e_i = 0, \qquad i = 1, \ldots, I \qquad (2.4)$$

which says that the demand functions are homogeneous of degree zero in $\mathbf{p}$ and Y. In general, the price and income elasticities depend on the level of prices and income. They inherit zero-degree homogeneity in $\mathbf{p}$ and Y from the demand functions. For a proof, differentiate (2.4) partially with respect to $\ln p_n$ and observe that $\partial e_{ij}/\partial \ln p_n = \partial e_{in}/\partial \ln p_j$ and $\partial e_i/\partial p_n = \partial e_{in}/\partial \ln Y$. Hence

$$\sum_j \partial e_{in}/\partial \ln p_j + \partial e_{in}/\partial \ln Y = 0, \qquad i, j, n = 1, \ldots, I \qquad (2.5)$$

The elasticities are constants in special cases only: $\mathbf{e} \equiv 1$ if and only if the consumer's direct utility function is homothetic to the origin, and $E \equiv -I$ (minus the identity matrix) if and only if the utility function has the Cobb-Douglas form.

The consumer's behaviour may equivalently be described by her inverse demand functions $p_i/Y = f_i(\mathbf{q})$, $i = 1, \ldots, I$, which give the prices that, at income Y, she is willing to pay for a given basket of goods. The local properties of these functions are summarised by a number of elasticities; the most important ones constitute the matrix D of quantity elasticities, $D \equiv \{d_{ij}\} \equiv \{\partial \ln f_i / \partial \ln q_j\}$, where $D = E^{-1}$. In general, the quantity elasticities depend on the levels of the quantities consumed of all products (which, in their turn, are homogeneous of degree zero in $\mathbf{p}$ and Y).

2.2.2 Producer behaviour

Every commodity is produced by a single firm. Each firm maximises its profits subject to the constraints imposed by its production technology and market environment. Factor markets are competitive so that each firm takes the factor prices as given. Product markets, in contrast, feature imperfect competition. Profit maximisation proceeds in steps. In the first step, the firm forms expectations on demand and factor prices and determines the expected cost-minimising factor quantities. At the beginning of the planning period, the firm adjusts capital to its desired level; during the period, capital is fixed, but the firm retains the flexibility to adapt the other inputs to surprises in their prices and in demand. In the second step, the firm minimises its variable costs at the realised prices of the variable factors and the predetermined quantity of capital. In the third step, the firm determines the profit-maximising quantity and price of its product.

2.2.2.1 Marginal cost

The firm uses three (groups of) production factors, capital k, labour l and materials m. For simplicity's sake, the model abstracts from technical change and non-constant returns to scale in production. Dual to the long-run production function $x = h(k, l, m)$ there is a cost function C

$$C = C(p_k, p_l, p_m, x) = c(p_k, p_l, p_m)x \qquad (2.6)$$

The unit cost function $c(\cdot)$ is linearly homogeneous in its arguments. It represents both long-run average cost ACL ($\equiv C/x$) and long-run marginal cost MCL ($\equiv \partial C/\partial x$). Logarithmic differentiation yields

$$\dot{ACL} = \dot{MCL} = c_k\dot{p}_k + c_l\dot{p}_l + c_m\dot{p}_m \qquad (2.7)$$

where $c_r \equiv \partial \ln c/\partial \ln p_r$, the cost share of factor r, and $\dot{a} \equiv d\ln a$.

The long-run factor demand functions follow from Shephard's Lemma, $r = \partial C/\partial p_r$. In rate-of-change form they read

$$\dot{r}^e = \zeta_{rk}^e \dot{p}_k^e + \zeta_{rl}^e \dot{p}_l^e + \zeta_{rm}^e \dot{p}_m^e + \dot{x}^e, \qquad r = k, l, m. \qquad (2.8)$$

The superscript e denotes the expected value of an exogenous variable or the expected optimal value of an endogenous variable. The ζ_{rs} correspond to the compensated price elasticities in the theory of consumer demand. They satisfy the following restrictions

$$\sum_s \zeta_{rs} = 0 \qquad (2.9)$$

$$\zeta_{rs} = c_s\sigma_{rs} \qquad (2.10)$$

$$\sigma_{rs} = \sigma_{sr} \qquad (2.11)$$

for $r, s = k, l, m$, with σ_{rs} the Allen partial elasticity of substitution between the factors r and s.

The dual of the short-run production function $x = h(k^e, l, m)$ is the variable-cost function $VC = VC(p_l, p_m, x, k^e)$. The variable-factor demand functions again follow from Shephard's Lemma, $n = \partial VC/\partial p_n \equiv VC_n, n = l, m$, while short-run marginal cost MCS is $\partial VC/\partial x \equiv VC_x$. In rate-of-change form

$$M\dot{C}S = \gamma_l \cdot \dot{p}_l + \gamma_m \cdot \dot{p}_m + \gamma_x \cdot \dot{x} + \gamma_k \cdot \dot{k}^e \qquad (2.12)$$

Consider first the elasticities γ_l and γ_m. Because $VC_l = l$, $\partial l/\partial x = \partial VC_x/\partial p_l$. Now, $\gamma_l \equiv \partial \ln VC_x/\partial \ln p_l = (p_l/VC_x) \cdot (\partial l/\partial x) = p_l \cdot (\partial l/\partial VC)$. Similarly, $\gamma_m = p_m \cdot (\partial m/\partial VC)$. So, γ_l and γ_m are *marginal shares of variable cost* and sum to one.[1]

In order to relate the elasticities of the MCS-function to those of the MCL-function, take the long-run capital demand function from (2.8) and substitute for $\dot{k}^e$ in (2.12)

$$\begin{aligned} M\dot{C}S = & \;\gamma_l \cdot \dot{p}_l + \gamma_m \cdot \dot{p}_m + \gamma_x \cdot \dot{x} + \\ & \gamma_k \cdot (\zeta_{kk}^e \cdot \dot{p}_k^e + \zeta_{kl}^e \cdot \dot{p}_l^e + \zeta_{km}^e \cdot \dot{p}_m^e + \dot{x}^e) \end{aligned} \qquad (2.13)$$

Suppose first that the firm's expectations were correct: $\dot{p}_r = \dot{p}_r^e, r = k, l, m$, $\dot{x} = \dot{x}^e$ and hence, $M\dot{C}S = M\dot{C}L^e$. The coefficients of corresponding variables in (2.7) and (2.13) must be equal. This yields

$$\zeta_{kk} = \frac{c_k}{\gamma_k}, \quad \gamma_x = -\gamma_k \qquad (2.14)$$

$$\zeta_{kr} = \frac{c_k - \gamma_r}{\gamma_k}, \quad r = l, m \qquad (2.15)$$

Solve for the elasticities of the MCS-function and use (2.10) to find

$$\gamma_k = -\gamma_x = \frac{1}{\sigma_{kk}} \qquad (2.16)$$

$$\gamma_r = c_r \frac{1 - \sigma_{kr}}{\sigma_{kk}}, \qquad r = l, m \qquad (2.17)$$

The superscript e has been deleted for notational ease.

The assumption that capital is fixed in the short run means that the producer may be rationed. In case the firm's expectations turn out to be correct, k^e is in fact optimal *ex post*; one might say that the ration 'just' bites. Neary and Roberts (1980) have studied the case that a ration strictly bites. They conclude that (2.14-2.15) holds good provided that the elasticities are evaluated at the *virtual prices*, that is, the prices that at production level x would induce the producer to voluntarily use the factor quantities k^e, l and m. They show that such prices exist and that for unrationed factors they coincide with the actual prices. The virtual price of capital services is given by $p_k = -\partial VC/\partial k^e$.

Substitution of (2.14-2.15), adorned with superscripts e, into (2.13) yields

$$\dot{MCS} = \gamma_l \cdot \dot{p}_l + \gamma_m \cdot \dot{p}_m - \sigma_{kk}^{-1} \cdot (\dot{x} - \dot{x}^e) + \\ \sigma_{kk}^{-1} \cdot \sigma_{kk}^e \cdot \{c_k^e \cdot \dot{p}_k^e + (c_l^e - \gamma_l^e) \cdot \dot{p}_l^e + (c_m^e - \gamma_m^e) \cdot \dot{p}_m^e\}$$

With approximately constant elasticities this may be simplified to

$$\dot{MCS} \approx \dot{MCL}^e + \gamma_l^e \cdot (\dot{p}_l - \dot{p}_l^e) + \gamma_m^e \cdot (\dot{p}_m - \dot{p}_m^e) - (\sigma_{kk}^e)^{-1} \cdot (\dot{x} - \dot{x}^e)$$
(2.18)

In words: the change in short-run marginal cost is approximately equal to the expected change in long-run marginal cost corrected for the extent to which production and the variable-factor prices deviate from their expected values. Capacity utilisation will be an excellent measure of the ratio x/x^e.

Smooth neutral technical change leads to a negative constant term in the equations for the rate of change of marginal cost. Increasing returns to scale adds a term in with a negative elasticity. With aggregate data, it would now be necessary to decompose changes in industry output into changes in the number of firms and in output per firm.

2.2.2.2 Marginal revenue

Firm i's profit is the excess of revenues over costs, $\Pi_i \equiv R_i - C_i$. Revenues are the product of the price and quantity of its product, $R_i \equiv p_i \cdot x_i$, while costs consist of VC_i and a fixed component FC_i representing capital costs:

$$\Pi_i (p_i, q_i) \equiv p_i \cdot x_i - VC_i (x_i) - FC_i, \qquad i = 1, \dots, I \qquad (2.19)$$

Production equals sales, which follow from the market demand functions $x_i = G_i(\mathbf{p}), i = 1, \dots, I$, which may be inverted to yield the inverse market demand functions $p_i = F_i(x), i = 1, \dots, I$. Consider the following set of maximum problems:

$$\max_{p_i \text{ or } x_i} \Pi_i \equiv p_i \cdot x_i - VC_i (x_i) - FC_i, \qquad i = 1, \dots, I \qquad (2.20)$$

subject to

$$x_i = G_i (\mathbf{p})$$

or , equivalently, $p_i = F(\mathbf{x})$. An important feature of this problem is that the $2I$ arguments of the maximands are interrelated through the constraints. Therefore, the I problems of profit maximisation are interdependent. Elimination of I variables shows that all maximands have the same list of arguments. Two cases merit special attention. Eliminating either the quantities or the prices from all maximands yields, respectively,

$$\max_{p_i} \Pi_i \equiv p_i \cdot G_i(\mathbf{p}) - VC_i(G_i(\mathbf{p})) - FC_i \qquad (2.21)$$

$$\max_{x_i} \Pi_i \equiv F_i(\mathbf{x}) \cdot x_i - VC_i(x_i) - FC_i, \qquad i = 1, \ldots, I \ (2.22)$$

As the elimination of variables does not really affect any problem, (2.20), (2.21) and (2.22) are mathematically equivalent and must yield identical solutions for **p** and **x**. Actually, if one properly accounts for the interdependence between the individual problems at the optimisation stage, one arrives at the *Contract Curve* (a cooperative solution), which is invariant under nonsingular transformations of the arguments.

It is common practice in economic theory, however, to ignore the simultaneity at the optimisation stage. The overall problem is split into a number of conditional problems, in each of which only firm i's own instrument is treated as an argument. The results thus obtained would apply if the firms behaved noncooperatively. (2.21) and (2.22) are two polar cases. Under *Cournot competition*, firms use their *quantities* as instruments, *taking the quantities of their rivals as given*. The 'first-order condition' for maximal profit of firm i states that (short-run) marginal cost equals marginal revenue:

$$MCS_i(x_i) = MR_i^C(\mathbf{x}) \equiv F_i(\mathbf{x})(1 + \delta_{ii}), \qquad i = 1, \ldots, I \quad (2.23)$$

where $\delta_{ii} \equiv \partial \ln F_i / \partial \ln x_i$. Under *Bertrand competition*, firms set *prices* for their products, *taking the prices of their rivals as given*. The 'first-order condition' is now

$$MCS_i(G_i(\mathbf{p})) = MR_i^B(\mathbf{p}) \equiv p_i\left(1 + \varepsilon_{ii}^{-1}\right), \qquad i = 1, \ldots, I \tag{2.24}$$

where $\varepsilon_{ii} \equiv \partial \ln G_i / \partial \ln p_i$.

Perfect competition on the market for a good prevails if it is produced by a large number of small firms, all of which take the market price p_i of the good as given. Marginal revenue is now equal to p_i, and profit maximisation implies that the firms produce up to the point where marginal cost equals p_i:

$$MCS_{is} = MR_{is} \equiv p_i, \qquad s = 1, \ldots, S_i \qquad (2.25)$$

with S_i the number of firms producing good i. These are the individual supply functions. One additional assumption suffices for the derivation of a market price equation. Multiply both sides of (2.25) by x_{is}, sum over s and divide by x_i:

$$\sum_s v_{is} \cdot MC_{is} = p_i \qquad (2.26)$$

where $v_{is} \equiv x_{is}/x_i$. Now, strong competition forces the firms to use similar techniques. In the limiting case where they use identical techniques,

$$p_i = MC_i \left(x_i / S_i \right) \qquad (2.27)$$

is the inverse market supply function or *price equation* for a competitively traded good.

2.2.3 Models of product markets

In a partial competitive equilibrium context, semi-reduced form equations for p_i and x_i follow from (2.27) and the market demand function $G_i(\mathbf{p})$. In a general competitive equilibrium model, the union of $G_i(\mathbf{p})$ and (2.27) for $i = 1, \ldots, I$, describes all product markets. An equivalent approach combines the inverse market demand functions $F_i(\mathbf{x})$ with the market supply functions (the solutions for x_i of (2.27)).

Likewise, the product markets in models featuring imperfect competition may be described in several equivalent ways. In the case of Cournot competition it seems natural to combine the inverse market demand functions with (2.23), whereas in the case of Bertrand competition it seems natural to combine the market demand functions with (2.24). However, our present concern is the comparison of models featuring alternative modes of producer behaviour in a world with price-taking consumers. As (2.23-2.25) show, the crucial differences reside in the form of the marginal-revenue functions. They stand out most clearly if the market demand functions describe the demand side and (2.23-2.24) are solved for the prices at which firms are willing to supply a given, as yet arbitrary quantity of its product. For this reason, the term *price equation* will denote the solution for p_i of the first-order condition $MC_i = MR_i$ *at a given level of marginal cost*. This deviates from much of the literature, where the term refers to the semi-reduced form equation alluded to in the opening sentence of this subsection.

(2.23-2.24) can easily be rewritten to an equation which expresses that the optimal price follows from marking-up marginal cost: $p_i = (1 + \mu_i) \cdot MC_i$. In the literature on price formation, a number of *ad hoc* assumptions on the mark-ups have been made, for example that they are constant or depend on the degree of capacity utilisation. The approach below deviates from this tradition in that it explicitly relates the properties of the mark-ups or, equivalently, the marginal-revenue functions to the microeconomic theory of consumer behavior. In particular, it acknowledges that the price and quantity elasticities depend on all prices, so that the rewritten versions of (2.23-2.24) are *not* price equations in the sense just defined. The true price equations are presented in Section 2.4, aggregation issues being dealt with first in Section 2.3.

2.3   Aggregation

Usually, one just ignores the hard problem of aggregating microeconomic relationships and applies micro results to aggregate data without much further ado (Bruno, 1979 is an exception). The best solution is to avoid aggregation. Approximation of this ideal requires the construction of a detailed, large-scale general-equilibrium model, in most other cases the problem must be faced. But our interest is not in aggregation itself. The justification for the assumptions below is that they enable us to obtain analytical results; weaker assumptions may suffice for other purposes. Unavoidably, approximations are involved. Their quality is enhanced by empirical regularities like the stability of income distributions and the similar movements of the prices of similar products.

2.3.1   Aggregation over consumers and over consumer items

For simplicity's sake, all demand is treated as consumer demand. All consumers are identical, so market demand is a multiple of an individual's demand and market prices depend on *per capita* quantities. Hence, the elasticities of the ordinary and inverse market demand functions coincide with their microeconomic counterparts.

Most empirical research deals with product groups but treats them as if they were elementary products, talking about their prices and quantities and juggling with them as if their existence is self-evident. This practice implicitly assumes that the utility function of the consumer is *homothetically weakly separable* with respect to some partitioning of the elementary products. It seems best to make the assumption explicit and to exhaust its implications. By a change of notation, the subscript i now denotes a product group. The function $q_i = q_i(q_{i1}, \ldots, q_{is}, \ldots, q_{iS})$ is a linearly homogeneous quantity aggregator for the group. The linearly homogeneous function $p_i = p_i(p_{i1}, \ldots, p_{is}, \ldots, p_{iS})$ is the dual price aggregator. In fact, p_i represents the minimal cost of 'producing' one unit of the composite product i from its constituents. The product of p_i and q_i equals Y_i, the expenditures on the group. Demand functions for composite products exist just like for elementary products. Homothetic weak separability implies a number of restrictions on the price and quantity elasticities. In particular, the own-price elasticities satisfy

$$e_{is,is} = w_s^i \cdot \left(\sigma_{ss}^i + e_{ii} \right) \tag{2.28}$$

with $w_s^i \equiv p_{is} \cdot q_{is}/Y_i$, σ_{ss}^i the Allen own partial elasticity of substitution of the product *within group* i and e_{ii} the own-price elasticity of group i. This relationship has a clear economic interpretation. First, a rise of p_{is} causes substitution between the 'factors' involved in the 'production' of composite i. The effect on q_{is} for a given level of q_i is given by a *conditional* compensated own-price elasticity, the term $w_s^i \cdot \sigma_{ss}^i$. Second, the unit cost p_i increases due to the rise of p_{is}, which causes substitution between the product groups at the given budget; the concomitant effect on q_{is} is measured by the term $w_s^i \cdot e_{ii}$. Typically, the elasticities $e_{is,is}$ will be much larger in absolute value than e_{ii}, to which our empirical knowledge refers. Unobserved competition between firms within the industry accounts for the difference. Therefore, inelastic industry demand is still compatible with realistically low mark-ups.

Duality implies an analogous expression for the own-quantity elasticities:

$$d_{is,is} = w_s^i \cdot \left(\tau_{ss}^i + d_{ii} \right) \tag{2.29}$$

with τ_{ss}^i the Hicks own partial elasticity of complementarity of the product *within group* i and d_{ii} the own-quantity elasticity of group i.

2.3.2 Aggregation over producers

Each good being produced by just one firm, the partitioning of goods into separable groups in the consumer's utility function induces an identical partitioning of firms. The firms in one group constitute an industry. Aggregation over the firms in an industry uses the expressions (2.28) and (2.29) for the own-price and own-quantity elasticities.

2.3.2.1 Aggregate marginal revenue under Bertrand competition

Multiply the first-order condition $MR_{i,s}^B = MC_{is}$ by q_{is}, sum over s and divide by q_i:

$$p_i \cdot \left(1 + \sum_s w_s^i \cdot e_{is,is}^{-1} \right) = \sum_s q_{is} \cdot MC_{is}/q_i \equiv MC_i \tag{2.30}$$

MC_i is a weighted *sum* (not a weighted *average*, unless the products are perfect substitutes) of the firms' marginal costs. Still, it is taken to represent the industry's marginal cost. The change of aggregate marginal cost may be approximated by substituting industry values for the variables in the expressions for marginal cost in Section 1.

Now, (2.28) is equivalent to $w_s^i/e_{is,is} = 1/(\sigma_{ss}^i + e_{ii})$. Sum over s to obtain

$$\sum_s w_s^i \cdot e_{is,is}^{-1} = \sum_s \left(\sigma_{ss}^i + e_{ii}\right)^{-1} \equiv S_i \cdot \left(\bar{\sigma}^i + e_{ii}\right)^{-1} \qquad (2.31)$$

where S_i is the number of firms in the industry and $\bar{\sigma}^i$ is implicitly defined as some average of the own Allen partial elasticities of substitution σ_{ss}^i. Use this result to rewrite the left-hand side of (2.30) to[2]

$$MR_i^B \equiv p_i \cdot \left(\left(\bar{\sigma}^i + e_{ii} + S_i\right)/\left(\bar{\sigma}^i + e_{ii}\right)\right) \qquad (2.32)$$

the industry's marginal-revenue function under Bertrand competition. In an approximate sense,[3] $\bar{\sigma}^i$ (which in general depends on S_i) and S_i characterise the internal competitive structure of the industry, while the external relations with the other industries have their impact through the own-price elasticity e_{ii}.

2.3.2.2 Aggregate marginal revenue under Cournot competition

Multiply the first-order condition $MR_{is}^C = MC_{is}$ by q_{is}, sum over s and divide by q_i:

$$p_i \cdot (1 + \sum_s w_s^i \cdot d_{is,is}) = \sum_s q_{is} \cdot MC_{is}/q_i \equiv MC_i.$$

Use (2.29) to rewrite the sum term in the expression on the left-hand side:

$$\begin{aligned}
\sum_s w_s^i \cdot d_{is,is} &= \sum_s \left(w_s^i\right)^2 \cdot \left(\tau_{ss}^i + d_{ii}\right) \\
&\equiv \left(\sum_s \left(w_s^i\right)^2\right)\left(\bar{\tau}^i + d_{ii}\right) \qquad (2.33) \\
&\equiv H_i \cdot \left(\bar{\tau}^i + d_{ii}\right) \equiv \left(\bar{\tau}^i + d_{ii}\right)/\bar{S}_i.
\end{aligned}$$

H_i is the Herfindahl-Hirschman concentration index of industry i; $\bar{S}_i \equiv 1/H_i$ may be called the 'effective' number of firms in the industry; $\bar{\tau}^i$ is implicitly defined as some average of the own Hicks partial elasticities of complementarity τ_{ss}^i. Use this result to write[4]

$$MR_i^C \equiv p_i \left(1 + H_i \cdot (\bar{\tau}^i + d_{ii})\right) = p_i(\bar{\tau}^i + d_{ii} + \bar{S}_i)/\bar{S}_i \qquad (2.34)$$

the industry's marginal-revenue function under Cournot competition. $\bar{\tau}_i$

(which in general depends on S_i) and $\bar{S}_i$ characterise the internal competitive structure of the industry, while the external relations with the other industries have their impact through the own-quantity elasticity d_{ii}.[5]

2.4 Aggregate price equations: theory

Our analysis focuses on marginal cost and on the interrelationship between industries, to the neglect of the influence of the industry's market structure on price. The results do convey, however, how the elasticities of demand and cost factors depend on market structure. Of course, one must bear in mind that in general all elasticities depend on the point of approximation in price-income space.

2.4.1 Present results

For better legibility, but without a real loss of generality, the price equations below refer to the case of two industries. In the equation for p_1, p_2 may be a weighted average of the prices of all other products. In a macro model, p_1 would be the price of domestic goods and p_2 the price of foreign goods. The general form of the aggregate price equation is

$$\dot{p}_1 = \pi_{MC} M\dot{C}S_1 + \pi_2 \dot{p}_2 + \pi_Y \dot{Y} \tag{2.35}$$

Under Bertrand competition,

$$
\begin{aligned}
\pi_{MC} &\equiv (\bar{\sigma}^1 + e_{11} + S_1)(\bar{\sigma}^1 + e_{11})/B \\
\pi_2 &\equiv S_1 \cdot (\partial e_{11}/\partial \ln p_2)/B \\
\pi_Y &\equiv S_1 \cdot (\partial e_{11}/\partial \ln Y)/B \\
B &\equiv (\bar{\sigma}^1 + e_{11} + S_1) \cdot (\bar{\sigma}^1 + e_{11}) - S_1 \cdot \partial e_{11}/\partial \ln p_1)
\end{aligned}
$$

Under Cournot competition,

$$
\begin{aligned}
\pi_{MC} &\equiv (\bar{\tau}^1 + d_{11} + \bar{S}_1)/C \\
\pi_2 &\equiv -(\partial d_{11}/\partial \ln p_2)/C \\
\pi_Y &\equiv -(\partial d_{11}/\partial \ln Y)/C \\
C &\equiv (\bar{\tau}^1 + d_{11} + \bar{S}_1) + \partial d_{11}/\partial \ln p_1)
\end{aligned}
$$

Some properties of the aggregate price equation hold independently of the type of competition.

Expression (2.35) is linearly homogeneous in marginal cost, competitors' prices and *per capita* income: $1 = \pi_{MC} + \pi_2 + \pi_Y$. This property derives from the linear homogeneity in industry prices and *per capita* income of the aggregate marginal-revenue functions.

- If the representative consumer's utility function is homothetic to the origin in the space of industry aggregates, $\pi_Y \equiv 0$.
- If utility is a Cobb-Douglas function of the industry aggregates, then $\pi_2 \equiv 0 \equiv \pi_Y$, so that $\pi_{MC} \equiv 1$.
- π_{MC} may exceed one (for an example, see Nieuwenhuis, 1986). Hence, π_2 and/or π_Y may be negative.
- If the industry's market structure approaches perfect competition, industry price approaches marginal cost, π_{MC} approaches one, and π_2 and π_Y approach zero.

Other properties do depend on the type of competition.

- Bertrand competition is more competitive than Cournot competition in the sense that the mark-up is lower for a given specification of the demand side. The level of the mark-up may be controlled by varying the parameters describing unobserved competition within the industry at given elasticities of industry demand.
- Under Bertrand competition, the denominator B in the expressions for the elasticities is large so that π_2 and π_Y tend to be small. With (nested) CES-type utility functions to describe the demand side, they are even negligible. However, a counterexample provided by Broer (1996) shows this is not generally true: in case of a homothetic Almost Ideal Demand System,[6]

$$\pi_2 = (1 + e_{11}) / (1 + e_{11} + e_{1s,1s}) \tag{2.36}$$

where e_{11} is the price elasticity of demand at the industry level and $e_{1s,1s}$ the price elasticity at the (representative-)firm's level. For $e_{11} = -3$ and $e_{1s,1s} = -9$ this yields $\pi_2 = 2/11 \approx 0.18$.
- For a given specification of the demand side, under Cournot competition the elasticities π_2 and π_Y may deviate farther from zero than under Bertrand competition because their denominator C is much smaller than B. However, the need to control the level of the mark-up reduces the scope for generating large values of π_2.

This analysis reveals that non-zero elasticities on competitors' prices and *per capita* income signal imperfect competition, that these elasticities are not likely to be large and approach zero if the industry's market structure approaches perfect competition.

2.4.2 Comparison with previous studies

Price equations occurring in the literature have much in common with our results, although there are differences of detail. In one respect, however, the similarity is often merely superficial, hiding an important conceptual difference. As mentioned above, most studies assume the elasticity of demand to be constant and thus obtain the result that price is proportional to marginal cost. The variables appearing in these price equations (for example, factor prices, productivity, capacity utilisation) thus enter through marginal cost. Still, several studies (for instance, Nordhaus, 1972, Maccini, 1978, 1981, and Bruno, 1979) present equations that include some aggregate price index and the level of aggregate income (or demand), which correspond to p_2 and Y, as explanatory variables. However, these are no structural price equations as defined above. The equation for price in these studies is the reduced form for p_1 of the two-equation model consisting of the demand function, $x_1 = g_1(p_1, p_2, y)$, and the first-order condition for maximal profits, $p_1 = (1 + \mu_1) \cdot MC_1(x_1)$. With non-constant returns to scale, marginal cost depends on the level of production, which, in turn, depends on all prices through the demand function. The elasticity on p_2 in the (semi-)reduced-form equation has quite a different interpretation from π_2 in (1). Indeed, even if π_2 is zero, the reduced-form elasticity may be quite large and even approaches one if the price elasticity in the aggregate demand function approaches minus infinity.

As mentioned in the introduction, Deardorff and Stern (1978), in summing up possible channels for the international transmission of inflation, suggested that price-setting firms may raise their profit margins if foreign competitors raise their prices. Calmfors and Herin (1979) incorporated this effect into their model by assuming that the mark-up homogeneous of degree zero in all prices. Nieuwenhuis (1986) and Zeelenberg (1986) appealed to the microeconomic theory of consumer behaviour and noted that it implies that the mark-up is homogeneous of degree zero in all prices and *per capita* income. Section 2.5.3 contains an extensive discussion of the role of competitors' prices. Until then, the focus is on the relationship between price and (marginal) cost to the neglect of the direct effect of p_2 and Y, like in the greater part of the literature.

It appears that many price equations occurring in the literature can be rationalised with a reference to marginal cost alone. The rate-of-change form of (2.2), the pricing formula of target-rate-of-return pricing, is very much like (2.7), the change of long-run marginal cost, with the target rate of return ρ replacing p_k. Of course, this formal similarity hides a conceptual difference. Whereas p_k is just the user cost of capital, ρ would represent a whole array of factors. According to Eckstein and Fromm (1968, pp. 1165),[7]

> The target rate of return is based on market structure and long-
> run economic conditions of the industry, including barriers to
> entry, international trade barriers, concentration, product dif-
> ferentiation, the necessary quantity and quality of managerial
> talent, long-run demand elasticities, the degree of risk attached
> to the profits, and the valuation placed on the firm's equity and
> debt instruments in the capital market.

However, this conceptual difference has had little practical consequences:
empirical researchers have tended to avoid using both ρ and p_k because they
are not directly observed and reliable proxies are hard to obtain.

This brings us to a comparison with full-cost pricing (2.3). It shares with
(2.12) the feature that only the prices of the variable factors appear explicitly
in the price equation. But whereas their weights are the *average* shares of
standard unit variable cost in (2.3), they are the *marginal* shares of *variable* cost in (2.12). A clear difference is the presence of the output-capital
ratio as an additional explanatory variable in (2.12). But again, these formal
differences have had little practical consequences. As to the coefficients of
the variable-factor prices, these have usually been estimated freely. As to
the output-capital ratio, whereas full-cost pricing in its pure form states that
price is independent of demand conditions, in applications invariably some
demand factor has been added to the equation in order to test this hypothesis;
the common rationale for this practice is that the mark-up would depend on
the state of demand. A popular demand factor is capacity utilisation, which
will be strongly correlated with the output-capital ratio for many industries.
With this modification, full-cost pricing comes close to short-run marginal-
cost pricing.

As a further test of the assumption that (oligopolistic) firms base their
prices on standard cost considerations, Eckstein and Fromm (1968) include
not only standard unit labour cost ULC^* but also the difference between
actual and standard unit labour cost into their equation, which in full reads:

$$p = \alpha_1 ULC^* + \alpha_2 p_m + \alpha_3(ULC - ULC^*) + \alpha_4(x/x^*) + \alpha_5 \quad (2.37)$$

The similarity to (2.18) is remarkable.

In conclusion, the early theories have not erred grossly in the specification of the price equations. Still, the somewhat insecure theoretical base has
contributed to a lack of clarity on several issues, in particular the impact of
imperfect competition on price formation.

2.5   Aggregate price equations: empirics

2.5.1   Dynamic specification

In empirical research, the dynamic specification of the equations to be estimated deserves attention. This is particularly true for the analysis of price formation, where an old and venerable theme is the rigidity of prices in oligopolistic industries as compared to more competitive environments. Several approaches exist.

The first and oldest approach takes the equilibrium relationship from a purely static model and modifies it in an *ad hoc* way by allowing for discrete and distributed lags. A popular practice has been to include the lagged value of the dependent variable in the right-hand side of the equation. This specification is the partial-adjustment mechanism. Its justification is that due to adjustment costs, each period only part of the gap between the actual and target value of the left-hand variable is closed.

The second approach is at the other extreme and specifies an intertemporal optimisation model for obtaining the behavioural equations. The advantage of this method is that it provides a firm theoretical footing for the equations to be estimated, ensuring consistency between short-run and long-run equilibria. Its main disadvantage is the inherent intractability due to the difficulty of finding closed-form solutions in other but the most simple cases. Applications of this method are still rare and mainly to be found in the analysis of factor demand. Maccini (1981, 1984) gives a theoretical analysis of price setting in an intertemporal optimising framework with adjustments costs on investments in inventories and capital, but he refrains from an empirical application.

The third, intermediate approach is called quasi-dynamic. It incorporates rigidities in a static model, but allows for changes of the fixed factors in applications of the model to time series data. In the present context, this leads to the distinction between long-run and short-run marginal cost; adjustment of the fixed factor, capital, according to its long-run demand function ensures consistency between both concepts of marginal cost. The results are qualitatively similar to those of a full-fledged intertemporal model. The equilibrium relationship now contains some medium-run elements in the form of deviations between the actual and 'normal' values of variables, as in (2.18).

In practice, the third approach is combined with the first approach. But recent advances in econometric methodology provide a preferable alternative. Short-run dynamics can be accounted for by specifying a relationship in *error-correction form*; cointegration techniques are used for retrieving both

the equilibrium relationship, provided by economic theory, and the short-run
dynamics, to be determined by the data. A simple example of the error-
correction mechanism applied to the price equation is

$$\dot{p}_t \approx \Delta \ln p_t = -\zeta e_{t-1} + \theta' \Delta \ln \mathbf{z}_t \qquad (2.38)$$

where $e_t \equiv \ln(p_t/p_t^e)$, with p_t^e the equilibrium price. Typical elements of
the vector $\mathbf{z}$ are the explanatory variables of the price equation (2.35), but
other variables may be added provided they do not affect the equilibrium.
The error term e_t is the residual of the equilibrium price equation in level
form. Insights from the statistical theory for this relatively new method are
helpful in putting into perspective some aspects of the empirical findings on
price formation to date.

2.5.2 Cost factors versus demand factors

With the present definition of 'price equation,' it is natural to split the set of
explanatory variables into *demand factors*, which stem from marginal revenue, and *cost factors*, which stem from marginal cost. However, this con-
flicts with the practice of labelling capacity utilisation as a demand factor,
which originates from the traditional rationale for its inclusion in the set of
explanatory variables. Empirical price equations have also featured other demand factors like the ratio of unfilled orders to sales or the ratio of unplanned
inventory building to sales, but these will be strongly correlated with capacity utilisation and measure essentially the same phenomenon. Cost factors
are the variables affecting unit cost, such as the factor prices. But many
studies have employed unit labour cost ULC and unit material cost UMC
instead, thus accounting for any changes in productivity due to factor substitution, technical progress, or scale economies.

The parent theories of the price equation, Walrasian tâtonnement and
target-rate-of-return pricing or full-cost pricing, suggested that demand factors would affect prices in competitive industries only. The early empirical
research on price formation focussed on tests of this hypothesis. In 1963,
Neild reported that in Britain, 'the relationship between trends in costs and
prices of manufactures shows considerable short-run stability and is not sensitive to cyclical variations in the pressure of demand' (quoted in Rushdy and
Lund, 1967). The hypothesis was rejected by Rushdy and Lund (1967) and
revived again by Godley and Nordhaus (1972). But most studies (for example, Eckstein and Fromm, 1968, de Menil, 1974, Gordon, 1975, Sahling,
1977) did find some influence of demand pressure on prices in concentrated
industries. Eckstein and Wyss (1972) reported evidence for stronger demand
effects in competitive industries than in concentrated industries, but even

this modified version of the hypothesis was contested by Straszheim and Straszheim (1976). In terms of the theory favoured in this paper, this last result means that the slopes of the short-run marginal-cost functions do not differ systematically between both classes of industry.

A general finding is that cost factors perform stronger than demand factors in the explanation of prices, both for concentrated and for competitive industries. Researchers have always had the intuitive understanding that this results from common trends in costs and prices, and so the recent theory of cointegration 'only' serves to formalise this intuition. Because (the log of) price is an integrated variable, the set of explanatory variables must contain an integrated variable like cost or else no cointegrating relationship will be found. Capacity utilisation, which by its very nature is a stationary variable, is not needed for cointegration to obtain. Its task is to pick up the cyclical component in a remainder term that is already stationary (if a cointegrating relationship between price and cost factors exists). It can never, on its own, provide an explanation for a trending variable (such a regression equation would be *unbalanced*).

A more subtle hypothesis seeks the difference between competitive and oligopolistic industries in the *speed* with which firms react to changes in cost and demand factors, rather than in the relative strength of their effects. This theory rests on two pillars. First, a policy of price smoothing will be optimal if firms face transitory demand shocks. Second, market power in concentrated industries extends the scope for price smoothing, whereas firms in competitive industries must take every opportunity to reap the profits of the day. Encaoua and Geroski (1986) give an overview of this branch of the literature and report evidence for five OECD countries that supports the theory.

According to the model of Section 2.4, the market structure of an industry affects the elasticities on marginal cost, competitors' prices and income. This link between market form and price formation has not been the object of investigation to date.

An old debate concerns the specification of the demand factor, that is, whether the *level* or the *rate of change* of capacity utilisation affects the *rate of change* of price. Walrasian tâtonnement appended to normal-cost or full-cost pricing in relative changes yields the level specification, which is analogous to the *strong* Phillips curve. However, a framework with optimising firms under certainty yields the rate-of-change specification, which is analogous to the *weak* Phillips curve; examples are this paper, but also de Menil (1974), Sahling (1977), Bruno (1979) and Maccini (1981).

For lack of a clear verdict of the early theory, many empirical researchers have taken an agnostic stance and tested both specifications. For example,

Eckstein and Fromm (1968) freely switch from the rate-of-change specification in their regressions for price levels to the level specification in their regressions for relative price changes without spending one word on the issue. On the whole, it seems fair to say that the empirical evidence favours the rate-of-change form; Gordon (1980) characterised a near- century of U.S. price behaviour with this specification. Probably, the increased reliance on microeconomic theory for the derivation of equilibrium relationships has also contributed to the predominance of this form. The Walrasian version is not dead, though. FKSEC, the currently operational macroeconometric model for the Netherlands of CPB, features price equations that contain both the rate-of-change and the level effect of capacity utilisation, continuing a long tradition.

Recent years have witnessed a revival of interest in this old issue. By analogy to the related literature on wage formation (Blanchard and Summers, 1986), the finding that the rate-of-change form characterises the data best has been interpreted as signalling the absence of strong, competitive, equilibrium-restoring market forces. Van Bergeijk *et al.* (1993) specify the following price equation,

$$\dot{p} = \alpha_1 U\dot{L}C + \alpha_2 \dot{p}_m + \alpha_3(x/x^e) + \alpha_4\Delta(x/x^e) + \alpha_5$$

and define the Market Inertia Criterium (MIC) as an indicator of the functioning of the price mechanism:

$$MIC \equiv 100\% \cdot \alpha_4/(\alpha_3 + \alpha_4) \tag{2.39}$$

A strong rate-of-change effect and a weak level effect would signal hysteresis on the goods market, which is a bad thing (Van Bergeijk *et al.*, 1993, p. 532)

> Hysteresis characterises situations in which the market mechanism does not function properly so that either excess capacity or excess demand can continue without a tendency to restore equilibrium.

Their estimates for nine OECD countries imply high values of 94% and over for the MIC of all countries except the USA and Canada, suggesting that hysteresis is a significant and widespread phenomenon.

The present paper questions this interpretation and even argues to the contrary. Our quasi-dynamic theory is quite clear on the specification of the equilibrium relationship: the influence of competitors' prices and income apart, prices move proportionally to marginal cost. The quasi-dynamic model will perform well empirically if prices react quickly to changes in marginal cost; in this case a rate-of-change effect will be found, its strength depending on the slope of the short-run marginal-cost function. If the quasi-dynamic model does not perform well, it may be adorned with additional dynamic elements through an error-correction mechanism. An appealing interpretation of the level effect is that it proxies the error-correction term. Now, fierce competition forces firms to react quickly to market changes and to maintain a close link between price and marginal cost in order to obtain maximal profits; this means that price change depends on the contemporaneous change of capacity utilisation (if the latter has any effect on marginal cost). With fast adjustment, little 'error' has to be corrected afterwards, meaning that the level term has no role to play. But when firms are shielded from competition, for example in an oligopolistic industry, they have greater discretion in their short-run pricing decisions, to the effect that the error-correction term has a more important role to play in restoring the long-run equilibrium relationship between price and marginal cost. This reasoning conforms well to the analyses of Iwai (1974) and Bruno (1979), who derive a level effect from a choice-theoretic model with uncertainty. In both models, the level term results from conscious error correction by firms; its coefficient is smaller the closer the model is to the competitive ideal and vanishes in the limit.

2.5.3   The role of competitors' prices

As to the part played by per capita nominal income, little can be said. So far, only Nieuwenhuis (1986) estimated price equations including this variable. His estimates suggest that in a number of cases its omission may be a serious misspecification, but no firm conclusion can be drawn from just one study. The remainder of this subsection assumes $\pi_Y = 0$ and discusses the role of (foreign-)competitors' prices.

2.5.3.1   Empirical evidence

Estimation results
Calmfors and Herin (1979), in a sectoral study for Sweden, distinguish three groups of industries, sheltered sectors, import competing sectors and export sectors. Estimates of π_2 for most industries in the first two groups and even for some in the last group fall in the range of -0.05 to 0.25. In the last group, however, several industries (those producing bulk goods) feature higher values (0.4 and over).

Zeelenberg (1986) constitutes the most extensive study of industrial price formation for the Netherlands. He does not treat the elasticities π_2 as constant but he presents the implied values at the sample means. The values generally exceed 0.4.

CPB's sectoral model ATHENA (CPB, 1990) of the Dutch economy contains price equations with low elasticities for foreign prices. In fact, these elasticities are zero for several industries.

In addition to sectoral studies, aggregate studies exist. Macroeconometric models of the Netherlands, from VINTAF-II (CPB, 1977) to FKSEC (CPB, 1992), feature price equations for final-demand categories with values for π_2 of about 0.5. Ormerod (1980) reports a value of 0.5 for manufacturing exports of the United Kingdom. For Dutch manufacturing, however, Draper (1985) obtained an elasticity of only just over one quarter. In a more recent endeavour, applying cointegration techniques to a simultaneous model of export volume and export price, he settled on a value of only 0.2 (Draper, 1996).

Interpretation

It thus appears that many empirical estimates of the foreign-price elasticity are higher than theory suggests. Several factors may be responsible for this. First, if the model is applied to the prices of final-demand categories, the cumulation of a number of small effects through intermediate deliveries raises the elasticity of the aggregate foreign price. However, this factor alone cannot account for the large elasticities in the macro models of CPB.

Second, estimation may fail to identify the long-run elasticities because equations are often estimated in rate-of-change form. Two notable exceptions are Ormerod (1980) and Draper (1996), who used the error-correction specification. Whereas Ormerod (1980) obtained a large long-run elasticity on foreign prices for U.K. manufacturing export prices (on which more will be said below), Draper (1996), studying the export price and volume of Dutch manufacturing, found that the short-run elasticity exceeds the long-run elasticity by a factor of two.[8]

Third, simultaneity may cause the estimates to be biased. The model underlying the industry price equations aggregates over domestic firms in the same industry, which compete on strongly interrelated markets. However, markets do not stop at national borders: domestic firms compete also with foreign firms producing similar goods. In this case, foreign prices are no longer exogenous. Hence, the strong correlation between domestic and foreign prices must be explained from common underlying causes like domestic and foreign costs. This remark applies to the results of Calmfors and Herin (1979) for the exporting industries producing bulk goods as well as to those of Zeelenberg (1986) for many industries. Marginal cost is another source of simultaneity: it is affected by capacity utilisation, which directly depends on the left-hand side variable (and on competitors' prices as well). One attempt to address this problem is to replace demand, the nominator of capacity utilisation, by the demand function. However, this changes the nature of the equation to be estimated and thus the interpretation of its elasticities. Indeed, this semi-reduced-form equation features a lower elasticity on the remaining elements of marginal cost and a higher elasticity on competitors' prices. Deleting capacity utilisation if found insignificant in single-equation estimation may yield similar effects as replacing demand by the demand function; this may explain the large foreign-price elasticity of Ormerod (1980) mentioned in the previous paragraph. Another attempt to address simultaneity bias is estimation by simultaneous-equation methods. Indeed, the low estimates of π_2 in Draper (1996) and ATHENA for the export price equations have been obtained by such methods.

Fourth, 'selectivity bias' may be part of the explanation. It is common practice to estimate a number of variants of an equation and to select one on the basis of not only statistical criteria (*e.g.*, goodness of fit) but also agreement with prior notions. Hence, the equations may have been selected that best fitted the prevailing view that a high elasticity of competitors' prices reflects strong competition. Indeed, 'price equations' of the form $p_1 = p_2$ have been incorporated in theoretical models of small open economies. In the limiting case of perfect substitution, the multiple equality $MC_1 = p_1 = p_2 = MC_2$ holds in equilibrium. However, one must distinguish between the *behavioural* relationships $p_i = MC_i$, $i = 1, 2$, and the *outcome* $p_1 = p_2$ from the full model that incorporates also demand functions for the products with a high (infinite) elasticity of substitution.

2.5.3.2 Relevance for macroeconomic modelling

Two models currently in use at the CPB are FKSEC and MIMIC. FKSEC is a traditional macroeconometric model with a firm empirical footing in time series analysis; for a description, see CPB (1992). MIMIC is an applied general-equilibrium model with a firm footing in microeconomic theory; for a description, see Gelauff and Graafland (1994). Both models contain price equations that fit the general form (2.35), and both models assume $\pi_Y = 0$. The elasticities on competitors' prices, however, differ greatly. As mentioned in the previous subsection, their values center around 0.5 in FKSEC. In MIMIC, their values are approximately 0.0 because this model assumes Bertrand competition and characterises the demand side with (nested) CES utility functions. Model comparisons revealed that this difference causes strongly divergent model outcomes. The issue may be illustrated with the help of a small-scale macro model, MINI.

MINI consists of fourteen equations, ten behavioural relationships and four definitions. Domestic absorption q is described by a system of two demand functions, one for domestic goods (q_1) and the other for foreign goods q_2, foreign absorption b is described by a traditional export equation with a (compensated) price elasticity of -2. The three long-run factor demand functions and the short-run marginal-cost function are consistent with an aggregate Cobb-Douglas production function. A *wage curve* featuring full price indexation describes the price of labour p_l, two price equations yield the price p_1 of domestic sales and p_b of exports, respectively. The definitions concern the price p of aggregate domestic absorption, the volume of gross product x, the value of national income Y and the price p of capital goods. The model contains five exogenous variables, labour supply þs and world trade m_w on the real side, and three foreign prices, the price of competing final imports p_2, the price of competing exports p_f, and the price of imported materials p_m. The model is linear in relative changes of the variables. The elasticities have been chosen in such a way that the model applies to the Dutch economy. MINI was in fact designed to mimic MIMIC.[9]

In the notation of MINI, the price equations read

$$p_1 = \gamma_{mc} \cdot mc + \gamma_2 \cdot p_2 \qquad (\gamma_{mc} + \gamma_2 = 1) \qquad (2.40)$$
$$p_b = \delta_{mc} \cdot mc + \delta_f \cdot p_f \qquad (\delta_{mc} + \delta_f = 1) \qquad (2.41)$$

Tables 2.1-2.3 present the matrices of reduced-form elasticities of several versions of MINI. The columns labeled (sumr) and (sumn) are the sum of the preceding two and three columns, respectively, and show that MINI is characterised by a number of homogeneities both in the real and in the nominal sphere. Table 2.1 presents the case $\gamma_2 = \delta_f = 0$, which corresponds to MIMIC. In fact, this version of MINI mimics MIMIC quite well. Note that the column corresponding to p_f is twice the column corresponding to m_w. Both variables appear only in the export equation in this version of the model. Hence, the corresponding columns of the matrix of reduced-form elasticities are proportional with a factor that equals the ratio of the two elasticities (here, the *Tinbergen two*). Most importantly, although the elasticities of foreign-competitors' prices in the *structural-form* price equations are zero, these variables do appear in the *reduced-form* price equations with quite high elasticities. Their relative magnitude depends on the price elasticities in the domestic and foreign demand functions. For example, increasing the price elasticity of exports (competing final imports) raises the reduced-form elasticity of p_f (p_2) of both prices (with a limiting value of one for ever increasing substitution elasticity).

Table 2.2 presents the matrix of reduced-form elasticities of MINI for the case $\gamma_2 = \delta_f = 0.5$, their approximate value in FKSEC. The changes from Table 2.1 are remarkable. For a discussion of the differences, it is best

Exogenous variable	l_s	m_w	(sumr)	p_2	p_f	p_m	(sumn)
Endogenous variable							
Wages, prices							
p_l	−0.40	0.40	0.00	0.23	0.80	−0.03	1.00
p_1, mc	−0.26	0.26	0.00	0.22	0.53	0.25	1.00
p	−0.19	0.19	0.00	0.45	0.38	0.18	1.00
p_b	−0.26	0.26	0.00	0.22	0.53	0.25	1.00
Production, final demand							
x	0.58	0.42	1.00	−0.29	0.83	−0.54	0.00
$q\,(\equiv Y - p)$	0.51	0.49	1.00	−0.51	0.99	−0.48	0.00
b	0.53	0.47	1.00	−0.45	0.94	−0.49	0.00
q_2	0.21	0.79	1.00	−1.38	1.58	−0.20	0.00
Factor demand							
k	0.51	0.49	1.00	−0.51	0.98	−0.47	0.00
l	0.72	0.28	1.00	−0.29	0.56	−0.27	0.00
m	0.32	0.68	1.00	−0.07	1.36	−1.29	0.00
Income							
Y	0.32	0.68	1.00	−0.07	1.36	−0.30	1.00

Table 2.1: Reduced-form elasticities of MINI, $\gamma_2 = \delta_f = 0$

Exogenous variable	l_s	m_w	(sumr)	p_2	p_f	p_m	(sumn)
Endogenous variable							
Wages, prices							
p_l	−0.45	0.45	0.00	0.37	0.61	0.02	1.00
p_1	−0.14	0.14	0.00	0.69	0.19	0.13	1.00
p	−0.10	0.10	0.00	0.78	0.13	0.09	1.00
p_b	−0.14	0.14	0.00	0.19	0.69	0.13	1.00
Production, final demand							
x	0.35	0.65	1.00	−0.56	0.88	−0.32	0.00
$q\,(\equiv Y - p)$	0.35	0.65	1.00	−0.85	1.17	−0.32	0.00
b	0.27	0.73	1.00	−0.45	0.63	−0.25	0.00
q_2	0.20	0.80	1.00	−1.19	1.38	−0.18	0.00
Factor demand							
k	0.17	0.83	1.00	−0.96	1.12	−0.16	0.00
l	0.53	0.47	1.00	−0.55	0.64	−0.09	0.00
m	0.08	0.92	1.00	−0.18	1.25	−1.07	0.00
Income							
Y	0.25	0.75	1.00	−0.07	1.30	−0.23	1.00

Table 2.2: Reduced-form elasticities of MINI, $\gamma_2 = \delta_f = 0.5$

to concentrate on the limiting case $\gamma_2 = \delta_f = 1.0$ presented in Table 2.3. Now, $p_1 \equiv p_2$ and $p_b \equiv p_f$, as in the textbook small-open-economy model. Contrary to conventional wisdom, an elasticity of one on foreign price in the structural form corresponds to perfect *collusion* between domestic and foreign firms, not to perfect *competition*. In this case, prices and marginal costs of domestic products are not linked at all. Lower domestic wages due to a positive labour supply shock do not feed into lower prices. In fact, domestic and foreign producers never change the relative prices of their products

Exogenous variable	l_s	m_w	(sumr)	p_2	p_f	p_m	(sumn)
Endogenous variable							
Wages, prices							
p_l	−0.51	0.51	0.00	0.57	0.36	0.07	1.00
p_1	0.00	0.00	0.00	1.00	0.00	0.00	1.00
p	0.00	0.00	0.00	1.00	0.00	0.00	1.00
p_b	0.00	0.00	0.00	0.00	1.00	0.0	1.00
Production, final demand							
x	0.10	0.90	1.00	−0.55	0.64	−0.09	0.00
$q (\equiv Y - p)$	0.18	0.82	1.00	−0.99	1.16	−0.17	0.00
b	0.00	1.00	1.00	0.00	0.00	−0.00	0.00
q_2	0.18	0.82	1.00	−0.99	1.16	−0.17	0.00
Factor demand							
k	-0.18	1.18	1.00	−1.01	0.84	0.17	0.00
l	0.33	0.67	1.00	−0.58	0.48	0.10	0.00
m	-0.18	1.18	1.00	−0.01	0.84	−0.83	0.00
Income							
Y	0.18	0.82	1.00	0.01	1.16	−0.17	1.00

Table 2.3: Reduced-form elasticities of MINI, $\gamma_2 = \delta_f = 1.0$

so that substitution between domestic and foreign products is completely absent. Accordingly, the (compensated) price elasticities in the aggregate demand functions do not affect the model outcomes so that they are not identified. Naturally, most entries of Table 2.2 have values in between those of Table 2.1 and Table 2.3.

The great differences between these three versions of MINI testify to the need of renewed empirical research into the true values of the elasticities γ_2 and δ_f.

2.6 Summary

This paper has reconsidered the implications of imperfect competition for the specification of aggregate price equations. More specifically, it has studied the two market forms of Bertrand oligopoly and Cournot oligopoly with differentiated products. Unlike most other studies, it has concentrated on the form of the marginal-revenue functions as the main difference between alternative market forms. Whereas in a perfectly competitive world a firm's marginal-revenue function is identically equal to price, it is a homogeneous-of-degree-one function of all prices and *per capita* income in imperfectly competitive markets. Consequently, price is a homogeneous-of-degree-one function of marginal cost, all other prices and *per capita* income. This result applies also at the industry level, with elasticities that depend on the industry's market structure. The fiercer competition in the industry is, the closer to one is the elasticity on marginal cost and the closer to zero are the elasticities on competitors' prices and income. Contrary to what economic theory suggests, many empirical estimates for the competing-price elasticities are quite large. The paper advanced several reasons for why these estimates may be biased upward.

The paper has also studied the short-run marginal-cost function, with capital as a fixed factor, and related its properties to those of the long-run marginal-cost function using the theory of rationing. It appears that short-run marginal cost is an increasing function of the rate of capacity utilisation, with an elasticity that is inversely related to minus the own elasticity of substitution of capital; the harder it is to substitute variable factors for capital in the long-run production function, the higher is the elasticity on capacity utilisation. Many price equations occurring in the literature can be rationalised with a reference to marginal cost alone, be it that the inclusion of capacity utilisation (or some other 'demand' factor) has been justified on different grounds. An old debate concerns the specification of the demand factor, that is, whether the *level* or the *rate of change* of capacity utilisation affects the *rate of change* of price. Here, theory and empirics do not conflict: Both favour the rate-of-change form. The paper contested the view that a weak level effect and a strong rate-of-change effect signal the absence of strong, competitive, equilibrium-restoring market forces and even argued to the contrary.

In summary, then, the view on price equations has long prevailed that the following two features signal fierce competition: *a*) a high elasticity on competitors' price and (hence) a low elasticity on marginal cost; *b*) a strong level effect and a weak rate-of-change effect of capacity utilisation. The main points this paper has tried to make are that on both scores, actually the reverse is true.

Acknowledgments

This is an abridged version of a paper presented at the workshop on 'Market Behaviour and Macroeconomic Modelling,' June 6-7, 1996, Groningen, The Netherlands. A recent result of Broer (1996) necessitated one substantive change (in Section 2.4.1). CPB Netherlands, Bureau of Economic Policy Analysis, Van Stolkweg 14, P.O. Box 80510, 2508 GM The Hague,The Netherlands, Tel. +31 70 3383380, Fax +31 70 3383350, Email ANH@CPB.NL.

68 *Ate Nieuwenhuis*

Notes

1. These marginal shares equal the corresponding average shares only if the long-run cost function satisfies a certain separability restriction
2. In a symmetric Bertrand equilibrium, (2.32) reduces to $MR_i^B = p_i \cdot (e_{is,is} + 1) / e_{is,is}$.
3. From the definition of $\bar{\sigma}^i$ by the last equality in (2.31),

$$\partial \bar{\sigma}^i / \partial e_{ii} = 1 - \sum_s \left((\bar{\sigma}^i + e_{ii}) / (\sigma_{ss}^i + e_{ii}) \right)^2 / S_i \approx 0,$$

   because the S_i terms of the sum are distributed araound $1/S_i$. Hence, ignoring the dependence does not seem a serious error and the assignment of separate roles to $\bar{\sigma}^i$ and S_i on one hand and e_{ii} on the other, as in the text, seems warranted.
4. In a symmetric Cournot equilibrium, (2.34) reduces to $MR_i^C = p_i \cdot (1 + d_{is,is})$.
5. This involves no approximation, as $\bar{\tau}^i$ does not depend on d_{ii}.
6. Note that (2.36) illustrates several of the above remarks.
7. Note that some of these factors appear as determinants of the industry's mark-up as derived above.
8. Much earlier, Artus(1974), *imposing* the restriction that the long-run elasticities are zero, found quite high short-run and medium-run elasticities.
9. A full decription of MINI is available from the author upon request.

References

Artus, J.R. (1974), "The Behavior of Export Prices for Manufactures," *IMF Staff Papers*, 21:583-604.
Bergeijk, P.A.G. van, R.C.G. Haffner and P.M. Waasdorp (1993), "Measuring the Speed of the Invisible Hand: The Macroeconomic Costs of Price Rigidity," *Kyklos*, 46:529-44.
Blanchard, O.S. and L.H. Summers (1986), "Hysteresis and the European Unemployment Problem," *NBER Macroeconomics Annual*, Cambridge (Mass.), pp. 15-78.
Broer, D.P. (1996), "The Effect of Competitor's Prices and Utilisation Rates on the Mark-up of Price over Cost," memo (*to appear as internal CPB memorandum*).
Bruno, M. (1979), "Price and Output Adjustment: Micro Foundations and Aggregation," *Journal of Monetary Economics*, 5:187-211.
Calmfors, L. and J. Herin (1979), "Domestic and Foreign Price Influences: A Disaggregated Study of Sweden," in: A. Lindbeck (ed.), *Inflation and Employment in Open Economies*, North-Holland, Amsterdam, pp. 269-306.
Centraal Planbureau (1977), "VINTAF-II, A Medium-Term Macro Model for the Netherlands" (in Dutch: VINTAF-II, een macromodel voor de Nederlandse economie op middellange termijn), *Occasional Paper No 12*, CPB Netherlands Bureau for Economic Policy Analysis, The Hague.
Centraal Planbureau (1990), "ATHENA, a Sectoral Model of the Dutch Economy" (in Dutch: ATHENA, een bedrijfstakkenmodel van de Nederlandse economie), *Monografie No 30*, CPB Netherlands Bureau for Economic Policy Analysis, The Hague.
Centraal Planbureau (1992), *FKSEC, a Macroeconometric Model for the Netherlands*, Stenfert Kroese, Leiden/Antwerpen.
de Menil, G. (1974), "Aggregate Price Dynamics," *Review of Economics and Statistics*, 56:129-40.
Deardorff, A.V. and R.M. Stern (1978), "Modeling the Effects of Foreign Prices on Domestic Price Determination: Some Econometric Evidence and Implications for Theoretical Analysis," *Quarterly Review of the Banca Nazionale del Lavoro* :333-53.
Draper, D.A.G. (1985), "Exports of the Manufacturing Industry: An Econometric Analysis of

the Role of Capacity," *De Economist*, 133:285-305.
Draper, D.A.G. (1996), "The Export Market" (in Dutch: De exportmarkt), *Research Memoran-dum No 130*, CPB Netherlands Bureau for Economic Policy Analysis, The Hague.
Eckstein, O. and G. Fromm (1968), "The Price Equation," *American Economic Review*, 58:1159-83.
Eckstein, O. and D. Wyss (1972), "Industry Price Equations," in: O. Eckstein (ed.), *The Econo-metrics of Price Determination*, Board of Governors, Federal Reserve System, Washington D.C.
Encaoua, D. and P. Geroski (1986), "Price Dynamics and Competition in Five OECD Coun-tries," *OECD Economic Studies*, 6:47-74.
Gordon, R.J. (1975), "The Impact of Aggregate Demand on Prices," *Brookings Papers on Eco-nomic Activity*, 6:613-62.
Gordon, R.J. (1980), "A Consistent Characterisation of a Near-Century of Price Behavior," *American Economic Review*, 70 (May, Papers and Proceedings):243-49.
Gelauff, G.M.M. and J.J. Graafland (1994), *Modelling Welfare State Reform*, North-Holland, Amsterdam.
Hall, R.L. and C.J. Hitch (1939), "Price Theory and Business Behavior," *Oxford Economic Papers*, :20-51.
Iwai, K. (1974), "The Firm in Uncertain Markets and Its Price, Wage and Employment Adjust-ments," *Review of Economic Studies*, 41:257-76.
Maccini, L.J. (1978), "The Impact of Demand and Price Expectations on the Behavior of Prices," *American Economic Review*, 68:134-45.
Maccini, L.J. (1981), "On the Theory of the Firm Underlying Empirical Models of Aggregate Price Behavior," *International Economic Review*, 22:609-24.
Maccini, L.J. (1984), "The Interrelationship between Price and Output Decisions and Invest-ment Decisions: Microfoundations and Aggregate Implications," *Journal of Monetary Economics*, 13(1):41-65.
Nieuwenhuis, A. (1986), "Sectoral Price Equations" (in Dutch: Prijsvergelijkingen per sector), *Research Memorandum No 14*, CPB Netherlands Bureau for Economic Policy Analysis, The Hague.
Neary, J.P. and K.W.S. Roberts (1980), "The Theory of Household Behavior under Rationing," *European Economic Review*, 13:25-42.
Nordhaus, W.D. (1972), "Recent Developments in Price Dynamics," in: O. Eckstein (ed.), *The Econometrics of Price Determination*, Board of Governors, Federal Reserve System, Washington D.C.
Nordhaus, W.D. and W. Godley (1972), "Pricing in the Trade Cycle," *Economic Journal*, 82:853-82.
Ormerod, P. (1980), "Manufactured Export Prices in the United Kingdom and the 'Law of One Price'," *The Manchester School*, 48(3):265-83.
Rushdy, F. and P.J. Lund (1967), "The Effect of Demand on Prices in British Manufacturing Industry," *Review of Economic Studies*, 34:361-73.
Sahling, L. (1977), "Price Behavior in U.S. Manufacturing: An Empirical Analysis of the Speed of Adjustment," *American Economic Review*, 67:911-25.
Straszheim, D.H. and M.R. Straszheim (1976), "An Econometric Analysis of the Determination of Prices in Manufacturing Industries," *Review of Economics and Statistics*, 58:191-201.
Zeelenberg, C. (1986), *Industrial Price Formation*, North-Holland, Amsterdam.

PART II
Imperfect Competition
(B)

3 A Critical Assessment of the Role of Imperfect Competition in Macroeconomics

Dennis W. Carlton

3.1 Introduction

Models of imperfect competition have proliferated in the macroeconomics literature.[1] These models have been used in both the business cycle literature, where the emphasis is on Keynesian type short-run movements in output, and in the growth literature, where long run properties of the economy are analyzed. These models involve firms having market power whereby a firm can profitably set price above marginal cost. Although these models have improved and made more realistic our understanding of certain economic forces, I claim that it is too much to expect these models based solely on market power to provide fundamentally new insights into how an economy works and that there are probably better areas to pursue to look for new insights into macroeconomics.[2]

One reason for my skeptical view on the ability of macro models with market power to add to our understanding of short-run movements in the economy is that such models give virtually identical theoretical insights as models with taxes and those models have been around a long time. Taxes create a wedge between price and marginal cost in the same way that market power creates a wedge. It is this 'wedge' that drives most of the interesting results in the macro literature dealing with imperfect competition. I will show that many of the Keynesian multiplier results derived from models using monopolistic competition are identical (after appropriate reinterpretation) to expressions of deadweight loss in the public finance literature. Since empirical estimates of these deadweight losses from market power are small, I am skeptical of the quantitative significant of these Keynesian-type results derived from market power.

Models with market power have been used to study growth. Although these models have yielded some valuable insights, I believe that other forces such as the existence of intellectual property laws may be at least as important a force as market power in understanding growth. The reason for the assumption of market power in models with endogenous innovation is to provide an incentive for innovative activity by creating a property right in the rents generated by any new discovery. But domestic and international laws protecting intellectual property are key determinants of whether such property rights can exist at all. There turns out to be a close relationship between property rights, human capital, and R & D and these relationships provide a much richer background to understand incentives for innovation than one with only market power.

Although I believe the 'standard' models of imperfect competition won't generate startling new insights into macroeconomics, models that depart from perfect competition in non-standard ways likely will. Specifically, models that focus on the frictions or costs of making a transaction or that recognise that firms choose 'policies' intertemporally which can differ from a sequence of static one period maximisation problems strike me as likely to provide promising new insights into how a macroeconomy works. I will explain why the usual notion of marginal cost loses its meaning in some realistic models and why this can help explain certain types of results regarding margins and intertemporal substitution in business cycles. I will concentrate on three areas in which to examine departures from the usual competitive assumptions: markets where sellers face a risk that they can't sell their goods always, durable good markets, and markets where the uncertain future affects timing. These last two cases belong to a general class of examples in which firms (or the market) choose policy functions over time.

Because I'm not a macroeconomist, I must begin with a caveat. I know enough international trade theory to recognise that it is not my comparative advantage to discuss macroeconomic models. Though I'm familiar with some of the literature, I am not familiar with it all. I therefore apologise in advance if I make a point that has already been made and I fail to cite the appropriate source. I also will try to avoid repeating my views in Carlton (1989) where I discussed much of the literature in industrial organisation, especially that dealing with price rigidity, and its relationship to macroeconomics.

This paper is organised as follows. First, I analyze the use of imperfect competition models to derive Keynesian multiplier results. Second, I will discuss growth theory and why understanding the forces leading to laws protecting intellectual property may well be more important for understanding growth than knowing the level of market power. I examine the close relationship between capital intensity, R & D intensity and intellectual property laws. Finally, I analyze three types of departures from the usual model of perfect competition that could yield interesting insights into market clearing and intertemporal behaviour .

3.2 Monopolistic competition and the Keynesian multiplier

3.2.1 Relation between deadweight loss and the Keynesian multiplier

In simple Keynesian models, we have the identity $Y = C + I + G$ where Y is income, C is consumption, I is investment, and G is government spending. If $C = a + bY$ with $b < 1$, and if I and G are exogenous, we obtain the Keynesian multiplier and the familiar result that $\frac{dY}{dG} = \frac{1}{1-b} > 1$, or that as G rises by \$1, Y increases by more than \$1. This result occurs because there is an implicit assumption that there is more labour that is willing to work holding wages and prices constant. The implication is that prior to an expansion of G by \$1, that there are unutilised resources – resources inefficiently idle. The economy was operating inside its production possibility frontier. Indeed, if labour is involuntarily unemployed, then its shadow value is zero (or lower than the value of leisure) and an expansion of output using these unutilised resources generates an enormous amount of income and welfare to the economy. Having otherwise useless resources produce valuable output is what makes Keynesian policy sound great and what leads to Keynesian multipliers of 1 and higher.[3]

Models of monopolistic competition usually posit that the monopolistic competition is at the final goods level and that all markets including the labour market clear.[4] This means that unlike the simple Keynesian model, the economy is always on its production possibility frontier.[5] The distortions that arise under monopolistic competition come from the gap between price and marginal cost for each good in the final good market. According to the reasoning in this literature, fiscal policy can generate marginal profit in the economy which then leads to increased demand for goods which leads to what looks like a Keynesian multiplier.

The analogy to Keynesian multipliers is more apparent than real. Although realistic and insightful, these models that produce what look like Keynesian multipliers have little to do with what I think of as Keynesian macroeconomics and more to do with the public finance analysis of taxation. Indeed, when I learned macroeconomics at M.I.T., I recall learning about the 'Okun gap,' the shortfall between potential GNP (achievable if people who

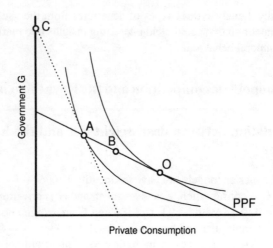

Figure 3.1: Equilibrium in a market with monopolistic competition

wanted to work could) and actual GNP. I also learned about 'Harberger triangles,' the loss to society from gaps between price and marginal cost caused
by taxes. The quip was that it takes many Harberger triangles to equal an
Okun gap. The public finance literature reflects the same sentiment when
it suggests that the loss from taxes pales in comparison to the loss from unemployed resources. (See, for instance, Musgrave and Musgrave, 1980, p.
319.) I view the current Keynesian literature based on models of monopolistic competition as equivalent to claiming that Harberger triangles[6] are the
same as an Okun gap.

Let me start by taking one of the clearest and most insightful papers in this
macroeconomic literature, Startz (1989). In that model, there are three types
of goods: private consumption goods available in the market, government
supplied consumption goods and leisure. All markets clear so there are no
idle resources and society is on its production possibility frontier. There is
market power in the pricing of all consumed goods. In the short run, all goods
have a constant return to scale production technology (there is a fixed cost
to produce, but, once the fixed cost is sunk, returns are constant). Therefore,
in the short run, the economy is on its production possibility frontier which
is linear among consumption goods.

In Figure 3.1, I oversimplify (Startz has more than 2 goods, as I already
described) to illustrate his key insights. The production possibility frontier
is linear with slope c and the economy is on it. No inefficiency from idle

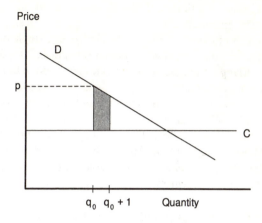

Figure 3.2: Deadweight loss

resources appear. The economy is at point A because the price ratio between government supplied goods (G) and private consumption is distorted by the presence of market power.[7] GNP is measured at point C which is based on the price ratio at point A. Welfare for infinitesimal movements from A to B can be measured by seeing what happens to the intercept of the line drawn through B, parallel to AC. As should be clear, GNP (and welfare) rises as the economy moves from A to B. The move from A to B expands output of private consumption, reducing output of G. This leads to an increased GNP because the relative price of private consumption is distorted causing the economy to be away from its optimal point O. (It may sound odd for fiscal policy to involve reducing G, but that is an artifact of my ignoring leisure in this simplified model. In Startz's model, an expansion of both G and private consumption occurs relative to leisure as fiscal policy improves welfare). The increase in GNP from a reduction in G (which I take as numeraire) that expands private consumption by 1 unit is ΔGNP = the lost value of G (which is $-c$) plus the value of the increased output of private consumption (which is just p). Hence, GNP rises by $p - c$, exactly the gap between price and marginal cost as a result of fiscal policy. More generally, with labour involved, there will be an increase in welfare, just as in the simple Keynesian case, but because the labour market is undistorted and clears, we do not have a zero (or low) shadow value for labour as is implicit in the usual Keynesian analysis and therefore should not expect as large a welfare gain as we are used to from Keynesian analyses.

This 'Keynesian multiplier' of Startz's model and the many models like his in the macro literature turn out to be nothing more than a typical Harberger triangle. In Figure 3.2, I show the Harberger triangle of deadweight loss associated with the gap between price and marginal cost and note that the change in the triangle from increasing private consumption from q_0 to $q_0 + 1$ is a reduction in deadweight loss of $p - c$, exactly as above. In Startz's and others' models, the increased GNP from expanding output in one market increases demand in other markets, which in turn increases demand elsewhere, and so on.[8] This is why the effect of fiscal policy is said to create a Keynesian multiplier. However, this multiplier is exactly the same as the Harberger loss 'triangle' plus what Harberger calls 'trapezoids' which represent the output shift in other markets (with distortions) caused by output shifts in the original market.

This result between the equivalence of the Keynesian multiplier from the macro literature using models of imperfect competition and the deadweight loss literature in public finance should in fact be exact even in more complicated models. Let t_i be the distortion between price and marginal cost for good i, q_i be quantity of good i consumed, and let Z be any government policy. Then, Harberger (1971) shows that the instantaneous change in welfare from altering Z is given by $\sum t_i \frac{dQ_i}{dZ}$ where $\frac{dQ_i}{dZ}$ is the total derivative of Q_i with respect to Z taking into account all general equilibrium effects that will alter Q_i. (Diamond and McFadden, 1974, derive similar formulas using duality theory.) I suspect that *every* demonstration of a 'Keynesian multiplier' in this market power literature can be reproduced by this formula for incremental deadweight loss.[9]

So what, you may ask? There are three main implications. First, the formula shows that these Keynesian results emerge because there is a gap between price and marginal costs, not because markets don't clear. These results are identical to those that emerge in a competitive economy with taxes. That is, if I know how a competitive model with taxes works, then I know all about 'Keynesian multipliers' from this new macro literature. Maybe that's perfectly sensible, but I doubt that's what Keynes thought or what most 'new' Keynesians believe.

Second, this new macro literature naturally focuses attention on market power as the reason why fiscal policy can improve welfare. Yet, the deadweight loss from taxes are likely as great or greater than the deadweight losses from market power.[10] Most attempts to measure the deadweight loss from market power in the U.S. economy produce small estimates as a percentage of GNP. Harberger's original estimate (1954) of loss was tiny (less than 0.1 percent of GNP), though subsequently there have been much higher estimates. Scherer and Ross (1990) summarise their best estimate based on the entire literature as around 1.3 percent of GNP. Of course, changes in this measure of deadweight loss as a result of government policy will be much smaller and it is changes that are of interest to Keynesians. In contrast, the deadweight loss from various taxes (corporate, income and the like) have been estimated in a higher range. For example, Harberger (1966) estimated the loss from corporate taxes alone to be about 0.5 percent of GNP (over 5 times his deadweight loss estimate from market power). Here too subsequent estimates have in creased Harberger's loss estimates. Boskin (1972) finds that taxes on capital (of which corporate taxes are the most important) amount to about 2 percent of GNP. Gravelle and Kotlikoff (1989, 1993) find values 7 to 15 times higher than Harberger. Aside from corporate taxes, there are income taxes. Harberger (1974) estimates the cost of labour taxes to be around .2 percent of GNP, double his estimate of loss from market power. A subsequent estimate by Browning (1979, p. 333) of the loss from the income tax is around 1 percent of GNP. Ballard, Shoven and Whalley (1985) estimate a cost of all taxes in the range 4 to 7 percent of GNP. Although there is consider able uncertainty regarding each of these calculations of deadweight loss, it is likely that the aggregate losses from taxes dwarf those from market power. Therefore, before worrying about market power and how it affects efficiency and output fluctuations, I would first worry about taxes which, unlike market power, are probably easier for a government to affect.

To provide some additional perspective on the importance of market power in the U.S. economy, I calculated the market share of the top four firms (four firm concentration ratio) for U.S. manufacturing in 1987.[11] Focusing on manufacturing, I find that about 60 per cent of U.S. industry has a four firm concentration ratio below 40 per cent, while only about 10 per cent of U.S. industry has a four firm concentration ratio over 70 percent. (Compared to 1935, this represents a less concentrated industry structure.) Moreover, industry sectors such as services, retail and wholesale trade are likely less concentrated than manufacturing. My assessment is that there is just not enough market power in the U.S. for it to be the key to understanding macroeconomics. Moreover, notice that the positive Keynesian effect on welfare would completely disappear if the firms with market power practiced perfect price discrimination for then there would never be a gap between price and marginal cost. Thus, in order to believe that there is a significant Keynesian effect on welfare, one must believe not only that there is sufficient market power to generate large deadweight loss but also that the ability to practice price discrimination doesn't seriously mitigate the distortionary effects of market power.

The third implication of the equivalence between Keynesian multipliers and deadweight loss deals with the relationship between the long and short run. In a model with taxes, the tax causes a distortion typically in both the short and long run. In Keynesian models with monopolistic competition, the Keynesian multiplier can vanish in the long run. The models that generate Keynesian multipliers with monopolistic competition distinguish between short and long run. In the long run, the fixed costs of entry of new firms dissipates any profit earned from expanding output. That is, in the long run, the extra gain from expanding a distorted market (that is, one with a gap between prices and cost) gets dissipated by the real cost of entry. This is one of Startz's (1989) most interesting results. This means that there is no Keynesian multiplier in the long run. All short run multipliers come from the extra profit from expanding production in a distorted market, and none from expanding employment of idle resources, since there are no idle resources. In the long run, not only are there no idle resources to use, there is no extra profit from expanding output. Hence, the multiplier vanishes. One crude way to understand this result is that each firm in the long run has constant average costs which become its long-run marginal cost. With constant long-run marginal costs, there is no pricing distortion between price and long run marginal cost, so there is no deadweight loss or alternatively no gain from expanding output.[12]

If the multiplier disappears in the long run, then it is only in the short run that government fiscal policy can positively affect income. But, this assumes that only the government and not firms can identify when fiscal policy is desirable. That is, suppose the economy is at some point A so that short-run fiscal policy can move the economy to B. If firms realise that the government will engage in such a program, firms will enter in anticipation of this action and turn the 'no long-run multiplier result' into a 'no short-run multiplier result.' This is, of course, the insight of Lucas (1972).

3.2.2 Empirical analysis

The fact that Keynesian multipliers in the new macro literature are just plain old deadweight loss is, in some sense, a detail. As I have stated earlier, taxes probably contribute as much if not more to gaps between price and marginal cost as does market power. Regard less of the source for the gap, the deadweight loss formula does tell us quite a lot about how short run fluctuations are likely to influence GNP. Economies with high taxes and lots of market power in sectors where output fluctuates a lot should have more pronounced business cycles than those in economies with no taxes and no market power, all else equal.

I have some snippets of suggestive evidence on this issue. (The analysis is highly preliminary.) I obtained annual information from OECD for about 25 years on real GDP and the importance of taxes (measured as taxes collected divided by GDP) for 9 countries.[13] I regressed the log of real per capita GDP on a time trend. The variability of the residual, all else equal, is a measure of the cyclic sensitivity of the countries' GDP. I examined whether that variability was related to the importance of taxes. Theory predicts a positive statistically significant correlation, but I do not find such a relationship in my admittedly small data set.

I would suggest at least three further directions for empirical work. First, the crude analysis could be applied to more countries and refined to analyze output movements by sector and relate those movements to the distortion between price and marginal cost caused by either taxes or market power. Second, one may wish to distinguish between output and input market distortions.[14] Taxes on inputs cause inefficient production, so that the economy operates within its production possibility frontier. My intuition is that 'Keynesian multipliers' and hence the value of output fluctuations will likely be larger if an economy is well within its production frontier than if the economy is already on its production frontier. Finally, the evidence suggests that the shocks hitting the different countries are very different. In order to test the theory, it is necessary to adjust somehow for the size of the shocks. One rough way to do this is to look at nearby countries, or countries amongst whom international trade is important, and see if the theoretical predictions are more likely to hold for those countries.

3.3 Growth and intellectual property

Macro models with market power have been used to understand growth. These models use market power to create an incentive for innovation. This section explains why intellectual property laws may be at least as important in understanding innovation as market power.

It has long been understood that no firm will spend resources to innovate if other firms can free ride by quickly and costlessly imitating the innovation. There are three main ways to create a property right in the revenue stream generated by the innovation in order to mitigate the free riding problem.

First, the firm can implement the innovation in such a way that it is hard for other firms to copy. This leads to trade secrets which firms can take steps to protect by controlling access of employees to information. However, protecting trade secrets is costly, can impede the efficient deployment of new technology and, many times, is not cost effective (Kitch, 1977). I have not seen models used in the macro literature explicitly based on protection of trade secrets.

Second, if the firm has market power so that it faces few equally efficient competitors, then competition will not drive price down to marginal cost after the innovation, and the innovating firm will be able to reap some of the benefits of the innovation.[15] The lower is the amount of competition faced by the innovator, the greater is the incentive to innovate. This approach, usually attributed to Schumpeter, has been used to understand growth in the recent macro literature (see for instance, Aghion and Howitt, 1992, and Grossman and Helpman, 1990, 1994, and Romer, 1990, 1994). These models have yielded impressive theoretical insights, though I'm less certain that any empirical link has yet been established between growth and market concentration. Indeed, at the microeconomic level, it has been difficult to reach consensus on what levels of industry concentration fosters the greatest innovative activity (see for instance, Cohen and Levin, 1989). Moreover, often the theoretical models take the degree of market power as given and stable, rather than modeling the dynamic probabilities of being leapfrogged by competing innovators. Still, these models have improved our under standing of growth.

My main question about this literature is how the theoretical insights that it provides regarding market power and growth can be used empirically. If market power is the driving force in these models, do these models have, as would appear, widespread implications for a country's antitrust policy towards mergers and joint ventures? Perhaps at some general level, but practically not really. The reason is that for most joint ventures or mergers and most industries, R & D is not very important nor is market power.[16] Moreover, the theory provides no guidance as to the most desirable levels of industry concentration to foster growth. Our ability to predict empirically how R & D will be affected as industry concentration rises through moderate ranges (which involve most joint ventures or mergers) is poor or nonexistent (Cohen and Levin, 1989). In any event, any prediction would require a detailed empirical analysis of the particular industry. Indeed, I think it unrealistic, except in unusual cases, to expect government agencies to be able to reliably predict the effect of joint ventures or mergers on R & D (see Carlton, 1995).[17] Nevertheless, there are some very important policy issues that these models of growth and market power raise but of which I have seen relatively little discussion in the macro literature – namely intellectual property rights.[18]

The third way to protect the investment by innovating firms is to grant them property rights in their intellectual property (for instance, patents) not only in their own country but in other countries. Indeed, the strength of the laws governing intellectual property will be a major determinant of the likely degree of competition and resulting market concentration in an industry where innovation is possible. Just like market concentration is the endogenous outcome of a market process, the laws governing intellectual property also will be the endogenous outcome of a political process. What is interesting is that there should be a close theoretical relationship between concentration, growth, intellectual property and government funding. Some initial investigations suggest that these relations are empirically valid and strong (Ginarte and Park, 1995, and Scalise, 1996).

Intellectual property laws create various benefits and costs for different groups within and across countries. Understanding the importance of these various interest groups helps to predict where intellectual property laws will develop.

There are several major beneficiaries of laws protecting intellectual property. They are:

1. The innovators for whom property rights allow them to earn a reward on their innovation.
2. The suppliers of factors of production to the innovating industry, especially when those factors generate rents.
3. The innovating country whose exports of innovating goods can better the terms of trade that the country faces.

Even in the complete absence of domestic laws preventing free riding on others' intellectual property, as long as free riding internationally is difficult, there could be an incentive for the government to support and subsidise R & D in order to encourage innovative products that get exported. For example, even if domestic competition is fierce and there are constant returns to scale so that, in the presence of free riding, any gain from a cost reducing innovation is passed on to consumers, the government could levy a tax on the industry allowing the government to capture in tax revenues the profits from the innovation. This amounts to taxing R & D exports when the innovating industry exports much of its output.[19] One group likely to benefit from such R & D subsidies is universities which can be expected to push for greater research money. We therefore can add as beneficiaries of stronger intellectual property laws:

4. Universities and research personnel.
5. Governments that subsidise R&D and tax industries that export R&D.

Offsetting these benefits are costs. The major one is that consumers will be deprived of the lower prices that free riding firms can charge in the short run.

To empirically investigate the relationship of intellectual property (IP) laws to other economic variables, I use the index system of Ginarte and Park

84 *Dennis W. Carlton*

(1995). They classify the strength of IP laws on a scale of 1 to 5 and provide an analysis of IP laws (see Appendix A for a detailed description). I also draw on some innovative work by Scalise (1996) who is investigating determinants of IP laws.

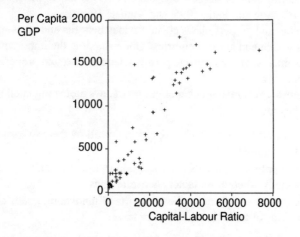

Figure 3.3: Per capita GDP vs. capital-labour ratio, 1990

It is well known (see Figure 3.3) that there is a relation between per capita GDP and the capital labour (K/L) ratio.[20] Figure 3.4 shows the relation between per capita GDP and the level of IP protection. The figure shows that there is the same type of positive correlation of per capita GDP with IP laws as with the K/L ratio. In fact, one should expect that with strong IP laws, firms should be willing to hold more capital since innovative capital can be better protected. Figure 3.5 confirms that there is a positive relationship between the K/L ratio and IP laws. The more capital per worker, the stronger are laws governing IP. Because there is a strong positive relation between human and physical capital across countries (Figure 3.6), it follows that IP laws are stronger where there is more human capital. It is well known that growth and human capital are related (Mankiw, 1995). Similarly, growth and IP laws are positively related. (Figure 3.7) Moreover, as Ginarte and Park (1995) have shown, there is a positive correlation between IP laws and R & D in a country. Finally, in Figure 3.8, I illustrate the relation between IP laws and government support of R & D. Figure 3.8 confirms the hypothesis that government spending on R & D and strong IP laws go together.

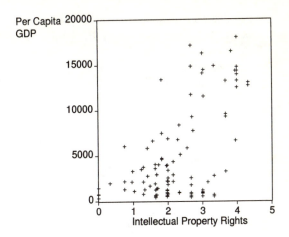

Figure 3.4: Per capita GDP vs. intellectual property rights, 1990

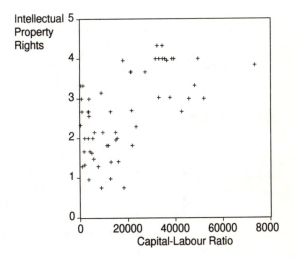

Figure 3.5: Intellectual property rights vs. capital-labour ratio, 1990

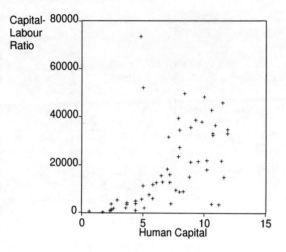

Figure 3.6: Capital-labour ratio vs. human capital per capita, 1990

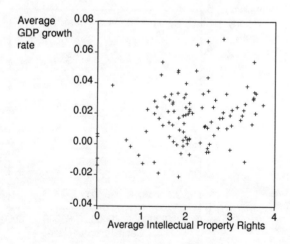

Figure 3.7: Per capita GDP growth vs. intellectual property rights 1960-90

I believe that understanding the relation between IP laws and income (or growth) is at least as important as understanding the one between market

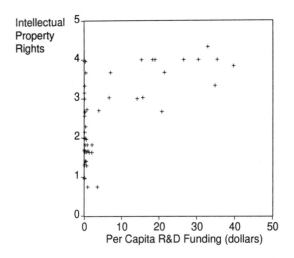

Figure 3.8: Intellectual property rights vs. per capita higher education R&D expenditures

power and growth for three reasons. First, IP laws are likely more important than market concentration in explaining income or growth. That is, if one redrew the figures using industry concentration instead of strength of IP laws, I would expect a weaker relationship. Second, even if one thought that industry concentration was the key determinant of innovative behaviour, one would still want to understand the determinants of IP laws because of their likely significant influence on concentration. Third, from a policy perspective, IP laws can often be more easily changed than industry concentration. Concentration can mainly be influenced by antitrust or industrial policy and neither policy has proven particularly adept at fostering growth (Carlton, 1995, and Klenow and Irwin, 1996).

3.4 The deviations from perfect competition not involving monopolistic competition

I am skeptical about the contribution to macroeconomics of models of imperfect competition because of *a*) their likely small quantitative significance based on deadweight loss calculations (especially compared to other distortions such as taxes), *b*) their assumption that all markets clear, and *c*) their ignoring of uncertainty. In this section, I explore three types of deviations from perfect competition that are likely to have greater ability to provide insight into macroeconomic phenomena. I will first focus on the transaction technology in the absence of market clearing. That is, if an auctioneer is not clearing markets for free, how costly and efficient are alternative mechanisms and what properties would an economy with such markets have? Second, I will focus on intertemporal substitution in the durable good sector. Third, I will focus on intertemporal substitution in the timing of investment. These three factors seem to me to contain the keys to questions that several macroeconomists pose such as why some resources appear not to be fully utilised and why there are business cycles. I do not mean to suggest that the topics I discuss are the only deviations from perfect competition important for understanding macroeconomics. For example, a key channel for understanding business cycles involves the credit market, a topic I do not discuss (see Bernanke, 1983).

3.4.1 Non-Walrasian market clearing

3.4.1.1 A simple model

With the same constant returns to scale production technology available to all and with perfectly divisible products, each consumer becomes self-sufficient and produces everything for himself. In this Robinson Crusoe economy, there is neither resource misallocation, resource underutilisation nor business fluctuations. Once we introduce comparative advantage, we obtain worker specialisation and trade. Economists have grappled with the notion that in the absence of costless Walrasian auction markets, the trading mechanism is what causes a macroeconomy to differ in performance from that of a Robinson Crusoe economy.[21] The idea of round-about production mattering is perhaps a reflection of this notion. There are many alternatives to an auction that have been explored, such as consumer search (Stigler, 1961, Diamond, 1982), fixed price models (Matsuyama, 1985, section 3) and a hybrid of the two (Prescott, 1975, Eden, 1990). I will focus on only one such model and then on only a stripped down version in order to highlight its salient features (see Carlton, 1977, 1978, 1989, 1991, Deneckere and Peck, 1995, Greely, 1996 for more detailed analyses).

As a general matter, economists have spent relatively little time studying the distribution sector – the sector responsible for allocating goods – even though wholesale and retail trade contribute about 17 percent to GNP, a figure close to manufacturing's share of about 22 per cent. To stress the role of the distribution sector in achieving efficiency, I note that distribution comprised only 7 per cent of GNP in the economies of the former Soviet Union in the early 1990s (Joskow and Schmalensee, 1994. See Carlton, 1991, for a discussion of the role of making markets).

Imagine a model in which a firm must set its price and choose its capacity before it knows its demand. A bakery could be a simple example, though more generally many firms set price and leave it unchanged for quite lengthy periods exceeding one year in many cases (see Carlton, 1986). In such a model, it is natural for buyers to be unable occasionally to purchase the good and for sellers to be unable occasionally to utilise fully their capacity. The likelihood that the good is available becomes an endogenous characteristic of the good that matters to buyers. The likelihood that the good goes unsold matters to sellers.

To set up one simple model, imagine that the number of customers that arrive is N and that each customer has a demand $x(p)$. Let N be a random variable with mean u and variance σ^2. Customers care about both the price and the probability of obtaining the good. For simplicity, assume that consumers go only to the firm that yields them the highest expected utility. (Allowing search would not alter the basic results.) An equilibrium consists of a price and capacity choice by each firm that yields zero expected profits and maximises a consumer's expected utility. In equilibrium, competitive firms take the utility level as given. (These models can be formulated as a game theoretic model in which the equilibrium I've described is the limit as the number of firms gets large. See Deneckere and Peck, 1995.)

To understand the firm's choices, write the firm's profits as

$$\Pi\left(p, S\right) = \left[\int_0^S p \cdot i \, \mathrm{d}F(i) + p \cdot S \cdot \int_S^\infty \mathrm{d}F(i) - c \cdot S\right] \cdot x\left(p\right)$$

where S is the maximum number of customers that the firm can serve, $F(i)$ is the cumulative probability that i or fewer customers arrive, and c is the constant cost of capacity. For simplicity, I assume that, if capacity is available, the firm can produce the good at no additional cost. There are several important implications of this model.

First, because it occasionally happens that goods go unsold – some capacity is purchased but is not ultimately used – the cost of those unsold goods (or the capacity to produce those goods) must still be paid for. Hence, the price of goods sold must exceed c in equilibrium! The gap between p and c is *not* a result of market power, but rather is a reflection of the transaction technology of the model. In equilibrium, each firm takes the utility level as given and so has no market power. Regardless of why the gap between p and c occurs, it follows that a certain expansion of output by 1 unit raises output by p and cost by c, adding to GNP by the amount of the gap.

One can use this model to assess the recent attempts to measure the gap between price and marginal cost and attribute the gap to market power.[22] For example, the attempts by Hall (1988) to measure market power by examining the Solow residual may correctly measure the gap between price and marginal production cost but incorrectly identify the source of that gap. Hall observes that marginal production cost, x, equals $\frac{\Delta L}{\Delta Q} W$, where W is labour and ΔL and ΔQ are changes in labour and output respectively. In competition, $p = x$, while with monopoly, $p \cdot u = x$ where u differs from one and equals $u = \left(1 + \frac{1}{E}\right)$ where E is the elasticity of demand. Hall can solve for u as $\frac{\Delta L}{\Delta Q} \frac{W}{P}$. Hall calculates u for several industries and obtains estimates for u significantly different from 1, leading him to conclude that significant market power exists in the U.S. economy.

To put the model that I have just sketched into Hall's frame work, imagine that if a customer shows up, then an additional cost, a, is borne for each additional unit sold as the cost of using some additional input L. Suppose that one additional unit of L is needed for each additional unit of output. Then in equilibrium, price equals $a + \gamma \cdot c$ where $\gamma > 1$ because price must cover the cost of unsold capacity. Hall would calculate u as $\frac{a}{\gamma \cdot c + a}$ and would attribute the deviation of u from unity to market power, even though there is no market power in the model in the sense that each firm takes the consumer's utility as given. Indeed, as a approaches 0, Hall's measured market power rises to infinity. The model under scores how tricky the concept of marginal cost can be. Although c is the cost of producing one more unit, c is not the expected marginal cost of producing *and selling* one more unit. Indeed, in this non-Walrasian model, that cost of selling depends on the characteristics of demand and the existing capacity.

The second implication of the model is that there are two separate decisions for the firm to make, price and capacity, even in competition. These two choices will influence how intensively the capacity is used. In contrast, in a Walrasian model, the firm has only one choice variable and there never is an inability to sell nor is there underutilisation of capital. In this model, unlike a Walrasian model, 'idle' resources some times exist (unsold capacity) but serve a (probabilistic) purpose of being there when they are needed. It would be wrong to characterise this 'idleness' as necessarily inefficient unless there is a reorganisation of production that, given the same transaction technology, can leave everyone better off.

Third, the gap between price and marginal production cost will depend on consumer preferences for obtaining certain delivery of the good and on the underlying characteristics of probability distribution of number of customers, N. Specifically, as the ratio of the mean (u) to the standard deviation (σ) of the random variable N rises, the transaction technology becomes more efficient in the sense that the gap between price and cost diminishes, for any probability of availability. The fact that σ influences cost creates a true technological externality in the model.[23]

One way to view a boom is as an increase in demand. The model tells us that if u/σ rises in a boom, as is likely, the transaction technology becomes more efficient and the gap between price and cost should narrow. The model would therefore predict countercyclical margins over the business cycles for those markets where auction markets aren't used – which would tend to be retail markets.[24] The empirical findings of Murphy, Shleifer, and Vishny (1989) support such an inference.

Finally, the model has implications for how shocks might get transmitted through out the economy and why positive demand shocks might get short-circuited in a recession. Consider how an increase in demand usually gets transmitted in a Walrasian economy. After hitting the initial market where a price increase mitigates the quantity shock, the quantity shock leads to increased demand for inputs. In the model just described, there is no mitigating price effect on quantity but there is a mitigating stock-out effect. If the firm in the output market that sees a positive demand shock is stocked out, then the positive demand effect is stopped dead in its tracks. Therefore, the inventory holding policy of each stage of production is critical to understanding how forcefully a shock gets transmitted. If u/σ is low in a recession, that will tend to lead to higher stock-outs,[25] which will tend to prevent positive demand shocks from being transmitted. This could explain why an economy finds it hard to get out of a recession.

3.4.1.2 Extensions of non-Walrasian models

The simple model discussed in the previous section, though helpful for understanding certain features of business cycles, is inadequate for providing a foundation for business cycles because it has no intertemporal linkages. There are at least three ways to create such linkages. First, the information at time t can influence predictions at time $t + 1$. Greely (1996) has successfully used this approach to develop a rich theory of business cycles possibilities. Second, a shock today can be dissipated over time at different rates depending on the willingness of consumers to wait.[26] As I have explained elsewhere, Carlton (1983, 1989), waiting is actually a Walrasian response that turns out to have enormous significance empirically, though I have rarely seen it used in empirical macro models. The simple idea is that in response to an increase in demand, either price could rise or consumers could wait for delivery or both. In general, the more willing consumers are to wait, the more stable is price. Finally, inventory holding can create intertemporal linkages.

The welfare properties of these models are complicated for two reasons. First, there is no insurance market to allow an individual to insure against the event of being unable to obtain the good. Second, there is a true technological externality in the model. The variability of one individual's demand raises the cost to everyone else. Both these reasons lead to market outcomes that lack the usual optimality properties of a Walrasian competitive equilibrium. See Carlton (1978) and Greely (1996).

3.4.2 Durables

Durable goods have received much attention by researchers in business cycles (for example, Murphy, Shleifer, and Vishny, 1989). The reason is simple, as Figure 3.9 shows. The production of new durable goods fluctuates enormously over the business cycle, plummeting in busts and skyrocketing in booms. Why, as Figure 3.9 shows, is the behaviour of durables so different from that of non-durables? I focus on two features of durables: intertemporal substitution, and the lack of an incentive to cut price in a bust for fear that it will adversely affect the firm's reputation for a price policy. The adverse

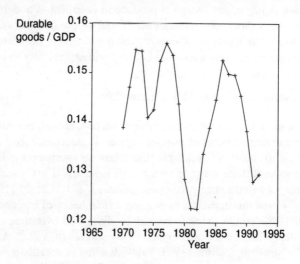

Figure 3.9: Relationship of durable goods to GDP

effect of a price cut on reputation impairs the firm's ability to earn a profit (Coase, 1974) and to obtain a steady stream of demanders. This reputation effect arises only where firms have some market power and so only applies to those durable goods industries with some market power. This effect from market power is different from the usual effect, discussed in Section 3.2, of pricing above marginal cost.

3.4.2.1 Durables and intertemporal substitution

Consumers of durable goods receive a flow of services from their stock of old goods plus their purchases of new goods. The demand curve for new goods is derived from the consumers' demand for a flow of services. Even if that flow demand curve reflects no intertemporal substitution (so that the flow demand $q_t[R_t]$ does not depend on R_i for $i \geq t+1$ where R is the implicit rental rate on capital goods), the demand for new goods will.[27] The demand curve a supplier sees, based on the price (not rental rate) of the new good will look highly elastic. Consumers can initially substitute away from new goods and let their old stock supply them. Through such substitutions, as well as keeping and maintaining old goods a little longer, consumers can respond to even a slight price increase in new goods by dramatic movement away from new goods. The elasticity for new goods will be especially high when the stock of old goods is high relative to the annual purchase of new goods.

Three implications follow from the characteristics of the demand for new durable goods. First, any economy wide shock gets reinforced by durables. For example, a slight positive shock can get magnified into a large positive quantity movement for new durables, because the demand is highly elastic.

Second, the price movements in both the new and used market should be small relative to quantity shifts between new and used goods. We have already seen the large quantity shifts in new goods in Figure 9. Figure 10 provides some very rough confirmatory evidence on new and used prices for one model of car. (Results for other models were similar.) It reveals that the price ratio of new to used cars does not show the large downward movements during severe recessions (for instance, 1975, 1981) that one might otherwise expect.[28]

Third, although the demand for new durables is highly elastic in the short run, in the long run as depreciation wears down the existing stock, the demand must become more inelastic. As Murphy, Shleifer, and Vishny (1989) explains, this provides a natural reason why a recession ends. Consumers, at some point, can wait no longer and must replace their durables.

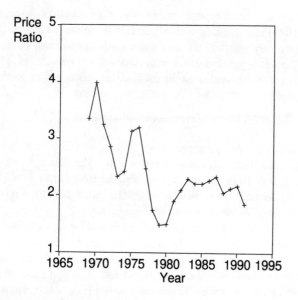

Figure 3.10: Ratio of new car price to four year used car price for the Cadillac Sedan Deville, 1969-1995

3.4.2.2 Durables and price reputation

The intertemporal substitution story just told made no mention of market power. Although I've explained why market power in general seems an unlikely source of new insight into macroeconomics, some of the work in durable goods monopoly may be an exception to that general rule.[29] Durable goods tend to be produced in industries that are more concentrated than non-durables in the sense that a greater fraction of output in durables is produced in industries where the four-firm concentration ratio exceeds 40 per cent.[30] Coase (1972) and a huge subsequent literature have explained why a monopolist of a durable good gains by following a stable price policy in order to create a reputation that he will not cut price. Indeed, in the United States, you see auto companies advertising how much of the original value a car maintains as it ages. The reason is that the amount a consumer is willing to pay for a durable good at time t depends on the asset's price in the future. Consumers will be willing to pay less today for a new good if prices are expected to fall in the future and thereby reduce resale value. Failure to develop a reputation that price will not be cut will erode the monopolist's ability to earn profits. In general, the monopolist concerned with his price reputation should follow a less variable price policy than one with no reputational concerns. This means that in a recession, the monopolist will be reluctant to cut price even though it may be profitable in the immediate short run to do so because to do so would destroy his reputation of following a stable price path.

This reasoning suggests that the margins in concentrated durable good industries should be countercyclical. It explains why firms should be reluctant to cut price to stimulate demand in a recession. (It is probably more accurate to say that the margins are more countercyclical than one would otherwise expect. I've already indicated that demand for durables is likely to be less elastic as a boom begins so that price should rise in booms, all else equal, in the presence of market power.) The evidence on margins in concentrated durable industries supports the empirical implications of the theory just sketched. Margins in concentrated durable good industries appear countercyclical (or acyclical) and, in any event, appear much less procyclical than margins in other industries.[31]

This example explaining why a monopolist's desire to establish a price policy can constrain the incentive to cut price is more general in its application than to just durable goods. Anytime either demanders' or suppliers' actions affect future demand or supply conditions, the behaviour can depart from simple static optimisation and can lead to systematic cyclic behaviour such as countercyclical margins. Let me provide an example.

Consider a group of consumers who wish to purchase a nondurable. Some consumers care little about whether they consume in the first or second period. Other consumers arrive only in the second period. A monopolist is trying to figure out how to price his product. There are two periods and the game repeats itself so the monopolist can develop a reputation for how he prices. Assuming that interest rates are zero, in a world of certainty P_1 must equal P_2 where P_i is the price in period i.

Now suppose that although P_1 is known, P_2 is uncertain because the monopolist's costs in period 2 are unknown or hard to predict because marginal cost is upward sloping and the number of customers is random in the second period. If P_2 is random, then the probability distribution of P_2 will affect purchase decisions in period 1. Even holding the mean of P_2 constant, the monopolist who puts a lot of weight in the tails of the probability distribution of P_2 will induce consumers in period 1 to wait until period 2 to purchase. The reason is that as long as the indirect utility function (or profit function in the case of a buyer who is a firm) is concave, Jensen's inequality creates a preference for randomness. But the monopolist may have good profit-based reasons to prefer not to switch consumers intertemporally. For example, spreading out demand over time could lower his costs[32] or could allow him to intertemporally price discriminate. The monopolist may choose not to lower price too much in period 2 in order not to create a price policy in the second period with a probability distribution with fat tails. This reputation for not having a wildly random price in period 2 allows him to maximise his profit even though it may be socially desirable to cut price. This discrepancy between the private and social incentives arises for the familiar one when there is market power – marginal revenue, not price, measures the gain to the monopolist.[33] These incentives can lead to countercyclical margins.[34]

3.4.3 Timing and intertemporal behaviour

The last example indicated how uncertainty can influence the timing of economic activity. Because intertemporal substitution is at the heart of business cycles, it is natural that understanding how uncertainty affects when people do things should be at the heart of business cycle research.

Substitution across time is analytically identical to substitution across commodities. What links prices intertemporally are interest rates at which consumers can borrow and lend. One fruitful line of business cycle research is studying the malfunction of credit markets and the effect of that malfunction on creating less intertemporal substitution than would otherwise occur, accentuating business cycles. (See for instance, Bernanke, 1983). Another line, and the one I will pursue here briefly, has to do with how uncertainty affects timing. Research on timing, exemplified by Dixit and Pindyk (1994), explains why what normally should matter in explaining behaviour – namely price which, in an intertemporal context, involves the interest rate – seems so impotent to describe investment or inventory behaviour .[35] I will focus on one element of timing, the resolution of uncertainty, which seems to have great potential, especially in conjunction with the analysis of durable goods, to explain business cycles.

The theory of uncertainty (for example, Raiffa, 1968) explains why it may pay to wait, rather than plunge ahead, even if the expected discounted present value of an investment is currently positive. The reason is that by waiting, one may learn information that would cause one to decide not to invest. By waiting, one can reduce or cut off the possibility of very bad outcomes. This means that, even if a project is desirable today, the investor will tend to wait, the more likely new information will arise to resolve future uncertainty. For example, suppose that the expected discounted present value of a project if done today is $100. There are two future equally likely outcomes – a big success ($1,200) or a big failure (-$1,000). Suppose that tomorrow, the investor will learn for certain whether the project will be a huge success or a failure. Then, even though the project is profitable to do today, it may pay for the investor to delay his investment to avoid the bad outcome and to invest tomorrow only if he learns that the project will be a success.

An uncertain environment can stop or slow down the economy as investors decide to wait to resolve their uncertainty. The decline in investment activity, reinforced by a cutback in durables, can lead to a recession. But, as an economy stays in a recession, two effects occur. First, without investment, the stock of durables declines, causing an expected upsurge in future demand, and thus the expected profitability of investment to increase. Second, the longer the economy is in a recession, the probability rises that the economy will escape it (because of the first effect). The combination of these two effects is ultimately to reduce the variance of future outcomes and the value of waiting. This causes investment to occur. Roughly speaking, if there are only two states of the world – boom and bust – then, as the probability of escaping the bust rises toward 1 with time, the variability of outcomes will diminish and the option value of waiting declines. (That is, if Y_1 and Y_0 are income in the boom and bust respectively, then the variance of income is $P \cdot [1 - P] \cdot [Y_1 - Y_0]^2$ where P is the probability that the boom will occur next period. Initially, P is low as the economy goes into recession, but P rises towards 1 as the economy uses up its durable stocks of goods, causing the variance to decline.) Therefore investment will increase, reinforced by an increase for durables, and the recession will end. Thus, the combination of the intertemporal substitution possibilities of durable goods and the option value from delaying investment work hand in hand to reinforce cyclical behaviour .

3.5 Conclusion

Models of monopolistic competition have sharpened our understanding of certain forces shaping macroeconomic activity. However, I am skeptical that these models will lead to fundamental new insights into macroeconomics. Public finance model with taxes yield the same insights and taxes are probably a more significant quantitative explanation for the gap between price and marginal cost. For growth theory, the determinants of intellectual property rights is related to market power and, I would suspect, of more important policy and quantitative significance. Departures from perfect competition that focus on imperfections in market clearing – real resources are used to clear markets – and the mechanisms of intertemporal substitution (especially under uncertainty) are areas in which research is likely to have large payoffs.

Since this is a conference on macroeconomic models, I will close by addressing what empirical recommendations, if any, flow from my analyses. Models where delivery lags appear in demand curves, where the price variance influences demand, and where the variance of key underlying information influences timing of investment strike me as fruitful. Models using market share to measure market power, with the possible exception of some durables, should not be a high priority.

Appendix A

OLS Regressions for figures (t-ratios in parentheses).
- Fig. 3.3: Per Capita GDP *vs.* Capital-labour Ratio (1990): 58 observations, $R^2 = .80$.

$$90gdp = \underset{(3.03)}{1520} + \underset{(15.20)}{0.304} \quad K/L90$$

- Fig. 3.4: Per Capita GDP *vs.* Intellectual Property Rights (1990)[36]: 106 observations, $R^2 = .37$.

$$90gdp = \underset{(-1.99)}{-1955} + \underset{(7.73)}{3047} \quad IP90$$

- Fig. 3.5: Intellectual Property Rights *vs.* Capital-labour Ratio (1990): 58 observations, $R^2 = .36$

$$IP90 = \underset{(11.90)}{1.94} + \underset{(5.64)}{0.000037} \quad K/L90$$

- Fig. 3.6: Capital-labour Ratio *vs.* Human Capital Per Capita (1990): 58 observations, $R^2 = .29$.

$$K/L90 = \underset{(-0.43)}{-2054} + \underset{(4.73)}{2936} \quad HC$$

- Fig. 3.7: Per Capita GDP Growth *vs.* Intellectual Property Rights (1960-90): 106 observations, $R^2 = .08$.

$$\Delta GDPC = \underset{(1.16)}{0.00539} + \underset{(2.93)}{0.00592} \quad \overline{IPR}$$

$\Delta GDPC$: Average per capita GDP growth rate (1960-90) from Summers and Heston (1991): $\Delta GDPC_{60-90} = (GDP_{90}/GDP_{60})^{1/30} - 1$. $\overline{IPR}$: Average IPR (1960-90): $\overline{IPR} = (IPR60+IPR65+IPR70+IPR75+IPR80+IPR85+IPR90)/7$.

- Fig. 3.8: Intellectual Property Rights *vs.* Per Capita Higher Education R&D Expenditures[37]: 57 observations, $R^2 = .37$.

$$IP90 = \underset{(17.27)}{2.18} + \underset{(5.64)}{0.0562} \quad High/pop$$

Acknowledgments

I thank Robert Gertner, David Greely, Janice Halpern, Anil Kashyup, Randy Kroszner, Kevin Murphy, Craig Scalise, Peter A.G. van Bergeijk, Arjen van Witteloostuijn, conference participants, and seminar participants at NBER. I am especially grateful to Julio Rotemberg and Richard Startz for helpful comments. I thank Colleen Loughlin, Greg Pelnar, Craig Scalise, Sam Smith and Thomas Stemwedel for their research assistance. This paper was prepared for the Conference on Market Behaviour and Macroeconomic Modelling Groningen, The Netherlands June 6-7, 1996. University of Chicago and National Bureau of Economic Research.

Notes

1. See Matsuyama (1995) and Silvestre (1993) for recent surveys and Mankiw and Romer (1991, vols. 1 and 2) for a collection of papers. Zarnowitz (1992) provides an excellent and insightful overview of all business cycle research.
2. I focus my analysis on models of monopolistic competition with Keynesian type multipliers and on models of monopolistic competition with long-run growth. Models of monopolistic competition may well be useful for analyzing other macro phenomena.
3. It is welfare, not GNP that matters to society. Practical considerations of national income accounting preclude the valuation of leisure. I will focus my analysis on welfare which GNP is supposedly designed to measure.
4. Hart (1982) is a notable exception.
5. When monopolistic competition occurs in intermediate goods, there is a distortion in factor usage which places the economy inside the production possibility frontier. This possibility seems not to have generated much interest in the macro literature.
6. In fact, Harberger 'triangles' become trapezoids in markets with existing distortions. See Harberger (1971).
7. In Startz's model, leisure rather than G is the undistorted good but that doesn't affect the point of the diagram.
8. A three-dimensional diagram would yield the same insight. The deadweight loss measures welfare changes. Changes in GNP would also measure welfare changes if leisure were appropriately valued. There is a simple relationship between changes in deadweight loss and changes in GNP when leisure is not valued. The change in welfare, as measured by the change in deadweight loss, equals the change in GNP plus the change in the value of leisure. Although the multiplier results in the literature focus on changes in GNP, I will focus on changes in welfare, since that is the more relevant measure.
9. Again, if GNP omits leisure, then changes in GNP differ from the formula just given by a term involving the change in the valuation of leisure. It also immediately follows that insights from public finance will be applicable to new Keynesian multipliers. For example, Proposition 3 of the Aloi, Dixon and Lawler paper in this volume follows im-

mediately from the public finance theorem that the incremental deadweight loss of taxes in an undistorted economy is zero.

10. Any regulation or policy that causes price to diverge from marginal cost would also cause a deadweight loss. As a technical matter, the comparisons that I present of the calculations of deadweight loss from monopoly and from taxes have the flaw that each calculation ignores other distortions. The precise calculation of interest is the relative contribution of taxes and market power to the incremental distortions between prices and marginal costs. Roughly, that relative contribution is given by the square root of the ratio of the relative deadweight losses reported in the economic literature for each distortion taken separately. Although it is possible for even small amounts of market power to generate significant distortions in the presence of existing distortions (for instance, taxes or regulations), that has been neither the theoretical nor empirical focus of the new Keynesian economics. Instead, that focus has been that market power alone produces Keynesian multipliers.

11. There are many drawbacks to using market shares at the four digit SIC Code level as reported by the Bureau of the Census to measure market power. See Chapter 9 of Carlton and Perloff (1994). For the purposes I use these numbers, I believe the drawbacks are not severe.

12. I ignore, as does most of the Keynesian literature, the subtle issue regarding optimal product variety. The use of the change in GNP as a welfare approximation to the change in utility is reasonable as long as consumers equate price ratios to their marginal rates of substitutions. New products change this analysis and their introduction means that welfare analysis based on changes in GNP can be inaccurate.

13. Australia, Canada, France, Italy, Japan, Netherlands, Sweden, U.K., and the United States.

14. Basu (1995) has explored some of these issues.

15. There is a subtle point here. Market power can be the result of an innovation that is hard to copy. Market power, by itself, does not guarantee that innovations can be protected.

16. Research joint ventures formed to achieve efficiencies, as distinct from simply achieving market power, raise issues that would require separate analysis.

17. Some empirical literature on innovation (for instance, Bernstein and Nadiri, 1988) suggests that the social rate of return to innovation exceeds the private rate. This would indicate that antitrust actions aimed at diminishing the ability of innovators to earn profits (*e.g.*, certain Section 2 actions under the Sherman Act in the United States) could be undesirable.

18. Helpman (1993) and Grossman and Helpman (1991) are exceptions in the theoretical literature and Ginarte and Park (1995) and Scalise (1996) are exceptions in the empirical literature.

19. Presumably, it is politically easier to levy a tax when most of the tax is to be paid by foreign consumers.

20. My data uses countries from the Summers and Heston data base (1990). Appendix A reports an OLS regression for each relevant figure in the text.

21. Before feeling forced to abandon the competitive paradigm, one should make sure that the competitive model is not consistent with the facts. The real business cycle literature exemplifies this approach. In Carlton (1983, 1989), I discussed how certain phenomena that look like non-Walrasian behaviour such as delivery lags and relatively inflexible prices are perfectly consistent with a competitive economy. (See also Zarnowitz, 1962.) Specifically, I stress the virtually ignored role of delivery lags as a (Walrasian) clearing device and show how important delivery lags are as a determinant of demand. However, I do go on to explain why market clearing models leave several facts unexplained. I return to the importance of delivery lags in Section 3.2.

22. See also Rotemberg and Summers (1990) and Eden and Griliches (1993).

23. A more sophisticated extension of the model (Carlton, 1991) examines how the externality can be controlled and how control of the externality can lead to a theory of marketing.

24. One reason why u/σ rises initially in a boom is that an initial demand surge raises average demand across products. The boom then encourages entry of new products which entails a sunk entry cost. When the economy falls back down to a recession, the new products remain causing u/σ in the recession to be lower than u/σ in the boom.

25. I have not examined empirical evidence on this matter. The relevant test would allow the number of products to differ between the beginning of the boom and the beginning of the recession. I suspect that stock-outs in some industries might be more of a problem during booms than busts.

26. The pathbreaking but often overlooked work by Zarnowitz (1962) is especially noteworthy.

27. Suppose Q_t = stock of goods at t, N_t = new goods at t, δ = depreciation rate, r = real interest rate and P_t = price of new goods at t. Suppose P_t is expected to remain constant. In equilibrium $N_t = Q_t = Q^*$ where Q^* = the consumption service demanded of durables at a rental price $(r + \delta)P_t$. If P_t rises, and is expected to remain high, $N_t = Q^* - (1 - \delta)Q_t$. As long as Q^* is large relative to N_t, the elasticity of N_t will be high. For many durable goods, the ratio of old to new goods is very high, leading to a very high demand elasticity for new goods.

28. It is beyond the scope of this paper to explain the shape of the curves in Figure 3.10, but that would appear to be a valuable line of research.

29. It is an exception in the sense that it provides a new theoretical reason why prices fail to fall. I am unsure of its empirical importance. Using data on U.S. concentration ratios and value of shipments for 1987, only about 5 to 10 per cent of durable output is from industries with four-firm concentration in excess of 70 per cent, though about 40 per cent is from industries with four-firm ratios above 40 per cent.

30. Surprisingly, the average four-firm concentration ratio is roughly the same for durables as nondurables.

31. The evidence on margins in general is not clear cut. See the discussion in Chapter 9 of Carlton and Perloff (1994), Domowitz, Hubbard and Peterson (1986, 1988), and Murphy, Shleifer and Vishny (1989). My interpretation of the evidence is that margins tend to be countercyclical based on the work by Mills (1936) for the Great Depression. Murphy, Shleifer and Vishny (1989) find countercyclical margins, using recent data. However, Domowitz, Hubbard and Peterson (1988) use recent data to find that with the exception of durables in concentrated industries, margins tend to be procyclical.

32. Why can't he induce consumers to consume in period 1 if it is cheaper? He can, but remember his savings are based on differences in marginal cost, while his inducements will affect his marginal revenue.

33. This point is related to Akerlof and Yellen (1985), Blanchard and Kiyotaki (1987), and Mankiw (1985).

34. Even if there is no market power, dynamic optimisation complicates the analysis. For example, in a model with adjustment cost, or a model with learning by doing, output today, q, affects costs above and beyond that measured by the partial derivative of cost today with respect to q. This is well known, but it creates serious measurement problems. If an act today will increase cost tomorrow, GNP accounting is unlikely to pick it up today. This tends to cause one to underestimate marginal cost in booms.

35. See Dixit and Pindyk (1994, p. 424) for a discussion of empirical attempts to explain investment timing. See also the recent successful attempt of Ghosal and Loungani (forthcoming).

36. 1990 Intellectual Property Protection: Park (1995a)

Each country's score is based on the following features of its patent protection: 1) availability of patent protection in various industries; 2) membership in international intellectual property agreements such as the Paris Convention (1883) and the Patent Cooperation Treaty (1970); 3) loss of protection through working requirements, compulsory licensing and conditions for revocation of patents; 4) availability of specific mechanisms for enforcement of patent protection; and 5) the duration of patent coverage in years.

A score of 0-1 is granted for each component based on the percent of 'important conditions or features' that are met for that component.

The unweighted scores are summed to give the country's level of Intellectual Property Protection.

0: no patent protection

5: maximum patent protection

37.   If one uses IP for the year 1980, instead of 1990, one obtains similar results.

References

Aghion, P. and P. Howitt (1992), "A Model of Growth through Creative Destruction," *Econometrica,* 60:323-51.
Akerlof, G. and J. Yellen (1985), "A Near-Rational Model of the Business Cycle, With Wage and Price Inertia," *Quarterly Journal of Economics,* 100(Supp.):823-38.
Aloi M., H. Dixon and P. Lawler, "The Multiplier in an Economy with Monopolistic Output Markets and Competitive Labour Markets," *This Volume.*
Ballard, C., J. Shoven and J. Whalley (1985), "The Total Welfare Cost of the United States Tax System: A General Equilibrium Approach," *National Tax Journal.*
Basu, S. (1995), "Intermediate Goods and Business Cycles: Implications for Productivity and Welfare," *American Economic Review,* 85:512-31.
Bernanke, B. (1983), "Non Monetary Effects of the Financial Crisis in the Propagation of the Great Depression," *American Economic Review,* 73:257-76.
Bernstein, J. and M. Nadiri (1988), "Interindustry R&D Spillovers, Rates of Return, and Production in High-Tech Industries," *American Economic Review,* 78(2):429-34.
Blanchard, O. and N. Kiyotaki (1987), "Monopolistic Competition and the Effects of Aggregate Demand," *American Economic Review,* 77:647-66.
Boskin, M. (1975), "Efficiency Aspects of the Differential Tax Treatment of Market and Household Economic Activity," *Journal of Public Economics,* 4:1.
Browning, E. and J. Browning (1979), *Public Finance and The Price System,* MacMillan, New York.
Carlton, D. (1977), "Uncertainty, Production Lags, and Pricing." *American Economic Review,* 67:244-9.
Carlton, D. (1978),. "Market behaviour With Demand Uncertainty and Price Inflexibility," *American Economic Review,* 68:571-87.
Carlton, D. (1983b), "Equilibrium Fluctuations when Price and Delivery Lag Clear the Market." *Bell Journal of Economics,* 14:562-72.
Carlton, D. (1986), "The Rigidity of Prices" *American Economic Review,* 76:637-58.
Carlton, D. (1989), "The Theory and Facts of How Markets Clear: Is Industrial Organization Useful for Understanding Macroeconomics?" in: R. Schmalensee and R. Willig (eds.), *The Handbook of Industrial Organization,* North Holland Press, Amsterdam.
Carlton, D. (1991), "The Theory of Allocation and Its Implications for Marketing and Industrial Structure," *Journal of Law and Economics,* 34:231-62.
Carlton, D. (1995), "Antitrust Policy Toward Mergers when Firms Innovate: Should Antitrust Recognize the Doctrine of Innovation Markets," Testimony before the Federal Trade Commission
Coase, R. (1972), "Durability and Monopoly," *Journal of Law and Economics,* 15:143-9.
Cohen, W. and R. Levin (1989), "Empirical Studies of Innovation and Market Structure," in: R. Schmalensee and R. Willig (eds.), *Handbook of Industrial Organization,* North Holland

Press, Amsterdam.

Deneckere, R. and J. Peck (1995), "Competition Over Price and Service Rate when Demand is Stochastic: A Strategic Analysis," *The RAND Journal of Economics*, 26:148-62.

Diamond, P. (1982), "Aggregate Demand Management in Search Equilibrium," *Journal of Political Economy*, 90:881-94.

Diamond, P. and D. McFadden (1974), "Some Uses of the Expenditure Function in Public Finance," *Journal of Public Economics*, 3:3-21.

Dixit, A. and R. Pindyk (1994), *"Investment Under Uncertainty,"* Princeton University Press, Princeton.

Domowitz, I., G. Hubbard and B. Peterson (1986b), "Business Cycles and the Relationship between Concentration and Price-Cost Margins," *Rand Journal of Economics*, 17:1-17.

Domowitz, I., G. Hubbard, and B. Peterson (1988), "Market Structure and Cyclical Fluctuations in U.S. Manufacturing," *Review of Economics and Statistics*, 55-66.

Eden, B. (1990), "Marginal Cost Pricing when Spot Markets Are Complete," *Journal of Political Economy*, 98:1293-1306.

Eden, B. and Z. Griliches (1993), "Productivity, Market Power and Capacity Utilization When Spot Markets Are Complete," *American Economic Review*, 83:219-24 (May).

Ginarte, J., and W. Park (1995), "Determinants of Intellectual Property Rights: A Cross-National Study," (*unpublished*), American University.

Ghosal, V., and P. Loungani, "Product Market Competition and the Impact of Price Uncertainty on Investment: Some Evidence from U.S. Manufacturing Industries," *Journal of Industrial Economics* (Forthcoming).

Gravelle, J. and L. Kotlikoff (1989), "The Incidence and Efficiency Costs of Corporate Taxation When Corporate and Noncorporate Firms Produce the Same Goods," *Journal of Political Economy*, 749-80.

Gravelle, J. and L. Kotlikoff (1993), "Corporate Tax Incidence and Inefficiency when Corporate and Noncorporate Goods are Close Substitutes," *Economic Inquiry*, 501-16.

Greely, D. (1996), "Economic Fluctuations in a Shopkeeper Economy," *mimeo*, University of Chicago.

Grossman, G. and E. Helpman (1991), *Innovation and Growth in a Global Economy*, MIT Press, Cambridge, Massachusetts.

Hall, R. (1988), "The Relationship Between Price and Marginal Cost in U.S. Industry," *Journal of Political Economy*, 96:921-47.

Harberger, A. (1971), "Three Basic Postulates for Applied Welfare Economics: An Interpretive Essay," *Journal of Economic Literature*, 9:109-38.

Harberger, A. (1974), *Taxation and Welfare*, Little Brown, Boston.

Harberger, A. (1954). "Monopoly and Resource Allocation," *American Economic Review*, 44:77-9.

Hart, O. (1982), "A Model of Imperfect Competition with Keynesian Features," *Quarterly Journal of Economics*, 97:109-38.

Helpman, E. (1993), "Innovation, Imitation and Intellectual Property Rights," *Econometrica*, 61:1247-80.

Joskow, P., R. Schmalensee and N. Tsukanova (1994), "Competition Policy in Russia during and after Privatization," Working paper no. 94-03, *World Economy Lab*, Massachusetts Institute of Technology, Cambridge.

Kitch, E. (1977), "The Nature and Function of the Patent System," *Journal of Law and Economics*, 20:265-90.

Klenow, P. and D. Irwin. (1996), "High-Tech R & D Subsidies: Estimating the Effects of Sematech," *Journal of International Economics*, 40:323-44

Lucas, R. (1972), "Expectations and the Neutrality of Money," *Journal of Economic Theory*, 31:197-99.

Mankiw, N.G. (1985), "Small Menu Costs and Large Business Cycles: A Macroeconomic Model," *Quarterly Journal of Economics*, 100:529-38.

Mankiw, N.G. (1995), "The Growth of Nations," *Brookings Papers on Economic Activity*, 257-310.

Mankiw, G., and P. Romer, eds., (1991), *"New Keynesian Economics"*, vols. 1 and 2, MIT Press, Cambridge, Massachusetts.

Matysuyama, K. (1995), "Complementarities and Cumulative Processes in Models of Monop-

104

olistic Competition," *Journal of Economic Literature,* 33:701-29.

Mills, F. (1936), *Prices in Recession and Recovery*, National Bureau of Economic Research, New York.

Murphy, K., A. Shleifer, and R. Vishny (1989), "Building Blocks of Market Clearing Business Cycle Models," in O.J. Blanchard and S. Fischer (eds.), *NBER Macroeconomics Annual 1989*, MIT Press, Cambridge.

Musgrave, R. and P. Musgrave (1980), "*Public Finance in Theory and Practice*," McGraw Hill, New York.

Prescott, E. (1975), "Efficiency of the Natural Rate," *Journal of Political Economy,* 83:1229-36.

Raiffa, H. (1968), *Decision Analysis: Introductory Lectures on Decision Making Under Uncertainty*, Addison Wesley.

Romer, P. (1990), "Endogeneous Technological Change," *Journal of Political Economy,* 98:S71-S102.

Romer, P. (1994), "The Origins of Endogenous Growth," *Journal of Economic Perspectives,* 8:3-22.

Rotemberg, J. and L. Summers (1990), "Inflexible Prices and Procyclical Productivity," *Quarterly Journal of Economics*, 103:851-74.

Scalise, C. (1996), "Human Capital Accumulation, Innovation and Intellectual Property Protection," *mimeo*, The University of Chicago Graduate School of Business.

Scherer, F. and D. Ross (1990), *Industrial Market Structure and Performance*, Houghton Mifflin, Boston.

Silvestre, J. (1993), "The Market Power Foundation of Macroeconomic Policy," *Journal of Economic Literature* 31:105-41.

Startz, R. (1989), "Monopolistic Competition as a Foundation for Keynesian Macroeconomic Models," *Quarterly Journal of Economics,* 104:737-52.

Stigler, G. (1961), "The Economics of Information," *Journal of Political Economy,* 69:213-25.

Summers, R. and A. Heston (1991), "The Penn World Table (Mark 5): An Expanded Set of International Comparisons, 1950-1988," *Quarterly Journal of Economics,* 106:327-68.

Zarnowitz, V. (1962), "Unfilled Orders, Price Changes, and Business Fluctuations," *Review of Economics and Statistics*, 44:367-94.

Zarnowitz, V. (1992), *Business Cycles: Theory, History, Indicators, and Forecasting*, National Bureau of Economic Research, University of Chicago Press, Chicago, IL.

4 An International Comparative Analysis of the State of Competition

Robert C.G. Haffner and
Peter A.G. van Bergeijk

1.1 Introduction

Since the OECD initiated its structural reform programme in the 1980s, the topic of structural change has been on the agenda of policy makers across the world. Motivated by the structural problems of stagflation in the 1970s, the deteriorating situation of public finance and the persistently high level of unemployment, it was increasingly recognized that 'open and efficient markets for goods and services, exposed to domestic and international competition, provide the crucial underpinnings for dynamic, high income economies' (OECD, 1994, p.7). Indeed, progress has been impressive in the areas of trade liberalisation and the deregulation of domestic and international financial markets, but the reform of markets for goods and services and, most notably, the labour market, was more tedious.

The Netherlands was no exception in this respect. During the 1980s and the beginning of the 1990s, the Netherlands had the reputation of being the 'cartel paradise' of Europe.[1] Partly this was because of the relatively admissive Dutch competition policy. For long, Dutch competition policy had been characterized by an abuse system, implying that agreements to restrain competition were allowed unless conflict with the public interest could be proven by the authorities. Formal agreements on restrictive practices were to be notified to the minister of Economic Affairs and were officially approved. The so-called Kartelregister listed more than five hundred cartel agreements with national effects on prices, supply and market division.[2] Thus, Dutch competition policy was out of line with the European antitrust policy which was based on the prohibition principle. Of the 'heavy' horizontal cases of cartel agreements dealt with by the European Commission during the period 1970-1990, almost 40 per cent were Dutch (De Jong, 1990).

At the same time, many policy analysts in the Netherlands felt that over-regulation, collusive private conduct, inefficient public monopolies and, in some cases, a lax attitude of competition policy authorities themselves, formed a major impediment to the long term economic growth potential (see, for example, Kremers, 1991). The Netherlands was supposed to lack economic dynamism compared to other countries especially in the sheltered sectors of the economy. However, this analysis was more based on 'gut-feeling' than on hard economic evidence. In the Netherlands, no quantitative economic research was available on the costs and benefits of the invisible handshake in the cartel paradise.[3] For the economics profession, the cartel paradise was more of a 'paradise lost.' It was not until the intended changes in competition policy were made public in the year 1992 that Dutch economists picked up this field of research. Crucial research questions in this respect were:

- how well do Dutch markets for goods and services function compared to other countries?
- what are the macroeconomic implications of a lack of competition?
- which Dutch markets are relatively rigid, and why is that so?

This paper will focus on answering the first question as we will discuss how competition on markets for goods and services can be measured. For this purpose, we will first provide an overview of the potential methodologies and indicators which can be used to measure the state of competition. This overview will be based on a brief discussion of the industrial economics literature. After that, we evaluate the suitability of each set of indicators for international comparisons. So, unlike most studies, we do not use perfect competition as a benchmark (at least not explicitly) but compare competition between countries. Each of the indicators discussed can also be used at the industry- or market level, but we will not do so here. Finally, we briefly discuss some of the macroeconomic implications of a lack of competition by presenting some simulation results. We conclude with a discussion of our findings and a brief evaluation.

4.2 The Structure-Conduct-Performance paradigm

The functioning of the market mechanism has received relatively little attention in the Netherlands both from policy makers and from economic scientists. The reason for this apparent gap in economic knowledge is not clear. Obviously, competition is one of those concepts which is intrinsically difficult to observe (Kühn *et al.*, 1992). However, we believe that the opportunities which an empirical operationalization of 'competition' provide would have justified more intensive research efforts. Given that a comprehensive set of indicators can be developed which provide a reasonable description of the state of competition, it would seem possible to assess empirically:

- the size of the policy problem, so that priorities can be set;
- the effects of current policies and changes in policies; and
- the relationship between structural policies and economic performance. Indicators of the state of competition are a precondition for any further analysis of the effects of reform policies, both *ex post* and *ex ante*.

In the current industrial organization literature, basically two approaches can be distinguished (see Scherer and Ross, 1990 or Martin, 1993 for overviews). The industrial economics or 'new industrial organization' approach is based on formal models of oligopoly markets. Tools of microeconomic theory, models of imperfect competition and game theory are used to rationalize observed outcomes. The focus is on the strategic interaction process between firms, the outcome of which is determined by a myriad of factors such as the type of game being played (repeated, simultaneous, sequential, for example), the choice variables of each firm (price, quantity, for example), the available information, the risk attitude and even notions like each firm's beliefs. Three interrelated reasons can be given why this approach is unsuitable for measuring the state of competition:

- Firstly, the analysis is directed at explaining observed outcomes by building (and sometimes testing) an internally consistent formal model which generates the observed behaviour. Thus the analysis tries to explain specific types of behaviour on specific markets. What we need is an approach which tells us what behaviour we are likely to see given a certain result (for example abuse of market power).
- Secondly, the large number of anecdotal theoretical models derived leads to an equally large number of equilibrium outcomes. According to Kreps and Spence (1985), '. . . the problem isn't that game theory is incapable of producing implicit collusion at all, but that it produces virtually every possible form.'
- Thirdly, the solutions are not robust, as they depend strongly on the set of *ad hoc* assumptions employed. A slight change in the game-theoretic context can completely change the policy implications of a model, which makes the number of factors influencing the outcome very high.

Thus, the high degree of formal rigour and determination of the outcome of the new industrial organization approach comes at the cost of applicability and relevance. This is why we resort to the industrial organization literature founded on the more traditional structure-conduct-performance paradigm (SCP paradigm). The main advantage of the paradigm is its high degree of generality, but it is relatively less successful in predicting actual market outcomes. First conceived by Mason in 1939 and extended by Bain (1956), the SCP-paradigm states that the performance in a particular market depends on the conduct of buyers and sellers, which in turn is dependant on the structure of the relevant market. Later refinements acknowledge the fact that the causality between structure, conduct and performance is not unidirectional and that market structure is in turn influenced by certain basic conditions (for example the price elasticity of demand, the nature of the relevant technology, the durability of the product, for example).

The three key elements of the SCP paradigm are (Scherer and Ross, 1990, pp. 4–7):

- *market structure*, that is, among others, the number and size distribution of sellers and buyers, the degree of product differentiation, and the extent to which entry and exit barriers exist;
- *conduct*, which covers the pricing policies, policies in the area of product differentiation such as advertising and research and development, and strategic or collusive behaviour;
- *performance*, which constitutes the productive and allocative efficiency, the dynamic efficiency in terms of economic growth and technological progress, and the profitability of the industry.

In terms of the goal of this paper, market structure indicates whether the preconditions to generate a dynamic competitive process are present. For example, the absence of barriers to entry or large dominant firms are preconditions which favour competition. When studying conduct, one tries to find out whether competition is actually taking place. This is done by looking at, among others, the pricing policies of firms and at company strategies in the fields of research and development and product differentiation. Finally, the 'performance' of firms or markets indicates whether the results of the competitive process are consistent with competition. For example, persistently high profits are generally not consistent with competition, as competition eliminates excess profits relatively quickly through entry of new firms.

In our analysis, the SCP-paradigm is merely a framework to group a number of indicators of the state of competition. This means that we do not propose any strictly deterministic links between structure, conduct and performance. Rather, the indicators presented in the next section should be seen as instruments for a first appraisal of the state of competition, so that time

Structure	Expected relation to competition
Market concentration	−
Market share mobility	+
Entry and exit	+
Import competition/FDI	+
Exports	+
Capital intensity	−
Consumer goods	−
Conduct	
Price flexibility	+
Collusion (Conjectural variation)	−
Advertising expenditures	+/−
R&D expenditures	+/−
Investment expenditures	+/−
Performance	
Price–cost margins	−
Productive efficiency	+
Persistence of profits	−
Output of innovations	+
Growth in sales	+

Source: based on Van Bergeijk and Haffner (1996), p. 65

Table 4.1: Summary of structure, conduct and performance indicators of competition

and manpower can more efficiently be devoted to a more detailed analysis of potentially collusive sectors.

4.3   Indicators of the state of competition

Table 4.1 summarizes the indicators of structure, conduct and performance that we will discuss here. Market structure variables are the most frequently used indicators of the state of competition at the industry or market level. In part, this is due to the view in the 1960s that structure determines conduct and performance, so that an analysis of market structure is sufficient.[4] Another reason for the frequent use of structure indicators is the fact that market structure is relatively easily observable.

Market concentration is a measure of the number and size distribution of sellers in a particular industry, which is a key factor distinguishing the theoretical models of perfect competition, oligopoly, and monopoly. High levels of concentration improve the leverage sellers have vis-à-vis buyers, especially when the buyers are large in number. In addition, market concentration improves the opportunities for operating a cartel effectively, because members will find it easier to detect secret price cutting and because the costs of negotiating a cartel are lower if concentration is high. Whether market power is also high in concentrated markets depends among other things on the presence of substitutes for the product concerned, the openness of the market for potential entrants and on the concentration of buyers.

Measures of concentration are often criticized for being static (Davies and Lyons, 1988, pp. 10–16). If competition is seen as a process by which competitors strive to get 'one step ahead' of their rivals, measures of concentration may not give an adequate description of the strength of competitive forces. In these cases, measures of 'dynamic concentration' or market share mobility may be helpful (Davies, 1988, p. 111). The basic idea is that because of the continuous thrust and counterthrust of competition, market shares will fluctuate accordingly. However, a problem with this measure of economic dynamism is that intensive rivalry need not always result in fluctuating market shares.

As long as **entry and exit** are costless and can occur very quickly, firms will not be able to exercise market power. According to the theory of contestable markets, hit-and-run entry ensures that prices cannot exceed average costs in a contestable market (Baumol *et al.*, 1982). Although perfect contestability depends on a number of crucial conditions which are quite restrictive (Paech, 1995), the degree of contestability can be assessed in three ways: by directly comparing entry and exit rates, by comparing the determinants of entry and exit rates and by looking at the impact of entry and exit rates on profitability.

First, entry and exit rates can be compared directly between industries or between countries. A relatively high level of entry and exit would indicate that entry and exit barriers are apparently not a problem, so that competition is fierce. However, when entry and exit rates are relatively low, it is difficult to ascertain whether barriers to entry are high. The reason is that in theory, the threat of entry is sufficient to discipline incumbent firms so that entry need not actually take place. The direct comparison of firm dynamics in a cross-section of industries or across countries may also be complicated by (sectoral or international) differences in the stage of the product life-cycle or the business cycle.

The second method relates entry and exit rates to variables which are expected to give incumbent firms an advantage over potential competitors such as product differentiation, economies of scale, excess capacity, absolute cost advantages or establishment regulations (see, for example, Lyons, 1988, pp. 34–51). This enables the researcher to determine which factors actually act as barriers to entry.

The third method relates entry and exit to profits in order to determine the strength of competitive forces (Kleijweg and Lever, 1994). The more entry by new businesses occurs at a given (industry-wide or firm-level) profit rate, the more competitive the market. While entry should respond to profits, profits should also respond to entry in competitive markets. The link between entry and profits is therefore a two-way causal relationship.[5]

Import competition and **foreign direct investment** (FDI) are a special kind of market entry by international competitors, which can have a substantial effect on existing firms. Competition is influenced positively when firms are exposed to foreign competition both abroad (through **exports**) and on the domestic market (through import penetration and FDI). International differences in relative factor prices and production processes may be quite large, and foreign entrants are often well-established firms in their home or other foreign markets. This may enable them to enter the market by significantly underpricing their domestic competitors or by offering superior quality. In addition, foreign firms may be less inclined or less able to participate in domestic cartels.[6]

Capital intensity is an indicator which may be heavily correlated with buyer concentration. Sectors with a high capital intensity are often characterized by high fixed costs, and consequently decreasing average costs, which favours large firms vis–vis smaller ones (Lyons, 1988, pp. 37–41). As a result, the minimum efficient scale is relatively high. In addition, new firms may have problems in raising sources of finance for entering these markets (because of imperfect capital markets), and the required capital goods are often too specific to be re-sold without loss of value. A high capital intensity may hence pose both entry and exit barriers.

Whether the product concerned is sold largely to other producers or to **consumers** determines several buying market characteristics, such as the number of buyers, the average size of buyers, the degree of professionality and expertise of the buyer and the degree in which middlemen are used to enable the sale (Kotler, 1980, pp. 267–9). Each of these characteristics influences the division of market power between buyers and sellers. Sellers will have relatively more market power in consumer goods industries, where the number of buyers is usually large, and the size and expertise of buyers is low. As such, the share of consumer goods in sales is often strongly correlated with the concentration of buyers: the higher the proportion of consumer goods sold, the more buyers, and therefore the lower the concentration of buyers (Dijksterhuis *et al.*, 1995, p. 653).

Conduct by firms which restricts competition is not easy to detect. This is especially true when one uses a macroeconomic or mesoeconomic approach. Above all, this is caused by the wide diversity of conduct that may or may not be the result of oligopolistic collusion. For example, just by looking at price changes in a cross-section of industries it is not possible to ascertain which industries engage in price-fixing agreements, because there simply may have been no economic reason to change prices. Collusion is also less

traceable in the case of price leadership in a transparent market, when fol-
lower firms are willing to co-operate with the leader's decisions, or when
all firms adhere to similar pricing rules. In general, the difference between
tacit collusion and forms of co-ordination which are more or less inherent
to the market concerned is very subtle. It is because of this complexity that
the links between indicators of firm conduct and competition should be interpreted with caution.

For our purposes, the most important indicator of firm conduct is **price
flexibility**. Theoretical explanations of price inflexibility are large in number. Prices in oligopoly may be less flexible because oligopolists fear the
breakdown of implicit or explicit collusion. Especially when products are
relatively homogeneous, small price changes can cause large fluctuations in
market shares. A price war may be the result (Ross and Wachter, 1975). In
addition, cartels try to make market behaviour as uniform as possible, with
price fixing as just one of the instruments in this respect. Cartel members,
however, often differ in their price preferences, for example because of differences in costs or different estimates of the price elasticity of demand. This
makes it difficult to reach an agreement. Once an understanding is reached,
parties to the arrangement may be reluctant to suggest a price increase for
fear of appearing weak to their 'partners'. As a result, there is a tendency to
avoid price changes once agreements are reached (Scherer and Ross, 1990,
p. 244). More tacit forms of collusion resort to several facilitating devices
as a way of communicating intentions to co-operate. Price changes may be
announced in advance, prices may be recommended by a central agency, or
some common pricing rule such as full-cost pricing may be adhered to.

Another indicator of firm conduct can be found in the way firms expect rivalsto respond to changes in their own output (Cowling and Waterson,
1976). This so-called **conjectural variation** is the (expected) change in the
output of all other firms in the industry with respect to the output of a single
firm i:

$$\lambda_i = \frac{d(Q - q_i)}{dq_i}$$

where λ_i is the conjectural variation of firm i, Q is total market demand
and q_i is the production of firm i. If the conjectural variation is zero, firm i
thinks that other firms will not react to changes in its output (the so-called
Cournot hypothesis). When the conjectural variation is equal to one, on the
other hand, firm i believes that when it restricts its output by one per cent,
rivals will do the same. If firm i maximizes its profits on the assumption that
$\lambda = 1$ and when the other firms react in the manner perceived, the monopoly
output will result. In this case, all firms in the market co-operate to restrict

output. When the conjectural variation is minus one, rivals will compensate any attempt of firm i to pull output off the market, so that total industry output will not change. This is what happens in the perfectly competitive case. The estimation of conjectural variations thus provides information about the degree of market power that individual firms have.[7]

Other conduct indicators are the **advertising** intensity, and the **research and development** and **investment** efforts of firms.[8] Advertising may have both positive and negative effects on competition (Scherer and Ross, 1990, pp. 405–6 and 435–7). On the positive side, advertising can provide valuable information to critical consumers about the existence of products, about prices, and so on. By expanding the available information about alternatives, demand curves may become more elastic so that market power may be reduced. On the other hand, advertising can influence economic dynamism negatively if it has characteristics of a fixed cost. If all firms in the market have to expend a certain amount on advertising, minimum efficient scale rises. In addition, advertising is a sunk cost, as it is generally not recoverable upon exit.[9] Advertising may also act as an entry barrier when entrants have to advertise more per unit sold compared to established firms, in order to convince prospective buyers of the quality of their product. Finally, advertising is a means of differentiating products and establishing brand loyalty, which makes demand less sensitive to price changes. Market power rises as a result. Empirical evidence in this area is still inconclusive.[10]

The R&D and investment expenditures of firms share many of the characteristics of advertising expenditures. Both types of expenditures create barriers to entry when they become necessary to enter the market successfully (Lyons, 1988, pp. 50–8). On the other hand, the R&D intensity is an indicator for the efforts that firms undertake to develop new products and processes. By analogy, the investment intensity serves to upgrade existing products and production facilities. With these new or better products and processes, firms hope to gain a competitive advantage. In other words, R&D and investment expenditures can be spurred by intense competition on dynamic markets, because firms which do not undertake these efforts will put themselves in an unfavourable competitive position. Many of today's most dynamic markets are characterized by large R&D expenditures, such as the semiconductor and aircraft industry, where rivalry is extremely strong. Whether or not R&D and investment expenditures promote competition is therefore unclear from a theoretical point of view.[11]

Firm or market performance refers to the outcome of the competitive process, and is usually analyzed jointly with indicators of firm conduct and market structure. Performance indicators of economic dynamism are the profitability of the firm or industry concerned, the productive and allocative efficiency which is achieved, the dynamic efficiency in terms of new products and processes and the growth in domestic or foreign sales.

A well-known indicator of profitability is the so-called Lerner index or **price–cost margin** $(P - C)/P$, which is a measure of the extent to which firms are able to raise their prices P above their marginal costs C. A Lerner index close to zero indicates intensive competition, while a relatively high value may indicate a lack of actual competition.[12]

The Lerner index can give a useful first impression of the degree to which possible abuses of market power have led to an increase in profitability.[13] In order to gain more insights into the causes of the high profits, profitability can be analyzed jointly with a market structure variable such as concentration. A positive correlation between the number of sellers and the price–cost margin is considered as evidence of collusive behaviour.[14] The consensus among industrial organization researchers, however, is that

> the correlation between industry concentration and either firm
> or industry profitability, if positive, is surely weak, and might
> very well be non-existent or even negative (Mueller, 1991, p.
> 3).[15]

One of the reasons for the weak relationship between concentration and profitability is that monopoly power may not only result in excessive profits, but could also lead to excessive costs. When competitive pressure is weak, managers may tolerate higher than necessary cost levels or may engage in 'rent-seeking'—which incurs substantial wasteful expenditures to obtain or strengthen monopoly positions. The degree of X-inefficiencies or **productive efficiency** is another possible indicator for the strength of competition, and could be measured by comparing the capital and labour inputs per unit of output among different plants within an industry and across nations. By fitting a 'best-practice' isoquant Q^* through the observations using the least inputs at varying capital–labour ratios, a ranking of relative productive efficiency can be obtained.[16] Empirical evidence suggests a positive correlation between productive efficiency and indicators of market structure such as the number of firms and the degree of import competition, thereby confirming the X-inefficiency hypothesis.[17]

Another problem with the Lerner index is that it assumes that markets are in equilibrium. If high profits in concentrated industries are only a temporary phenomenon, because competition becomes fiercer in response to profit opportunities, economic dynamism is high. This means that not only the size of (excess) profits at a particular moment in time should be considered, but also the duration of excess profits.

A way around this problem is provided by the persistence of profits liter-ature, which tries to measure how fast excess profits erode (Brozen, 1971). The idea is that in a dynamic and competitive environment, high profits can-not persist long because other firms will want 'a piece of the cake.' High profits will trigger entry by new firms and increased competition between existing firms. As a result, profits approach their long-run equilibrium value relatively quickly. On the other hand, competition may be hampered by bar-riers to entry or collusive behaviour of incumbent firms so that profit levels are permanently high.

The dynamic performance of firms and markets is usually measured by output indicators such as the number of new products, the number of new patents for **innovations**, or the number of quotations in scientific journals. In dynamic high- growth sectors, firms need to invest continuously in new products and processes to maintain market share and profitability. The global consumer electronics industry is an example. Competition is fierce because of short product life-cycles, critical and demanding consumers and large dif-ferentials in costs and quality between different producers. Firms must con-tinuously develop new products and processes in order to survive. Thus, a high innovative output of a sector can be an indication that competition is strong.

Dijksterhuis *et al.* (1995) also consider **growth in sales** (domestic or foreign) as an indicator of economic dynamism. Given that no large over-capacity exists initially, growth in sales creates room for new competitors to enter the market. In these markets, existing firms must continuously upgrade capacity through investment, because otherwise new competitors will try to enter the market. For the same reason, existing firms must keep prices in line with costs. However, it is important to keep in mind that growth in sales can be caused by many things. At the mesoeconomic level, a sector which grows strongly may be profiting from trade liberalization or an expansion in the demand for the product concerned. This does not have to go hand in hand with intensive competition *within* the sector. At the level of the individual firm, a dominant firm with superior products may have substantial market power, but at the same time experience a strong rise in sales.

4.4 Observing the invisible hand

The indicators and methodologies of the preceding section can be used to analyze the state of competition at the microeconomic and mesoeconomic level. How well are they suited to assess the state of competition for an economy as a whole? In general, they can give only very crude indications of the problem at hand because the high level of aggregation conceals many of the underlying dynamics. In addition, the usual international differences in statistical and accounting practices warrant a serious caveat with regard to the comparability of the various figures. All in all, we see the empirical evidence to be presented more as an illustration of the applicability and the kind of results that can be achieved with each methodology than as adequate representations of the state of competition at the macroeconomic level. We do not present figures on R&D, investment and advertising expenditures and on the growth in sales because of their theoretically unclear and, in any case, indirect relationship with competition.

Before we turn to presenting some of the available empirical evidence on conduct and performance, a remark on market structure seems appropriate. We have a number of reasons for not presenting evidence on market structure variables. As explained in the preceding section, market structure indicates whether the preconditions to generate a dynamic competitive process are present. In principle, therefore, observing conduct and performance is suffi-cient to assess the functioning of the competitive process. Moreover, market structure is influenced by many other factors. For example, exposure to for-eign competition and market concentration are strongly influenced by the size of the domestic market, as countries with a relatively large domestic market tend to have a relatively low degree of import competition and a rel-atively large number of firms in an industry. Likewise, capital intensity is influenced by the relative price of labour compared to capital and the cost structure (decreasing or increasing average costs) of the market. This makes international comparisons of market structure variables especially difficult. However, we do see a role for market structure variables in explaining dif-ferences in conduct and performance at the industry level within a country because here the interpretation of market structure is more straightforward.

Table 4.2 shows two indicators of conduct. The first column shows an in-dication of the time it takes for current cost and demand changes to be passed on into prices. More specifically, it presents the time period in which half the adjustment of prices to current cost and demand changes is completed (the half-life of cost and demand changes). Current cost and demand changes take relatively long to be passed through in prices in the Netherlands, Greece and the US (1958–80), while in Japan there seems to be virtually no lag in price responses. This result for Japan is in sharp contrast with the estimate of the product market inertia coefficient (PMIC). This coefficient gives an in-dication of the degree in which prices are able to clear markets. A low value of the PMIC indicates that prices react quickly to excess capacity, thereby reducing the level of excess capacity in the next period. If the value of the PMIC is high, neither excess capacity nor excess demand influences the rate of inflation. In Japan and the European markets for goods and services, the price mechanism does not seem to function well as the PMICs are all at the 94 per cent level and higher. Taiwan, the US and Canada do relatively well in this respect. The finding that the estimates of the adjustment speed of Japanese product markets differ so widely may be due to the fact that the price rigidity parameter relates to both demand and cost changes whilst the PMIC relates to demand (changes) only. This implies that in Japan, prices adjust relatively quickly to cost changes but relatively slowly when excess demand or excess supply occurs.[18]

Average speed of price adjustment (months, various estimation periods)		Estimated product market inertia coefficients 1974–92	
high adjustment speed			
Japan 1971-79	1.6	Taiwan[c]	23
Austria 1974-88[a]	1.5/3.2[b]	U.S.A.	42
Sweden 1970-70	3.8	Canada	83
Canada 1940-80	3.8	Australia	90
U.K. 1963-74[a]	3.9	Denmark	94
Australia 1949-74	4.7	U.K.	95
U.K. 1970-79	6.2	Germany[d]	95
U.S.A. 1958-80	6.3	France	95
Greece 1963-77[a]	7.0	Netherlands	97
Netherlands 1970-92	7.1	Italy	97
		Japan	99
low adjustment speed			

[a] Quarterly data used
[b] Speed of price adjustment in response to cost and demand changes, respectively
[c] 1976–94
[d] 1973–91
Sources: Encaoua and Geroski (1986), Dixon (1983), Weiss (1994), Wijnstok (1995), Domberger (1979), Bedrossian and Moschos (1988), Van Bergeijk and Haffner (1996)

Table 4.2: Conduct indicators for selected OECD economies

The performance indicators of Table 4.3 show that the mark-ups of prices over marginal costs are all significantly larger than one at the usual confidence intervals, with the UK and the US on the lower end of the scale and France and Japan on the upper end.[19] Finally, the spread in long run profits indicates the degree of persistence in the profit differentials of the most profitable and least profitable categories of firms. A low spread in long run profits indicates that competition eliminates any profit differentials to a large extent. Japan and Germany perform well according to this indicator of competition, while the Netherlands, France and Canada display a high degree of profit persistence.

Table 4.4 summarizes the above results. The table is based on 16 studies for 17 countries and shows only those countries for which three out of four methods of measuring the state of competition have been applied. The numbers in parentheses indicate the score of each country on any indicator; an index score of 100 is assigned to the best performing country (and 0 to the worst performing country). The table highlights the difficulties of obtaining a clear ranking in this respect.[20] The United States has the highest overall score, even though it does not perform well in terms of the flexibility of prices to demand and cost changes. A relatively clear picture also emerges for Australia and France, which have a relatively low score on all four indicators.

Estimated mark-up ratios for total manufacturing 1971-90		Spread in long-run profits (various estimation periods)	
high adjustment speed			
U.K.	1.14	Japan 1964-82	1.5
U.S.A.	1.15	Germany 1961-82	1.6
Denmark	1.16	U.K. 1951-77	2.6
Norway	1.16	U.S.A. 1964-80	3.7
Belgium	1.17	Sweden 1967-85	6.1
Netherlands	1.18	U.S.A. 1950-72	7.5
Canada	1.18	Netherlands 1978-91	8.6
Germany	1.18	France 1965-82	9.7
Italy	1.20	Canada 1968-82	18.7
Sweden	1.20		
Japan	1.22		
Australia	1.25		
Finland	1.25		
France	1.26		
low adjustment speed			

Sources: OECD Secretariat, Jenny and Weber (1990), Khemani and Shapiro (1990), Kleijweg and Nieuwenhuijsen (1995), Odagiri and Yamawaki (1990), Schwalbach and Mahmood (1990), Cubbin and Geroski (1990) and Mueller (1990).

Table 4.3: Performance indicators for selected OECD economies

4.5 The macroeconomic implications

The preceding section established that in many economies, distortions of competition seem to exist. However, the question is what the consequences of these distortions are. Even large deviations from the best practice benchmark should not be qualified as alarming if the costs in terms of foregone economic growth and employment are negligible. One possible route to follow is to use the indicators of the state of competition as scenario variables for simulations with applied models. Table 4.5 shows two of such simulations with computable general equilibrium models for the Dutch situation.[21]

Country	Profit persistence		Mark-up		Price flexibility		PMIC		Avg. score
1 U.S.A.	3.7	(92)	15%	(92)	6.3	(15)	42%	(100)	73
2 Sweden	6.1	(73)	20%	(50)	3.8	(60)	n.a.		61
3 Germany	1.6	(99)	18%	(67)	n.a.		95%	(7)	58
4 Japan	1.5	(100)	22%	(33)	1.6	(100)	99%	(0)	58
5 U.K.	2.6	(94)	14%	(100)	6.2	(16)	95%	(7)	54
6 Canada	18.7	(0)	18%	(67)	3.8	(60)	83%	(28)	39
7 Netherls.	8.6	(59)	18%	(67)	7.1	(0)	97%	(4)	32
8 Australia	n.a.		25%	(8)	4.7	(44)	90%	(16)	23
9 France	9.7	(52)	26%	(0)	n.a.		95%	(7)	20

Sources: See tables 2-3

Table 4.4: Overall ranking of countries according to four different indicators (index score between brackets, 100 = best practice)

Both simulations are counterfactuals: they show what the development of endogenous variables would have been if competition had been stronger.

The second column of Table 4.5 presents a historical simulation with the model MESEM over the period 1984-1990 where the PMIC is used as a scenario variable. The calculations are made under the assumption that the labour market clears in the long run (except for structural unemployment).The macroeconomic benefit of increasing competition is assumed to be equal to the difference between two model simulations over this period: one where the PMIC is equal to 100 per cent (which seems reasonable as the PMIC is equal to 97 per cent in The Netherlands) and one where the PMIC is only 50 per cent (which roughly corresponds to the situation in the United States). The effects of improving the functioning of the market mechanism are substantial as average annual growth of production increases by 0.5 percentage points while the growth rate of employment rises by 0.1 percentage points. Investment declines due to an improved allocation and better use of capital which causes capital productivity to rise.

In MIMIC, the long-run effects are simulated of a reduction in the mark-up in two sectors of the economy: the construction sector and the sheltered (service) sector. It is assumed that due to rent-sharing mechanisms, the reduction in the mark-up leads to wage moderation in these two sectors. Because of the relatively uniform distribution of wage claims across sectors in the Netherlands, it is assumed this wage moderation also spills over to the other sectors of the economy. Other sectors also benefit from the lower price of goods and services from the construction and sheltered sectors. The long-run effects of increasing competition are quite considerable as a 1 percentage point reduction in the mark-up increases aggregate production and employment by about 0.5 percentage points.

All in all, these simulations illustrate that policies aimed at increasing

	Reduction PMIC in MESEM from 100% to 50%[a]	Reduction mark-up ratio in MIMIC by 1%-point[b]
Private production	0.5	0.5
Private employment	0.1	0.6
Private consumption	0.0	0.5
Consumption price level	-0.6	-0.4
Private investment	-0.9	0.5
Exports	0.9	0.2

[a] Annual growth rates in percent, 1984-1990
[b] long run deviations from baseline in percent.
Sources: Van Bergeijk, Haffner and Waasdorp (1993) and Nieuwenhuis and Terra-Pilaar (1996)

Table 4.5: Macroeconomic benefits of increasing competition

competition in the Netherlands may have large potential benefits.[22] However, it is important to keep in mind that the effects of structural adjustment policies also depend on the functioning of the markets for labour and capital, especially in the short run. Correcting large market distortions may involve a substantial reallocation of labour and capital which can only take place quickly and efficiently when these markets function adequately.

4.6 Discussion

An important lesson from our discussion is that 'the invisible hand' can take many different forms, and each of the methodologies discussed here may measure a different aspect of competition. For example, prices may react quickly to changes in costs and demand, but if firms maintain excess capacity for strategic reasons, the assessments based on the product market inertia coefficient and the speed of price adjustment will not be the same. Thus, conclusions about the functioning of the market mechanism based on one indicator alone should be treated with caution. Future research should be directed at investigating the relationship between the various aspects of competition and how they could be operationalized.

Basically, the difficulties in identifying the state of competition are due to three categories of problems which are of a statistical, methodological and analytical nature. The statistical problems are the well known issues of comparability (differences in definitions, statistical sources and reliability) and availability. These problems are especially relevant when attempting international comparisons. A methodological problem is that trying to derive broad generalizations always involves a certain degree of aggregation over industries, firms and products. This hampers the interpretation of any indicator of competition. For example, consider any indication of weak competition. If this indication was derived by aggregating over different products, it is possible that competition is actually very fierce because monopoly-like positions can only be retained temporarily for each product and are then taken over by new or other products. The more you aggregate, the worse this problem becomes. The analytical problem is essentially caused by a lack of what Fisher (1989) calls 'generalizing theory.' Generalizing theory provides propositions that apply under fairly wide circumstances. In this case, theory should show how market structure, conduct and performance influence each other. With respect to the conduct and performance indicators, it is difficult to distinguish between[23]:

- slow/quick adjustment of prices or profits. This is due to the joint estimation of the speed of price or profit adjustment and a model for this adjustment itself. In this way, the level to which prices or profits should supposedly adjust to may not be correct because of an incorrect adjustment model;
- market power and efficiency. Market power supposedly leads to high profits, but this may also be due to superior efficiency. The important question in this respect is whether or not the rewards to efficiency improvements are of a temporary nature;
- competition and X-inefficiencies. Low profits can both be caused by fierce competition and by inefficiency;
- innovativeness and a lack of competition. Profits should be the (temporary) reward for innovativeness, because otherwise the incentive to innovate would diminish. Profits are also a necessary condition for innovation. So high profits may both be due to a lack of competition and to innovativeness.

In our view, these problems can only be (partly) circumvented by the simultaneous study of structure, conduct and performance at a level of aggregation which is as low as possible. When several indicators point into the same direction, this may be a reason for a further investigation. The indicators and methodologies described in this paper will provide useful material for such an approach. Note however that many of the conduct and performance indicators have already been tested for correlation with market structure variables. The results were generally consistent with theoretical expectations.[24]

Moreover, the macroeconomic implications of a lack of competition seem to be considerable. Both simulation exercises show significant benefits in terms of competitiveness, employment and economic growth of policies aimed at stimulating competition on markets for goods and services. However the results are sensitive to the ability of labour and capital markets to adjust to the changes in resource allocation.

Finally, our analysis shows that significant departures from perfect competition could be assessed in almost every single case. This calls into question many current day analyses in which product and factor markets are assumed to be competitive and clear instantaneously. As Okun (1981, p. 19) wondered:

> But why should one cling to the maintained hypothesis that product markets are always in equilibrium in the short run? Perpetuating the hypothesis preserves the principle of conservatism about, and respect for, this classical model that has served for so long in so many dimensions. But it is difficult to find any other virtue. In particular, empirical evidence pulls the analyst in the opposite direction.

Acknowledgments

The authors thank Henk Jager, Jarig van Sinderen and participants of the workshop for useful comments. Correspondence: Economic Policy Directorate, Research Unit, Ministry of Economic Affairs, P.O. Box 20101, 2500 EC The Hague, The Netherlands. The views expressed in this paper do not necessarily reflect the opinion of the government of the Netherlands or the Netherlands Central Bank.

Notes

1. This expression was first coined by De Jong (1990).
2. See OECD (1993) for an overview of the restrictive agreements in the Cartelregister.
3. This caused C. van Gent of the Dutch ministry of Economic Affairs to call the recent overhaul of the Dutch competition policy a 'revolution without revolutionaries' (Van Gent 1996).
4. The theoretical justification for these analyses was not given until the publication of the influential paper of Cowling and Waterson (1976).
5. See Geroski and Schwalbach (1991) and Geroski (1991) for an overview of the theoretical and empirical literature.
6. See Haffner (1993) for a survey of a number of investigations on the relationships between market structure variables and competition in the Netherlands. Import competition has a significantly positive effect on competition in four out of five cases (in one case the effect

    was insignificant).

7.   Martin (1993) and Bresnahan (1989) provide overviews of this literature. See also Maggi (1996) for a recent variation on this theme where the mode of competition is measured by a structural parameter which is potentially easier to observe.

8.   Legal strategies are also considered as an indicator of firm conduct. An example is the patenting strategy of firms, which may help firms to appropriate a larger part of the benefits of their research (Cohen and Levin 1989, pp. 1090–95). However, legal strategies are difficult to measure. Whether advertising is considered an aspect of market structure or firm conduct, is rather arbitrary. As far as product differentiation through advertising is an intrinsic characteristic of the product concerned, it may be considered an aspect of market structure. However, because advertising is nowadays an integral part of company strategy and it is often consciously used both to create market niches for the own product and barriers to entry for competitors, we rank advertising under 'conduct.' The same holds for research and development efforts of firms, which may be classified both as 'conduct' and as 'performance.' Here, we make the distinction between the R&D input (conduct) and the results of these efforts, the output of new products and processes (performance).

9.   This depends on how successful the product and the advertising campaign concerned have been. No one would attach value to the advertising campaign of a flunked product, but a brand name of a successful product may have potential to be (re-)sold.

10.   See Das *et al.* (1993) for a recent overview of the empirical evidence. In this paper, a measure of market share instability is used to test for the effects of advertising on competition. The empirical results show a statistically significant positive correlation between advertising intensity and market share instability in 163 US manufacturing industries during the period 1978–88. According to this investigation, therefore, advertising generally stimulates competition.

11.   See Cohen and Levin (1989) for an overview of the—still inconclusive—empirical evidence concerning R&D expenditures.

12.   Price–cost margins usually vary over the business cycle, as a better utilization of productive capacity results in lower unit costs during business cycle upswings. Any empirical investigation into the relation between price–cost margins and competition should therefore control for the effects of the business cycle.

13.   See Hall (1986), Roeger (1995) and Felder (1995) for three more sophisticated versions of the Lerner index. See also Carlton and Perloff (1994), chapter 9 on some of the measurement issues.

14.   Another explanation is the hypothesis that efficient firms achieve both high market shares and high profits. The observed positive correlation between concentration and profit levels can therefore also be due to the superior efficiency of large firms. See Eckard (1995) and Bennenbroek and Harris (1995) for recent investigations of this topic.

15.   See also Schmalensee (1989), p. 976 on this conclusion.

16.   Another method relies on a direct comparison of productivity levels using industry specific conversion factors to make the different sectoral productivity levels internationally comparable. See Van Ark (1995).

17.   See for example Scherer and Ross (1990, pp. 668–78) for an overview of the empirical evidence, and Donni and Fecher (1994) for a recent application to the insurance industry in the OECD countries. See also MacDonald (1994), who shows that import competition led to large increases in labour productivity growth in highly concentrated industries in the US during the period 1972–87.

18.   Not all of the differences in PMICs are statistically significant, however. For example, the PMIC for Japan is indistinguishable from the PMICs of Italy and the Netherlands at

a 95 per cent confidence rate. However, the PMICs of the European countries do differ significantly from the PMICs of Canada, the United States and Taiwan.

19. Again, not all estimated mark-ups differ in a statistical sense. For example, the point estimate of the United States differs significantly from the estimates for Japan, France, Italy, Australia, Finland and Sweden (but not from the other estimates).

20. Of course, this is only one out of many possible rankings.

21. See van Sinderen (1993) and Gelauff and Graafland (1994) for more details on MESEM and MIMIC, respectively.

22. See Van Bergeijk and Haffner (1996), chapter 6 for a survey of macroeconomic analyses of the effects of privatization, deregulation and a more vigorous competition policy. These analyses confirm the above conclusion.

23. Competition policy authorities also have to cope with the problem of distinguishing between tacit collusion and strategic interdependence. In the latter case, collusive equilibria can result even if the conduct of the market participants is noncooperative (Phlips 1996). This result is the more probable, the more transparent the market concerned is.

24. See Van Bergeijk and Haffner (1996), chapter 4 and Haffner (1993), p. 49 for an overview.

References

Ark, B. van (1995), "Manufacturing Prices, Productivity and Labor Costs in Five Economies," *Monthly Labor Review*, Bureau of Labor Statistics, July:56–72.

Bennenbroek, N. and R.I.D. Harris (1995), "An Investigation of the Determinants of Profitability in New Zealand Manufacturing Industries," *Applied Economics,* 27:1093–101.

Bain, J.S. (1956), *Barriers to New Competition*, Harvard University Press, Cambridge, Mass.

Baumol, W.J., J. Panzar and R. Willig (1982), *Contestable Markets and the Theory of Industry Structure*, Harcourt Brace Jovanovich, San Diego.

Bedrossian, A. and D. Moschos (1988), "Industrial Structure and the Speed of Price Adjustment," *Journal of Industrial Economics,* 26:459–75.

Bergeijk, P.A.G. van, and R.C.G. Haffner (1996), *Privatization, Deregulation and the Macroeconomy*, Edward Elgar, Cheltenham.

Bergeijk, P.A.G. van, R.C.G. Haffner and P.M. Waasdorp (1993), "Measuring the Speed of the Invisible Hand: The Macroeconomic Costs of Price Rigidity," *Kyklos,* 46:529–44.

Bresnahan, T.F. (1989), "Empirical Studies of Industries with Market Power," in: R. Schmalensee and R. Willig (eds.), *Handbook of Industrial Organization*, North-Holland, Amsterdam:1011–58.

Brozen, Y. (1971), "Concentration and Structural and Market Disequilibria," *Antitrust Bulletin* 16 (2):241–8.

Carlton, D.W. and J.W. Perloff (1994), *Modern Industrial Organization*, Harper Collins, New York.

Cohen, W.M. and R.M. Levin (1989), "Empirical Studies of Innovation and Market Structure," in: R. Schmalensee and R. Willig (eds.), *Handbook of Industrial Organization*, North-Holland, Amsterdam: 1060–107.

Cowling, K. and M. Waterson (1976), "Price-Cost Margins and Market Structure," *Economica* 43:267-74.

Cubbin, J.S. and P.A. Geroski (1990), "The Persistence of Profits in the United Kingdom," in: D.C. Mueller (ed.), *The Dynamics of Company Profits: An International Comparison*, Cambridge University Press, Cambridge:147–69.

Das, B.J., W.F. Chappell and W.F. Shughart (1993), "Advertising, Competition and Market Share Instability," *Applied Economics,* 25:1409-12.

Davies, S. (1988), "Concentration," in: S. Davies and B. Lyons (eds.), *The Economics of Industrial Organisation*, Longman, London:73–126.

Davies, S. and B. Lyons (eds.) (1988), *The Economics of Industrial Organisation*, Longman,

London.

Dijksterhuis, G.B., H.J. Heeres and A.J.M. Kleijweg (1995), "Indicatoren voor dynamiek" (Indicators of Economic Dynamism; in Dutch), *Economisch-Statistische Berichten*, 80: 652–7.

Dixon, R. (1983), "Industry Structure and the Speed of Price Adjustment," *Journal of Industrial Economics*, 23:25–37.

Domberger, S. (1979), "Price Adjustment and Market Structure," *Economic Journal*, 98:96–108.

Donni, O. and F. Fecher (1994), "Efficiency and Productivity of the Insurance Industry in the OECD Countries," *mimeo*, University of Liège, Liège.

Eckard, E.W., (1995) "A Note on the Profit–Concentration Relation," *Applied Economics*, 27:219–23.

Encaoua, D. and P. Geroski (1986), "Price Dynamics and Competition in Five OECD Countries," *OECD Economic Studies* (6), Paris, OECD:47–74.

Felder, F. (1995), "The Use of Data Envelopment Analysis for the Detection of Price above the Competitive Level," *Empirica*, 22:103–13.

Fisher, F.M. (1989), "Games Economists Play: a Noncooperative View," *RAND Journal of Economics*, 20 (1):113–24.

Geroski, P.A. (1991), *Market Dynamics and Entry*, Basil Blackwell, Cambridge.

Geroski, P.A. and J. Schwalbach (1991), *Entry and Market Contestability*, Basil Blackwell, Cambridge.

Gent, C. van, (1996) "New Dutch Competition Policy: A Revolution without Revolutionaries," Paper presented at the Conference *Economic Science: An Art or an Asset?*, The Hague, January.

Gelauff, G.M.M. and J.J. Graafland (1994), *Modelling Welfare State Reform*, Elsevier, Amsterdam.

Haffner, R.C.G. (1993), "De meting van dynamiek: een onderzoek naar de marktwerking op goederenmarkten in nederland" [Measuring Market Dynamics: A Survey for Product Markets in the Netherlands], *mimeo*, Erasmus University, Rotterdam and Ministry of Economic Affairs, The Hague.

Hall, R.E. (1986), "Market Structure and Macroeconomic Fluctuations," *Brookings Papers on Economic Activity*, 2:285–322.

Jenny, F.Y. and A.P. Weber (1990), "The Persistence of Profits in France," in: D.C. Mueller (ed.), *The Dynamics of Company Profits: An International Comparison*, Cambridge University Press, Cambridge:123–9.

Jong, H.W. de (1990), "Nederland: het kartelparadijs van Europa?" [The Netherlands: the Cartel Paradise of Europe?; in Dutch], *Economisch-Statistische Berichten*, 75:244–248.

Khemani, R.S. and D.M. Shapiro (1990), "The Persistence of Profits in Canada," in: D.C. Mueller (ed.), *The Dynamics of Company Profits: An International Comparison*, Cambridge University Press, Cambridge:77–105.

Kleijweg, A.J.M. and M.H.C. Lever (1994), "Entry and Exit in Dutch Manufacturing Industries," *EIM/Fundamental Research* 9409/E, Zoetermeer.

Kleijweg, A.J.M. and H.R. Nieuwenhuijsen, (1995) "Winstpersistentie in de Nederlandse Industrie (1978–91" (Persistence of Profits in Dutch Manufacturing 1978–91; in Dutch) *EIM Fundamental Research*, Zoetermeer.

Kremers, J.J.M. (1991), "Naar een sterkere binnenlandse groeidynamiek," (Towards stronger interior dynamism of growth; in Dutch) *Economisch-Statistische Berichten*, 76:1228–32.

Kotler, P. (1980), *Principles of Marketing*, Englewood Cliffs, New Jersey.

Kreps, D.M. and M.A. Spence (1985) "Modeling the Role of History in Industrial Organization and Competition," in: G.I. Feiwel (ed.), *Issues in Contemporary Microeconomics and Welfare*, MacMillan, London:340–78.

Kühn, K.-U., P. Seabright and A. Smith (1992), "Competition Policy Research: Where Do We Stand?," *CEPR Occasional Paper 8*, Centre for Economic Policy Research, London.

Love, J.H. (1995), "The Measurement of Entry Rates: Reconsideration and Resolution, *Empirica*, 22:151–157.

Lyons, B. (1988), "Barriers to Entry," in: S. Davies and B. Lyons (eds), *The Economics of Industrial Organisation*, Longman, London:26–72.

MacDonald, J.M. (1994), "Does Import Competition Force Efficient Production?," *Review of*

126   *Robert C.G. Haffner and Peter A.G. van Bergeijk*

Economics and Statistics, 76:721–7.

Maggi, G. (1996), "Strategic Trade Policies with Endogenous Mode of Competition," *American Economic Review*, 86:237–58.

Martin, S. (1993), *Advanced Industrial Economics*, Blackwell Publishers, Cambridge.

Mueller, D.C. (1990), "The Persistence of Profits in the United States," in: D.C. Mueller (ed.), *The Dynamics of Company Profits: An International Comparison*, Cambridge University Press, Cambridge:35–59.

Mueller, D.C. (1991), "Entry, Exit and the Competitive Process," in: P.A. Geroski and J. Schwalbach (eds), *Entry and Market Contestability,* Basil Blackwell, Cambridge:1–23.

Nieuwenhuis, A. and P.A. Terra-Pilaar (1996), "Aspecten van marktwerking in beschutte bedrijfstakken" (Aspects of Market Functioning in Sheltered Sectors; in Dutch), *mimeo*, CPB Netherlands Bureau of Economic Policy Analysis.

Odagiri, H. and H. Yamawaki (1990), "The Persistence of Profits: International Comparison," in: D.C. Mueller (ed.), *The Dynamics of Company Profits: An International Comparison*, Cambridge University Press, Cambridge:169–85.

OECD (1993), *Economic Surveys* (1993)/(1994), *The Netherlands*, OECD, Paris.

OECD (1994a), *Assessing Structural Reform: Lessons for the Future*, OECD, Paris.

OECD (1994b), *Employment Outlook*, OECD, Paris.

Okun, A.M. (1981), *Prices and Quantities: A Macroeconomic Analysis*, Basil Blackwell, Oxford.

Paech, N. (1995), *Die Wirkung potentieller Konkurrenz auf das Preissetzungsverhalten etablierter Firmen bei Abwesenheit strategischer Asymmetrien* (The Impact of Potential Competition on Pricing Policies of Established Firms in the Absence of Strategic Asymmetries; in German), Duncker & Humblot, Berlin.

Philips, L. (1996), "On the Detection of Collusion and Predation," *European Economic Review* 40:495–510.

Roeger, W. (1995), "Can Imperfect Competition Explain the Difference between Primal and Dual Productivity Measures? Estimates for US manufacturing," *Journal of Political Economy* 103:316–30.

Ross, S.A. and M.L. Wachter (1975), "Pricing and Timing Decisions in Oligopoly Industries," *Quarterly Journal of Economics,* 89:115–37.

Scherer, F.M. and D. Ross (1990), *Industrial Market Structure and Economic Performance*, Third edition, Houghton Mifflin, Boston.

Schmalensee, R. (1989), "Inter-industry Studies of Structure and Performance," in: R. Schmalensee and R. Willig (eds.), *Handbook of Industrial Organization,*: North-Holland, Amsterdam:951–1009.

Schwalbach, J. and T. Mahmood (1990), "The Persistence of Profits in the Federal Republic of Germany," in: D.C. Mueller (ed.), *The Dynamics of Company Profits: An International Comparison*, Cambridge University Press, Cambridge:105–23.

Sinderen, J. van (1993), "Taxation and Economic Growth," *Economic Modelling,* 10:285–300.

Weiss, C.R. (1994a), "Market Structure and Pricing Behaviour in Austrian Manufacturing," *Empirica,* 21:115–31.

Wijnstok, J.C. (1995), *De snelheid van prijsaanpassing in Nederlandse sectoren* (The Speed of Price Adjustment in Dutch Industries; in Dutch), Ministry of Economic Affairs and Erasmus University Rotterdam, The Netherlands.

PART III
Financial
Market
Imperfections

5 The Importance of Credit for Macroeconomic Activity: Identification through Heterogeneity

Simon Gilchrist and Egon Zakrajšek

5.1 Introduction

Recent work in macroeconomics emphasizes the role of credit in the transmission mechanism for monetary policy and as a propagation mechanism of business cycle shocks.[1] While much evidence has been gathered, not all researchers agree on the relevance of credit for the transmission of monetary policy nor for its relevance as a propagation mechanism of business cycle shocks. For the most part, every one agrees on the facts at hand but differs on their interpretation. In short, the argument is over identification. The primary purpose of this paper is to clarify the identification issues involved; to highlight those identification schemes that are promising avenues for measuring the importance of credit in aggregate fluctuations; and to provide a discussion of both previous evidence and new evidence in light of the identification schemes proposed.

The role of credit in the monetary transmission mechanism can be divided into two separate phenomena. The first has been dubbed the 'credit channel' of monetary policy. The second has been called the 'financial accelerator.' Both phenomena rely on credit frictions that are absent in the standard neoclassical models that economists typically use to explain business cycle fluctuations. Both, however, are complimentary to the standard 'money channel' described in textbook treatments of monetary transmission. As such, they provide an additional rather than competing mechanism for the propagation of monetary policy shocks.

The credit channel emphasizes the importance of bank lending in the monetary transmission mechanism. The existence of the credit channel presumes that capital markets are imperfect, owing to information asymmetries between borrowers and lenders. As a consequence, some borrowers are unable to borrow on the open market without paying large premiums on external finance. Banks specialise in information intensive loans and are able to reduce the premium for bank-dependent borrowers. Monetary policy has real consequences because of its effect on banks' ability to lend. Open market operations lead to a contraction in reserves and a decrease in funds available for lending. As long as banks face imperfections in issuing certificates of deposit (CDs) to offset the contraction in reserves, bank lending must fall. Bank-dependent borrowers, consequently, are forced to seek funds at a much higher cost on the open market—to the extent they are able to obtain funds at all. As a result, spending by bank-dependent borrowers contracts.

The financial accelerator emphasizes the importance of balance sheet conditions in propagating shocks to the economy. As with the credit channel, the existence of the financial accelerator depends on the assumption that capital markets are imperfect, and that external and internal finance are not perfect substitutes. The crucial point for the financial accelerator is that the size of the premium on external funds depends on the firm's balance sheet condition. As balance sheets deteriorate following a contractionary monetary policy—regardless of whether the initial effect comes through interest rates or the initial spending decline of bank-dependent borrowers—premiums on external finance rise, exacerbating the overall decline in spending.

Because the financial accelerator relies only on the assumption of credit market frictions and not on the additional assumption that a contraction in reserves limits banks' ability to lend, it is both a broader phenomenon than the credit channel, and a necessary condition for the existence of the credit channel. Thus evidence in favour of the financial accelerator is crucial for proving the existence of the credit channel, while the converse is not true.

All convincing evidence in favour of either a credit channel or the financial accelerator comes from studies that focus on differential behaviour of agents. This is the premise of our paper. The focus on differential behaviour is important for two reasons. First, models that incorporate financial frictions are more relevant for certain types of agents, certain classes of borrowers, and certain sectors of the economy. The propagation mechanisms generated by these models are more relevant at certain points in the business cycle, namely when cash flows are dropping and balance sheets are deteriorating. Second, because of the difficulties associated with formulating and estimating true structural models, empirical exercises seeking to establish the validity of either a credit channel or a financial accelerator must make comparisons against benchmarks where such credit effects are less likely to be relevant. By observing and measuring the differential behaviour of economic agents under consideration, one can potentially attribute some, if not all, of the difference in behaviour to frictions caused by credit markets. We elaborate on this premise in the next section. We then turn to a discussion of the existing evidence and provide some new evidence on the relevance of credit for monetary policy and macroeconomic fluctuations.

To limit the scope of the discussion, we only address firm behaviour, although all of the identification issues apply equally well to consumers; see Attanasio (1994) for a recent discussion of credit issues on the consumer side. To further limit the scope of the paper, we center our discussion around evidence generated from one data set: the *Quarterly Financial Reports* (QFR). Thus a secondary purpose of our paper is to provide a progress report on research using the QFR data. Because it is available at high frequency and at various levels of aggregation, the QFR data set is uniquely suited for analyzing credit issues and how they relate to macroeconomics. As we discuss below, the QFR data has already provided valuable insights into the identification issues raised by credit market imperfections. In addition, the QFR data has provided substantial evidence in favour of a credit mechanism, especially through the financial accelerator described above.

5.2 Identification through heterogeneity

To understand the essential role that heterogeneity plays in any identification scheme used to measure the importance of credit in the economy, it is useful to review briefly the theoretical underpinnings that motivate the existence of a premium on external funds, and how such a premium would respond to changes in interest rates and aggregate demand conditions. We then turn to a discussion of financial intermediaries and the role of monetary policy.

As a starting point, consider the implications of neoclassical investment theory. According to this theory, firms make investment decisions to maximise the net present value of profits. If interest rates rise, the net present value of profits falls, making investment less attractive. If expected future profits fall, net present value also falls, once again leading to a drop in investment spending. It is important to note that the firm's investment decision depends entirely on the future returns of the specific project under consideration and not on the current or past financial position of the firm. If the firm must borrow to complete or undertake the investment project, creditors are willing to lend the necessary funds at the current open-market interest rate. Thus we have the celebrated Modigliani-Miller (1958) result that real and financial decisions of the firm are completely separable.

In the presence of capital market imperfections, the separation of real and financial decisions no longer occurs. Balance sheet conditions affect the firm's ability to borrow at current interest rates. The theoretical motivation for this link can be found in the vast literature on asymmetric information and moral hazard in credit markets. An important insight from this literature is that such credit market imperfections create a wedge between the cost of external and internal finance. This wedge exists to compensate lenders for the risk that a borrower may either *ex ante* misrepresent the value of a given investment project or may *ex post* behave in a manner that expropriates value from the lender. To mitigate such risk, the lender must monitor the borrower, incurring costs in the process.

In general, the premium on external funds will be highest where information asymmetries are the most severe, and where the risk of opportunistic behaviour is hardest to mitigate. Thus small firms with idiosyncratic projects that are more difficult to value than those of large firms will face higher premiums. Younger firms with returns less known to the market will face higher premiums. By the same token, firm with projects backed by collateral will face lower premiums. More generally, the lower the collateralisable net-worth of the firm, the greater the premium on external funds.

An example of such a situation is displayed in Figure 5.1.[2] The dd line represents the demand for funds by the firm. It is a downward sloping function of the cost of funds. The ss line represents the supply of funds. Up to the point W (the firm's net-worth), lenders face very little risk of opportunistic behaviour and are willing to lend at the open market interest rate. Beyond W, however, lenders charge a premium over the open market rate to compensate for the increased probability of opportunistic behaviour on the part of borrowers. Because of the premium on external funds, the supply of funds curve for the individual firm is upward sloping, leading to an investment level I^*, below the perfect markets level I^P.

While the under-investment result is interesting in its own right, what matters for understanding the effects of monetary policy shocks is how the premium on external funds varies with both the state of aggregate demand and the risk-free interest rate. We consider both in turn. Figure 5.2 shows the effects of a rise in demand. In the perfect markets case, as demand increases, expected future profits rise, making firms more willing to invest. This shifts out the demand curve to dd' and raises investment spending for a given cost of funds. The credit market frictions amplify this effect. At higher profit levels, net-worth has increased for a given project size, and the benefits of reneging on contractual obligations are lower. Borrowers are less likely to default, and lenders need not monitor as often. With less monitoring, the required premium on external funds falls (a rightward shift in the ss curve), and the effect of the demand shock on investment spending is magnified.

A rise in risk-free rates has a similar magnification effect as shown in Figure 5.3. At higher interest rates, default probabilities rise, causing lenders

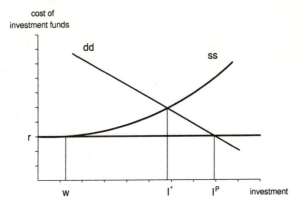

Figure 5.1: The premium for external funds

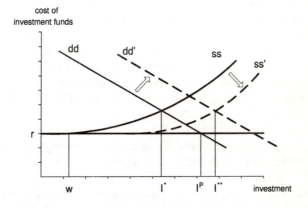

Figure 5.2: The premium for external funds. Impact of a rise in demand.

to increase the premium on external funds. An increase in the premium puts firms at even greater risk of default. This leads to an increase in the required premium that is much larger than the rise in the open-market interest rate. Once again, the initial shock is magnified through its effect on the premium for external funds.

The connection between the financial accelerator and the credit channel is easily understood once one recognises the special role that banks play in

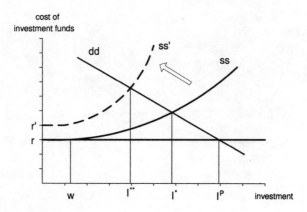

Figure 5.3: The premium for external funds. A rise in the riskless interest rate

the credit intermediation process. Given the high cost of obtaining information for certain classes of borrowers, it is natural to expect certain institutions to specialise in information gathering activities. Traditionally, banks have performed this role, in part because of the information advantage they obtain through observing would-be-borrowers' deposit flows.[3] Over time, institutions such as banks develop knowledge specific to their class of borrowers in general and to their own customers in particular. By reducing the information asymmetry, financial intermediaries can lower the premium on external funds. Because such knowledge is difficult to convey to third party lenders, disruptions in the supply of credit available through these intermediaries can have immediate and large consequences on spending.[4] With traditional borrowing relationships destroyed, many bank-dependent firms and consumers will be forced into the market where they face stiff premiums on external funds. At the prevailing rates, many may simply forgo planned investment projects, leading to large sudden drops in spending by certain classes of borrowers.

The link between theory and empirical work is established by determining factors that are likely to influence the size of the premium on external finance, the degree of the magnification effects, and the extent to which a firm must rely on bank loans rather than some other form of finance less subject to supply shocks through open market operations. Both the size of the premium on external finance, and the degree to which a firm is tied to the bank loan market rather than other forms of external finance are heavily influenced by the firm's size, age and previous financial track record. Industry specific characteristics such as the riskiness of projects, the degree to which investments are collaterisable, and the difficulty associated with evaluating borrowers' claims are also likely to be important determinants of the premium on external funds. By comparing the behaviour of firms with such characteristics relative to the behaviour of firms which have little difficulty obtaining funds at the open market rate, we can test for the presence of financial frictions and measure the extent to which these frictions distort firm hiring and investment patterns.

For the magnification effect, a key determinant of the premium on external funds is W—the net-worth of the firm, or the level of unencumbered assets or future earnings available as collateral for new investment projects. Given the complexity of present day financial contracts, it is difficult to quantify precisely such a concept. Nonetheless, some basic indicators of financial health commonly used by market analysts seem informative. Firms with high leverage ratios are likely to face greater difficulties obtaining new funds on the market, as are firms with low coverage ratios—that is, firms with a high level of current interest payments relative to their earnings. By using disaggregated data to compare behaviour of firms in different financial positions, we can potentially measure the distortions created by financial frictions.

In addition to being useful indicators of firms with imperfect access to credit markets, both of these variables are intuitively appealing in understanding the asymmetric nature of the financial accelerator discussed above. In good times, as profits increase and firms become flush with cash, borrowers have little trouble financing new investment projects and making existing debt payments. Under such conditions, a shock to earnings or interest rates will have very little magnification effect through the premium on external funds. As the economy turns down and balance sheet positions are weakened, however, a greater number of firms find themselves saddled with large debt, high interest payments, and low cash flow. In such precarious financial positions, these firms will face high premiums on external funds, either through direct price effects, credit rationing, or more severe non-price contract terms such as restrictive debt covenants. With only aggregate time-series, however, one cannot identify the extent to which firms are moving from one class to another and, therefore, how important is a credit mechanism in creating business cycle asymmetries.

The identification strategy of comparing one class of firms to another to measure the extent and importance of financial frictions takes advantage of the inherent heterogeneity underlying most aggregate time-series data. In addition, it emphasizes the limitations of models that focus on representative agents. The limitation of representative agent models does not come from our inability to formulate representative agent models with important credit frictions.[5] Rather, the limitation comes from our inability to distinguish such models from business cycle models with alternative propagation mechanisms that do not rely on credit frictions. Only by relying on the fact that some firms, at least some of the time, do not face the adverse consequences associated with limited access to credit markets, can we identify how other firms, at other times, are seriously affected by such restricted access.

5.3   Evidence from financial data

We now turn to a discussion of identification through the use of financial data. We focus on the comparison between aggregate lending data and disaggregated lending data. We argue that only disaggregated lending data can provide convincing evidence of either the presence of a credit channel or the financial accelerator. We also wish to emphasize that evidence from lending data alone is not sufficient to establish the relevance of credit frictions in propagating and amplifying business cycle shocks. We also need supporting evidence from non-financial data. This is discussed in the next section.

5.3.1   Interpreting the existing evidence

Prominent studies that attempt to gauge the importance of credit in the macroeconomy have focused on the following criteria: To what extent do movements in aggregate credit or aggregate bank loans either explain or lead movements in real side variables? Both King (1986) and Ramey (1993) show that total bank lending has no marginal forecasting power for either industrial production or other macro real-side variables. Romer and Romer (1990) show that monetary aggregates fall immediately following a shift to tight monetary policy and 9 months prior to the ensuing drop in output, whereas bank loans fall only coincidently with the resulting decline in output. These results are taken by the authors as strong evidence against a credit channel for monetary policy, and at least in King's case, as strong evidence against credit mattering at all for the transmission of monetary policy shocks.[6]

As Bernanke and Blinder (1992), Gertler and Gilchrist (1993), and Bernanke, Gertler, and Gilchrist (1994) all emphasize, however, such empirical exercises do not provide information on either the relevance of a credit channel for monetary policy or the presence of a financial accelerator. First, banks liquidate securities rather than contract loan volume immediately following a tightening of monetary policy. Banks do so in part to offset the effects that tight money will have on their ability to lend to valued customers. Thus the fact that bank loans only fall with a 9 month lag rather than immediately following a switch to tight monetary policy provides no information about banks' ability to obtain funds through CD issuance and, consequently, cannot be considered as a relevant test for the existence of a credit channel.

Second, the notion that bank lending should have marginal predictive power once one controls for monetary policy through either a monetary aggregate or an interest rate instrument such as the Federal Funds rate assumes that credit disruptions provide an important independent source of shocks. According to the general equilibrium theories linking balance sheet conditions to real activity, however, no such shocks need exist.[7] The financial accelerator is an amplification device, not an independent source of variation. Although disruptions to credit supply—through independent shocks to bank lending such as changes in regulatory policy, for example—may have large effects, such shocks need not be empirically important for credit to matter in conditioning the economy's overall response to either changes in monetary policy or other sources of variation.

Finally, once one recognises that the effects of credit frictions on debt quantities are most likely identified through the differential behaviour of certain classes of borrowers, one must question the relevance of any exercise that focuses only on aggregate lending patterns. Kashyap, Stein and Wilcox (1993) (KSW hereafter) were the first to make this point empirically. They argued that a credit channel for monetary policy could be more readily identified through the differential behaviour of bank loans relative to commercial paper movements rather than by looking only at total lending. According to KSW, bank loans would shrink following tight money, whereas commercial paper would expand, in part because, customers shut out of the bank loan market would naturally turn to commercial paper.

The evidence supports this contention. Bank loans drop relative to commercial paper following tight money, though the mechanism is not quite what KSW described. Using even more disaggregated data for the manufacturing sector, both Oliner and Rudebusch (1992) and Gertler and Gilchrist (1993) show that movements in the aggregate mix between commercial paper and bank loans are not driven by bank *vs.* non-bank lending but by small- *vs.* large-firm borrowing. In particular, following tight money, all types of borrowing by small firms fall, whereas borrowing by large firms actually expands in the first few quarters following a monetary contraction.

The expansion of credit to large firms following tight money is not often recognised and is worth emphasising in the context of identifying the effects of monetary policy shocks through aggregate lending behaviour. As Gertler and Gilchrist (1993) point out, many firms have a strong countercyclical demand for short-term credit as inventories rise, and cash flows fall in the first few quarters following a tightening of monetary policy or at a business cycle turning point. If funds were available at prevailing open market rates, all firms would increase their borrowing to smooth the effect of declining cash flows. Only those firms with relatively unimpeded access to credit, however, are able to obtain the desired funds. Thus following a shift to tight money, 'high quality' firms with access to the commercial paper market expand their credit (Calomiris, Wachtel and Himmelberg, 1994); firms with bank commitments draw down their lines of credit (Morgan, 1992); and the 'high quality' bank customers receive the funds obtained through the banking system's security liquidation (Lang and Nakamura, 1992). Those left out in the cold are the smaller, riskier, less-valued bank customers which, once shut out of the bank loan market, have no recourse but to curtail operations, liquidate inventories, cut investment spending and reduce their workforce. Their reductions in spending further exacerbates the downturn, leading to an even greater contraction than before.

Once one recognises the countercyclical demand for credit generated by an adverse shock to monetary policy, it becomes immediately obvious that important credit frictions may be at work with very little observable effect on aggregate credit quantities, especially over the first few quarters following a switch to tight monetary policy. If this were the case, we would not necessarily expect any observable relationship between aggregate credit movements and future output movements. We would expect, however, an observable relationship between the differential borrowing rates of high- *vs.* low-quality borrowers and future output movements, especially in a framework that does not control for the original source of the shock. In addition, we would expect monetary policy to have a strong effect on the relative borrowing patterns of these two types of firms.

5.3.2 Some new evidence from financial data

In this section, we test the proposition that the borrowing rates of 'low quality' firms relative to 'high quality' firms have predictive power for aggregate variables. We also test the proposition that differential movements in such borrowing rates are influenced by monetary policy. To obtain a debt measure for 'low quality' and 'high quality' firms, we follow Gertler and Gilchrist (1993) and use the ratio of short-term debt of small manufacturing firms relative to short-term debt of all manufacturing firms, constructed from the published QFR data. We call this ratio the small/all mix.[8] We view the first test as complementary to Ramey (1993) who uses the QFR data to examine the predictive power of short-term debt issued by small firms relative to that of large firms for total industrial production. We view the second test as complementary to the evidence presented in Gertler and Gilchrist (1993) who characterise the behaviour of small firm and large firm borrowing in response to monetary policy shocks using impulse response functions. In addition, both tests complement the analysis of small- *vs.* large-firm borrowing patterns provided by Oliner and Rudebusch (1992).

We test the first proposition by examining the predictive power of the small/all mix for the following measures of aggregate economic activity: real GNP, manufacturing industrial production, manufacturing inventories, and manufacturing employment.[9] We also examine the predictive power of the small- and large-firm debt series separately, as well as the relative behaviour of bank and non-bank debt for small and large firms. Table 5.1 reports the results of this exercises in the context of a bivariate VAR system. The top panel reports probability values from the exclusion tests for each debt variable across the various measures of real economic activity.[10] The bottom panel reports the t-statistics for the sums of coefficients on the debt variables and thus provides an indication of the sign of the effect that each debt variable has on real activity.

The bivariate results provide strong support for the hypothesis that credit flows between small and large firms predict real economic activity. The probability values from the exclusion test for the small/all mix are less than 0.01 in three out of the four cases. In addition, the t-statistics on the sums of coefficients indicate that an increase in the small/all debt ratio leads to a highly significant increase in the growth rates of GNP, manufacturing industrial production, and manufacturing employment, as one would expect. The effect on inventories is ambiguous and probably reflects the dynamics associated with unexpected inventory buildup following a slowdown in economic activity.

Separating these variables by small firms *vs.* large firms and bank *vs.* non-bank debt indicates a number of interesting patterns. First, differences in the predictive power of debt variables arise through differences in class of borrower and not through differences in class of debt. Thus bank debt and non-bank debt behave in a similar manner, but small *vs.* large firm debt does not. In particular, for large firms increases in either bank or non-bank debt predict *declines* in real economic activity while for small firms the opposite occurs. Although, the similarity in predictive power of bank *vs.* non-bank debt provides potential evidence against a 'direct credit channel' for monetary policy, it is important to be cautious with this interpretation. As we argued above, we would expect banks to continue to make loans to their larger, more valued customers as they sell off securities. In addition, for small firms, the non-bank category is very small and is not as reliable of an indicator of credit behaviour.[11]

Second, the t-statistics and p-values for exclusion tests of the first three lending variables confirm that the predictive power of the mix between bank loans and commercial paper for manufacturing comes entirely through the ratio of small firm borrowing relative to total borrowing (that is, the small/all mix) and not through differential movements between bank loans and commercial paper by firms with potential access to both markets.[12] This finding further supports the evidence presented in Oliner and Rudebusch (1992), Gertler and Gilchrist (1993), and Calomiris *et al.* (1994) that differential movements between bank loans and commercial paper reflect differential movements in debt by type of borrower and not by type of debt.

Finally, it is worth noting that much of the predictive power of the small/all mix can be captured by looking only at large firm behaviour in the bivariate regressions. When we augment these regressions to include other variables such as the Federal Funds rate and inflation, as in Table 5.2, the predictive power of large firm debt variables vanishes, while the small/all mix still retains significant predictive power for the growth rates of GNP, manufacturing inventories, and manufacturing employment. Overall, we find that the ratio of small firm borrowing relative to total borrowing has both the predictive power and sign one would expect based on credit theories.

The second prediction to be tested is whether monetary policy has any effect on the borrowing patterns of small and large firms. To test this hypothesis, we regress each debt variable on four lags of itself and four lags of the Federal Funds rate. We also consider a multivariate specification that includes the growth rate of industrial production for the manufacturing sector. We report the probability values from exclusion tests and t-statistics for sums of coefficients on the Federal Funds rate in Table 5.3. The results from both the exclusion tests and the tests of sums of coefficients provide strong support for the hypothesis that monetary policy significantly affects the differential growth rates of short-term debt between small and large firms. In particular, an increase in the Federal Funds rate leads to a contraction of small firm borrowing relative to large firm borrowing.

In conclusion, the evidence from financial data disaggregated by size class in manufacturing confirms the fact that the differences in short-term borrowing behaviour between small and large firms have substantial predictive power for real economic activity. In addition, the data is consistent with the view that monetary policy plays a crucial role in determining the pattern of such differences in borrowing behaviour, in the direction suggested by credit-based propagation theories.

5.4 Evidence from non-financial data

While consistent with a role for credit in the economy, the evidence using financial data alone cannot solve the identification problems posed by the literature. We must ask why funds flow from one class of borrower to another, and why such flows might have forecasting power for aggregate economic activity. One alternative explanation that does not rely on credit is that small, bank-dependent firms are subject to a different set of shock processes and adjustment mechanisms than large firms, or firms which are identified as having free access to credit markets. If, for example, small firms are on the fringes of the industrial process as suppliers or niche market producers, they may well be subject to more rapid and deeper contractions than their large firm counterparts. If this were true, we may indeed expect the ratio of small- to all-firm borrowing to respond to monetary policy shocks and lead the business cycle as the above evidence suggests.

There are two approaches to solving the identification problem here. The first approach is to rely on additional time-series evidence that is also consistent with a credit interpretation but much harder to explain with an alternative non-credit related phenomenon. The other approach is to go directly to micro data and control for as many of the alternative shock and adjustment processes as possible, using both reduced form and structural techniques. We discuss both in turn.

5.4.1 Evidence from the aggregate QFR data

To identify credit effects through time-series data one must have a data set that provides both a long time-series dimension and enough heterogeneity to form a basis of comparison between agents with differential access to capital markets. By providing balance sheet and income statement data over the period 1959:Q1-1991:Q4 across different size classes of manufacturing firms, the QFR data is uniquely suited to the task. Using this data, Gertler and Gilchrist (1994) provide substantial evidence on the differential behaviour of small *vs.* large firms over the business cycle and in response to monetary policy shocks.

Regardless of the source of this differential behaviour, the Gertler-Gilchrist evidence is striking. Following a shift to tight monetary policy, the contraction of small manufacturing firms—defined as firms in the bottom 30th percentile of the manufacturing size distribution—is 2.5 times greater than that of large firms over a 12 quarter horizon. This contraction can be seen across a wide variety of variables, but it is most noticeable in sales, inventories, and short-term borrowing. While one could potentially explain the differential response of sales with an alternative demand story, it is much harder to explain the differential response of the inventory/sales ratio and debt/sales ratio with such a story. The evidence clearly suggests that large firms obtain additional funds to finance inventories as sales are declining, whereas small firms do not. In addition, Bernanke, Gertler and Gilchrist (1994) show that controlling for industry-specific demand conditions does not substantially reduce the differential response between small and large firms, as one would expect if a demand based alternative were the true explanation.

Perhaps the most compelling evidence from a sceptic's point of view is the finding that small firm spending is highly responsive to current credit conditions, even after controlling for the lagged dynamics normally associated with spending equations. Two pieces of evidence are relevant here. The first piece is the finding by Gertler and Gilchrist (1994) that small firm inventory investment is highly responsive to a coverage variable that measures the ratio of income to short-term debt payments, whereas large firm inventory investment is not. Thus balance sheet conditions affect real decisions for small firms but not for large firms. This is true even after one controls for alternative sales processes and inventory adjustment speeds. A related piece of evidence is the finding by Oliner and Rudebusch (1994) that small firm business fixed investment is highly responsive to cash flow shocks during recessionary periods. Such an asymmetric response arises naturally from a model with credit frictions but is much more difficult to reconcile with a model that assumes perfect capital markets.

5.4.2   Evidence from the firm-level QFR data

While the time-series evidence based on small *vs.* large manufacturing firms paints a compelling picture of the process one would expect to observe if credit conditions play an important role in both the monetary transmission mechanism and business cycle fluctuations, the fact that the data is aggregated by size rather than by a more direct indicator of capital market access is a major limitation. Additional problems are posed by trying to control for industry effects and other aggregation issues. Fortunately, the underlying firm-level data set used to construct the published—aggregated by size class—QFR data is available through the Center for Economic Studies at the U.S. Bureau of the Census for the period 1977:Q1-1991:Q3. Unlike Compustat or other firm-level databases more commonly used in micro studies that seek to identify the effect of credit frictions on real behaviour, the firm-level QFR data set is comprehensive for all manufacturing, covering all corporations not just publicly traded ones.[13] The fact that it is comprehensive implies that one can correctly aggregate results to obtain macroeconomic implications. The quarterly frequency of the data allows one to consider directly issues at a business cycle frequency. While work on this database is preliminary, some interesting results have already emerged. We once again discuss these results in the context of identifying the role of credit in the macroeconomy.

The principal identification problem posed at the micro level is how does one separate a firm's response to a change in its financial position from its response to new profit opportunities? This identification problem can be easily understood in the context of standard firm-level investment regressions. As discussed in Section 5.2, a positive shock to profits has two effects. First, to the extent that high profitability today signals high profitability tomorrow, firms will want to invest more. This is the standard neoclassical response. Second, higher profits today signal greater net-worth and an improved financial position. The improved financial position lowers the premium on external funds and boosts the investment spending of constrained firms. In this manner, investment expenditures are more responsive to innovations in current earnings than the neoclassical model would suggest.

To test this hypothesis of 'excess sensitivity' of investment to cash flow, past researchers regressed investment on Tobin's Q and either current or past earnings.[14] The identifying assumption of this approach is that Tobin's Q adequately proxies for future profit opportunities through the forward looking behaviour captured by the stock market.[15] To the extent that current or past earnings still had explanatory power for investment—even after controlling for future profit opportunities through Tobin's Q—they did so because of credit market frictions. This identification scheme, however, was called into serious doubt by researchers who found either little observable relationship between investment and Q or implausibly high adjustment cost estimates (low Q coefficients). These results suggest that Tobin's Q is not an adequate proxy for future profit opportunities. Since, in principal, Tobin's Q measures the present value of future earnings streams attributable to new investment, cash flow might help predict this stream, in which case, one could not attribute the large, positive coefficient on cash flow solely to financial effects.

Gilchrist and Himmelberg (1994) (G-H hereafter) formalise this point by using a VAR forecasting framework to decompose the effect of cash flow on investment into two separate components—a component that forecasts future profitability under perfect capital markets, and a residual component that may be attributable to financial frictions. The results of their methodology provide the following insights for identification of credit effects. By relying on Tobin's Q to control for future profit opportunities, rather than a VAR based alternative that controls for predictive power of cash flow, one dramatically overstates the effect of cash flow on investment. This is especially true for firms classified as financially 'unconstrained' and for which Tobin's Q is a particularly bad proxy for future profit opportunities. Thus without properly controlling for such profit opportunities, even large firms with commercial paper ratings appear overly responsive to cash flow shocks, relative to the perfect markets benchmark. Once one controls for the forecasting power of cash flow, however, all evidence of excess sensitivity disappears for 'unconstrained' firms, and it is reduced for 'constrained' firms.

The other lesson for the identification of credit effects provided by G-H is that although the level of response of investment to cash flow differs substantially with and without controlling for the forecasting component, the difference in the response of investment to cash flow across constrained and unconstrained subgroups is actually greater, once one controls for the predictive content of cash flow for future profit opportunities. This result is encouraging since it suggests that even if we cannot correctly identify the underlying investment model, by making comparisons across subgroups of firms, we are still likely to obtain a reasonably correct answer for the degree of excess sensitivity of constrained firms relative to unconstrained firms.[16]

While the G-H results are informative, it is not clear how robust they are to alternative time frames, data sets, and forecasting rules. The latter is particularly important since G-H rely on a VAR forecast that is restricted to be common across all manufacturing firms, once they control for fixed firm and time effects. To the extent that such a restriction is invalid, we may not obtain a good proxy for future profit opportunities, and we may seriously bias the parameter estimates on cash flow.

Gilchrist and Zakrajšek (1995) investigate this point using the quarterly frequency, firm-level QFR data. The application of the VAR based measure of profit opportunities is particularly important for the QFR data set, since many of the firms in the sample are not publicly traded and, therefore, do not have a stock market based measure of profit opportunities available. Using the G-H methodology, Gilchrist and Zakrajšek (1995) compare a variety of forecasting rules, including firm-specific, industry-specific (2-digit SIC), sector-specific (durables vs. non-durables) as well as aggregate forecasting equations and find the G-H results robust to these alternatives. In fact, of all forecasting systems considered, the G-H restriction that the forecasting equation is common across all firms in manufacturing, after controlling for fixed firm and time effects, provides the smallest residual sensitivity of investment to cash flow for both financially constrained and unconstrained subgroups. In addition, Gilchrist and Zakrajšek (1995) confirm the G-H result that unconstrained subgroups show no excess sensitivity of investment to cash flow, once one controls for cash flow's forecasting power of future profit streams. Nonetheless, there remains a large residual correlation between investment and cash flow for constrained subgroups, even after controlling for the predictive power of cash flow for future profit opportunities. In fact, Gilchrist and Zakrajšek (1995) find that the investment of constrained firms is just as responsive to cash flow shocks as it is to future profit opportunities, with an elasticity around 0.12. This latter finding combined with the fact that, by their definition, financially constrained firms account for over 30 per cent of the capital stock, strongly suggests that financial frictions are an important determinant of business fixed investment in the manufacturing sector.

The methodology used by Gilchrist and Himmelberg (1994) and Gilchrist and Zakrajšek (1995) follows that of numerous other researchers who start with a well-specified investment equation and then look for departures from this equation that are consistent with a model based on financial frictions. The alternative model under imperfect capital markets is neither specified nor estimated. Therefore, although these exercises are useful in providing evidence against the null hypothesis of perfect capital markets, they do not provide an alternative set of parameter estimates which can be used to identify the decision rule of the financially constrained firms. A major limitation is the difficulty involved in specifying the alternative, since theoretical models incorporating credit frictions are either too simple or too intractable to take to data.

Some headway has been made in constructing investment models with financial frictions that can be taken to data. One approach is to specify *ad hoc* but realistic rules that govern a firm's ability to obtain external funds and then solve the model using numerical techniques. Gross (1994) is a recent example of this approach. Not only does Gross (1995) specify the alternative to the perfect capital market case, but he also takes the model to data by estimating the reduced form of the decision rule using non-parametric methods. While the reduced-form results provide additional evidence in favour of a credit mechanism for firm-level investment, we still do not obtain the underlying parameter estimates that are necessary to fully evaluate the decision rule and thus to quantify the overall importance of credit for investment spending. For this type of exercise, we must still rely on reduced-form interpretations of the data.

Identification of capital market frictions using reduced-form equations at the micro level depends on similar methods to those used with time-series data. The point is to develop empirical evidence that is consistent with a model based on capital market frictions but is much more difficult to explain in a world where such frictions are absent. For example, although we would expect cash flow to be an important explanatory variable for investment, even in the absence of capital market frictions, this is not so obviously true for inventories, since nearly all structural models of inventory behaviour rely on sales rather than profits as the principal determinant of optimal inventory investment. In addition, to the extent that cash flow has greater explanatory power for firms that are likely to be constrained, we have further evidence in favour of a financial markets imperfection story.[17]

This identification scheme is used by Zakrajšek (1994) to measure the importance of financial frictions in the retail sector, using a sample constructed from the firm-level QFR data. The advantage of the data set that provides information on non-publicly traded firms is particularly important for the retail trade, where a much lower proportion of total assets is held by publicly traded firms. The focus on inventory investment in retail trade is motivated by the fact that inventory investment in this sector is the most volatile component of aggregate inventory investment (see Blinder and Maccini, 1991). Both the cross-sectional and time-series results from Zakrajšek (1994) are consistent with the presence of a financial accelerator in the retail trade sector. First, cash flow is both, statistically and economically, a significant predictor of inventory investment for firms with 'weak' balance sheet conditions—that is, firms with large debt burdens and no access to the commercial paper market.[18] Second, the predictive power of cash flow for inventory growth of firms with weak balance sheet conditions is highly asymmetric over the course of the business cycle, increasing considerably in recessions relative to normal times.

5.4.3 Some new results from firm-level QFR data

While the results of Zakrajšek (1994) are interesting, it is important to examine their robustness across broader sectors of the economy. Accordingly, we apply the identification scheme used in Zakrajšek (1994) to a similarly constructed firm-level QFR data set for the manufacturing sector. We estimate an inventory regression that includes both lags of inventories and sales to capture desired inventory behaviour and lagged cash flow to capture a financial effect. We rank observations based on last period's net-leverage ratio and split the data into four quartiles based on this ranking.[19] The details of the exact econometric specification and data construction are provided in the appendix.

Table 5.4 provides the first set of estimation results. It compares the response of inventory investment to cash flow shocks across the four subsets of firms, classified by leverage. All four categories show a positive response of inventory investment to cash flow, with the coefficient on cash flow increasing monotonically across the four leverage categories. The monotonic increase in cash flow coefficients is consistent with the view that internal funds are an important determinant of inventory investment for financially constrained firms. Nonetheless, the fact that inventory investment of firms in the lowest quartile of the net-leverage distribution responds to cash flow is difficult to interpret and highlights the costs of eschewing structural models even as benchmarks. An extreme interpretation would attribute all of the explanatory power of cash flow to the effect of capital market frictions. An alternative interpretation would attribute the explanatory power of cash flow for low net-leverage firms to an underlying perfect capital markets model and attribute the differential effect across different quartiles to capital market frictions. Even with this more restrictive interpretation, we have substantial evidence of excess sensitivity of inventory investment to earnings for high net-leverage firms.

While Table 5.4 provided information on the average response of inventories to cash flow across the full sample, Table 5.5 provides information on the cyclicality of the response over the business cycle. We do this by reestimating the inventory equation across two year subintervals for firms divided into two categories–low *vs.* high net-leverage.[20] In Table 5.5, we report the cash flow coefficient and its standard error for both the low and high net-leverage categories in each sub-sample. We also report the differential response of inventories to cash flow shocks across the two subgroups as well as the associated standard errors. Table 5.5 clearly shows the cyclical nature of the importance of internal funds for inventory investment. The highest differential responses occur in the 1980-1982 downturn and following the 1989 monetary contraction that preceded the 1990 recession. In these episodes, the differential effect of cash flow on inventory investment is nearly twice as high as in 1985-86, the period with the lowest differential response.

It is worth emphasising that the findings here come not from small *vs.* large firm comparisons as in Gertler and Gilchrist, but from a sample of large firms classified by financial policy. As such, one cannot easily explain away the differences in inventory investment response by attributing them to unmodeled industry or size effects. In addition, besides providing independent support for the idea that the differential inventory behaviour between firms documented by Gilchrist and Gertler (1994) is driven by financial factors, this evidence confirms the findings of Kashyap, Lamont and Stein (1994) on the cyclical nature of credit effects in inventory equations.

While these regressions provide strong micro evidence in support of a financial accelerator for inventory investment, it is important to ask do these differential effects matter in the aggregate. While a complete answer to this question is beyond the scope of this paper, we provide two pieces of evidence to suggest they would indeed matter. We proceed by calculating the share of inventories that would fall into the two quartiles with high net-leverage (*i.e.* firms above the 50th percentile of the net-leverage distribution). This

percentage is plotted in Figure 5.4. There are two noteworthy aspects to
this figure. First, high net-leverage firms account for a significant share
of inventories—at least 30 per cent. Second and more striking, the share of
inventories held by high net-leverage firms rises dramatically following tight
money episodes and during recessions, with at least a 15 per cent increase
during the 1981–82 recession, and a steady climb after the onset of tight
money in 1989 through the 1991 recession. Based on this evidence, it is easy
to see how one could obtain asymmetric responses of inventory investment
to financial conditions in the aggregate data. As more firms become highly
indebted throughout the economic downturn, and as the responsiveness to
earnings of highly indebted firms increases throughout the downturn, the
amplification effects associated with the financial accelerator become more
relevant for aggregate economic activity.

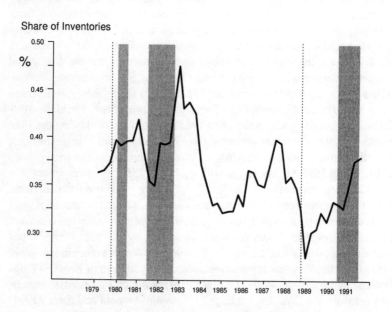

Figure 5.4: The relative importance of financially constrained firms

5.5 Conclusion

This paper addresses the issues surrounding the identification and quantification of the effects of financial market imperfections on firm behaviour. The paper emphasizes the essential role that heterogeneity plays in assessing the importance of credit market frictions, and the need for data sets that accurately reflect such heterogeneity when measuring the relevance of financial frictions. The paper also provides some new empirical evidence using time-series debt data. In particular, we find substantial evidence that small-*vs.* large-firm borrowing has predictive power for a variety of measures of aggregate economic activity. We also find that monetary policy has a substantial influence over the differential behaviour of these debt variables, with a tightening of monetary policy leading to a drop in small firm debt relative to large firm debt.

Using firm-level data, we find substantial evidence that inventory investment is highly responsive to the availability of internal funds for firms that find themselves in weak balance sheet positions. In addition, the percentage of manufacturing inventories held by such firms increases dramatically during economic downturns, making overall inventory investment much more sensitive to balance sheet conditions during such periods of low economic activity. Overall, these results provide substantial support for the view that a credit mechanism plays an important role in conditioning the macroeconomy's response to underlying economic disturbances.

Appendix

From the certainty component of the firm-level QFR data base, we selected an *unbalanced* panel of firms from 1979:Q1 through 1991:Q3.[21] From this unbalanced panel, we dropped all firms that had tenure of less than 8 quarters or had any discontinuities in their time-series record.

Inventories: The QFR data reports the book value of total inventories. Firms in the sample were required to hold strictly positive inventories at each point of their tenure in the panel. In order to eliminate the inflation bias from the inventory growth rate, inventory stocks were deflated by the implicit GNP deflator prior to constructing growth rates. We let N_{it} denote the real value of inventories of firm i in period t.

Sales: To construct a real measure of sales, the reported nominal value of sales was deflated by the implicit GNP price deflator. As with inventories, firms were required to have strictly positive sales at each point of their tenure in the panel. We let S_{it} denote the real value of sales of firm i in period t.

Internal Finance: The measure of internal funds in this paper is defined as cash flow relative to last period's total assets. Cash flow is defined as income (or loss) from operations plus depreciation, depletion, and amortisation of property plant, and equipment. Both cash flow and the book value of total assets are deflated by the implicit GNP price deflator prior to constructing the internal finance ratio. We let Π_{it}/TA_{it-1} denote the ratio of real profits of firm i in period t to its real assets in period $t-1$.

Net-Leverage: Financial leverage (the ratio of total debt to total assets) is normally thought of as a measure of a firm's balance sheet condition. A potential problem with identification strategy that classifies firms into 'constrained' and 'unconstrained' subgroups according to leverage ratio is that total assets consist of a variety of different assets. In particular, a highly liquid component of total assets, in addition to cash stocks, includes time deposits, CDs, and other readily marketable securities that can be quickly and at little cost converted to cash-on-hand and used to finance inventory investment if internal funds are low and external credit is unavailable; see Kashyap, Stein, and Lamont (1994) for the evidence of this phenomenon during the 1982 recession. A more comprehensive measure of a firm's overall balance sheet condition is, according to Sharpe (1994), the net-leverage. The net-leverage is constructed by subtracting a firm's net short-term assets from both the numerator and the denominator of a firm's leverage ratio. Net short-term assets consist of cash stock, all short-term investment, and trade receivables, minus trade payables.

Quartiles: In each period t, a firm is assigned to one the 4 quartiles based on its $t - 1$ period's net-leverage. In particular, the first quartile contains all firms with net-leverage below the 25th percentile; the second quartile contains all firms with net-leverage between the 25th and the 50th percentile; the third quartile contains all firms between the 50th and 75th percentile; and, finally, the fourth quartile contains all firms with net-leverage above the 75th percentile. We allow the cutoff points (*i.e.* the 25th, 50th, and 75th percentile) to vary over time by computing them for each *year* of our sample separately. For example, a firm in 1980:Q2 is assigned to the first quartile, if its net-leverage in 1980:Q1 is less than the 25th percentile of the net-leverage distribution computed over all four quarters of the year 1980.

We use the following econometric specification to measure the effect of credit frictions on inventory investment:

$$
\Delta \ln N_{it} = \beta_1 \ln \left(\frac{N}{S} \right)_i + \beta_2 \ln \left(\frac{N_{it-1}}{S_{it}} \right) + \beta_3 \Delta \ln N_{it-1}
$$
$$
+ \beta_4 \Delta \ln S_{it-1} + \beta_5 \left(\frac{\Pi_{it-1}}{TA_{it-2}} \right) + d_t + \epsilon_{it}.
$$

The dependent variable is the growth rate of inventories. The first two terms reflect the effect of the deviation of the current log inventory/sales ratio from its firm-specific target, $\ln(N/S)_i$. A consistent estimate of the target inventory/sales ratio for firm i is computed according to

$$
\ln \left(\frac{N}{S} \right)_i = \frac{1}{T_i} \sum_{t=1}^{T_i} (\ln N_{it} - \ln S_{it}),
$$

where T_i denotes the number of quarters that firm i is in the sample (a minimum of 8 quarters and a maximum of 51 quarters). The lag of Π_{it}/TA_{it-1} is meant to capture the effect of financial frictions on inventory investment; the fixed time effect, d_t, is added to control for aggregate shocks such as overall price movements or interest rate shocks, while lags of $\Delta \ln N_{it}$ and $\Delta \ln S_{it}$ are included to capture any additional short-run dynamics.

p-values from Exclusion Tests on Debt Variables and
t-statistics on Sums of Coefficients for Debt Variables

Debt Variable	Dependent Variable							
	GNP		IP		Inv.		Empl.	
	p-val	t-stat	p-val	t-stat	p-val	t-stat	p-val	t-stat
Manufacturing Mix[a]	0.04	2.75	0.01	2.60	0.08	1.75	0.07	2.11
Large Firm Mix[b]	0.76	0.73	0.32	0.73	0.79	0.30	0.56	0.42
Small/All Mix[c]	0.00	4.31	0.00	3.87	0.07	1.37	0.00	3.34
Small Firm S-T Bank	0.84	-0.12	0.28	-1.06	0.74	0.29	0.62	-0.11
Small Firm S-T Debt	0.80	-0.09	0.19	-1.05	0.82	0.41	0.47	-0.07
Large Firm S-T Bank	0.02	-3.17	0.00	-3.96	0.05	-1.08	0.02	-3.37
Large Firm S-T Debt	0.01	-3.52	0.00	-4.35	0.07	-0.83	0.01	-3.52

Table 5.1: Short-term debt and aggregate economic activity, results from a bivariate system

p-values from Exclusion Tests on Debt Variables and
t-statistics on Sums of Coefficients for Debt Variables

Debt Variable	Dependent Variable							
	GNP		IP		Inv.		Empl.	
	p-val	t-stat	p-val	t-stat	p-val	t-stat	p-val	t-stat
Manufacturing Mix[a]	0.40	1.61	0.17	1.43	0.21	1.95	0.36	1.20
Large Firm Mix[b]	0.91	0.15	0.52	0.11	0.98	0.16	0.73	-0.12
Small/All Mix[c]	0.05	2.69	0.13	2.27	0.02	2.59	0.06	2.31
Small Firm S-T Bank	0.98	0.17	0.36	-1.04	0.55	0.27	0.67	-0.38
Small Firm S-T Debt	0.95	0.33	0.35	-0.91	0.77	0.43	0.65	-0.22
Large Firm S-T Bank	0.32	-1.85	0.11	-2.70	0.02	-1.97	0.13	-2.49
Large Firm S-T Debt	0.28	-1.70	0.11	-2.68	0.06	-1.66	0.11	-2.28

Table 5.2: Short-term debt and aggregate economic activity, results from a multivariate system

Notes: The bivariate system includes four lags of the growth rate of the dependent variable and four lags of the growth rate of the debt variable. The multivariate system also includes four lags of the change in the Federal Funds rate. Sample range: 1975:Q1–1991:Q4. [a] Ratio of short-term bank loans to commercial paper plus short-term bank loans for all manufacturing firms. [b] Ratio of short-term bank loans to commercial paper plus short-term bank loans for firms above the 30th percentile in sales. [c] Ratio of short-term debt for firms below the 30th percentile in sales relative to short-term debt of all firms.

152   *Simon Gilchrist and Egon Zakrajšek*

p-values from Exclusion Tests on the Federal Funds Rate
and t-statistics on Sums of Coefficients for Federal Funds Rate

Debt Variable	Bivariate System[a]		Multivariate System[b]	
	p-val	t-stat	p-val	t-stat
Manufacturing Mix[c]	0.12	-2.35	0.31	-1.90
Large Firm Mix[d]	0.82	-0.50	0.95	-0.69
Small/All Mix[e]	0.00	-3.81	0.06	-2.70
Small Firm S-T Bank	0.01	1.59	0.13	0.05
Small Firm S-T Debt	0.02	1.53	0.13	0.14
Large Firm S-T Bank	0.00	3.29	0.26	2.07
Large Firm S-T Debt	0.00	3.61	0.10	2.44

[a] The bivariate system includes four lags of the growth rate of the debt variable and four lags of the change in the Federal Funds rate. [b] The multivariate system includes four lags of the growth rate of the debt variable, four lags of the change in the Federal Funds rate, and four lags of the growth rate of manufacturing industrial production. [c] Ratio of short-term bank loans to commercial paper plus short-term bank loans for all manufacturing firms. [d] Ratio of short-term bank loans to commercial paper plus short-term bank loans for firms above the 30th percentile in sales. [e] Ratio of short-term debt for firms below the 30th percentile in sales relative to short-term debt of all firms.

Table 5.3: The effect of the federal funds rate on S-T debt

Dependent Variable: $\Delta \ln N_{it}$

	Quartile I	Quartile II	Quartile III	Quartile IV
$\ln (N/S)_i$	0.145	0.124	0.127	0.161
	(0.004)	(0.004)	(0.004)	(0.004)
$\ln (N_{it-1}/S_{it})$	−0.144	−0.124	−0.127	−0.169
	(0.003)	(0.004)	(0.004)	(0.004)
$\Delta \ln N_{it-1}$	−0.145	−0.085	−0.081	−0.028
	(0.006)	(0.006)	(0.006)	(0.006)
$\Delta \ln S_{it-1}$	−0.081	−0.067	−0.072	−0.069
	(0.005)	(0.005)	(0.005)	(0.005)
Π_{it-1}/TA_{it-2}	0.282	0.396	0.493	0.582
	(0.031)	(0.027)	(0.035)	(0.040)
R^2	0.098	0.077	0.074	0.088
Obs	29,612	29,869	29,575	58,484

Notes: Standard errors in parentheses. All equations included fixed time effects (not reported) and are estimated with OLS. A minimum of 8 quarters and up to 51 quarters of data were used to compute a consistent estimate (*i.e.* the sample mean) of the firm-specific inventory-sales target ratio, $(N/S)_i$. The log of this variable is included to control for fixed individual effects. Sample range: 1979:Q3–1991:Q3.

Table 5.4: The effect of cash flow on inventory investment

Dependent Variable: $\Delta \ln N_{it}$

Time Period	Unconstrained[a]	Constrained[b]	Difference[c]
79:Q3-80:Q4	0.384	0.612	0.288
	(0.052)	(0.065)	(0.083)
81:Q1-82:Q4	0.320	0.649	0.329
	(0.055)	(0.067)	(0.087)
83:Q1-84:Q4	0.410	0.552	0.142
	(0.055)	(0.070)	(0.089)
85:Q1-86:Q4	0.395	0.526	0.131
	(0.056)	(0.058)	(0.081)
87:Q1-88:Q4	0.357	0.540	0.183
	(0.051)	(0.065)	(0.083)
89:Q1-90:Q4	0.247	0.564	0.317
	(0.056)	(0.076)	(0.094)

Notes: Standard errors in parentheses. All equations included fixed time effects (not reported) and are estimated with OLS. A minimum of 8 quarters and up to 51 quarters of data were used to compute a consistent estimate (*i.e.* the sample mean) of the firm-specific inventory-sales target ratio, $(N/S)_i$. The log of this variable is included to control for fixed individual effects. Sample range: 1979:Q3–1990:Q4. [a] Point estimates on cash flow for firms in Quartiles I and II. [b] Point estimates on cash flow for firms in Quartiles III and IV. [c] Difference in point estimates on cash flow between constrained and unconstrained firms.

Table 5.5: The effect of cash flow on inventory investment: asymmetric effects over the business cycle

Acknowledgments

This paper was originally prepared for the conference entitled 'Is Bank Lending Important for the Transmission of Monetary Policy?' sponsored by the Federal Reserve of Boston and held in North Falmouth, MA, June 11-13, 1995. We thank Mark Gertler for valuable comments and the CES staff at the Bureau of the Census for providing the firm-level QFR data. The second author thanks the Federal Reserve Board for their gracious hospitality during the project.

Notes

1. See Gertler (1988) and Bernanke (1993) for exhaustive reviews of the literature.
2. This example is based on the costly state verification (CSV) model presented in Gertler and Gilchrist (1991). Early examples of CSV models are Townsend (1979), Gale and Hellwig (1985), and Williamson (1987). Additional models that incorporate financial frictions in various guises include Jensen and Meckling (1976), Jaffe and Russell (1976), Leland and Pyle (1977), Stiglitz and Weiss (1981), and Myers and Majluf (1984).

154   *Simon Gilchrist and Egon Zakrajšek*

3. Fama (1980) outlines the special role of banks; Himmelberg and Morgan (1995) provide a more recent discussion.

4. A central issue in the literature, of course, is the link between monetary policy and credit supply disruptions. According to the traditional credit view, contractions in monetary policy drain reserves from the system and force a contraction on both the asset and liability side of the balance sheet. As long as banks do not face a perfectly elastic supply of funds schedule in the CD market, monetary policy contractions reduce the supply of bank loans. Romer and Romer (1990) forcefully argue against any such link between reserve contractions and loan supply, owing to the ability of banks to issue certificates of deposit not subject to reserve requirements. Kashyap and Stein (1994), on the other hand, argue in favour of such a link.

5. Bernanke and Gertler (1989) present one example of such models.

6. Both Ramey (1993) and Romer and Romer (1990) are careful to point out, however, that their findings do not provide evidence against the importance of a broader credit mechanism such as the financial accelerator.

7. Examples of dynamic general equilibrium models that incorporate a financial accelerator include Greenwald and Stiglitz (1993), Bernanke and Gertler (1989, 1990), Calomiris and Hubbard (1990), Gertler (1992), and Kiyotaki and Moore (1993).

8. Other lending variables that reflect potential differences in borrower quality are available, though generally not for as long a time period. In addition, the QFR data is disaggregated by size as well as by type of debt (*e.g.* bank *vs.* non-bank, commercial paper, *etc.*). The data is therefore well-suited for making additional comparisons of bank *vs.* non-bank debt. We focus on short-term rather than long-term debt to avoid the measurement problems associated with disentangling stocks and flows when only stocks are observable. As Gertler and Gilchrist (1993) show, the overall conclusions regarding disaggregated debt movements do not depend on the use of short-term rather than long-term credit quantities. In addition, Oliner and Rudebusch (1992) provide similar evidence based on total debt.

9. We focus on the manufacturing variables since the lending data are constructed using manufacturing firms only. Ramey (1993) provides a similar test of predictive power for aggregate industrial production alone, although, she uses the ratio of small- to large-firm borrowing rather than small- to all-firm borrowing. Since these variables are simple transformations of each other, it makes very little difference which one is used.

10. We start the estimation in 1975:Q1 to allow for the fact that credit conditions may have changed following the regulatory change that allowed banks to issue large time deposits without being subject to reserve requirements.

11. It is also possible that some firms are pushed out of the of the commercial paper market into the bank loan market as their credit quality deteriorates. Such an effect would mute any differential response for large firms.

12. Gilchrist and Gertler (1994) show that manufacturing firms with total assets less than $250 million have virtually no commercial paper outstanding. In addition, 90 per cent of manufacturing commercial paper is issued by firms with total assets greater than $1 billion.

13. In fact, the firm-level QFR data is the only known U.S. database that systematically provides either high-frequency or firm-level information about corporations that are not publicly traded. In addition, the firm-level QFR data provides information for the retail, wholesale, and mining sectors of the economy. Unlike the manufacturing sector, the sampling of these sectors is not comprehensive; income and balance sheet statements are provided only for firms with assets above $50 million; see Long and Ravencraft (1993) and Zakrajšek (1994) for detailed descriptions of the firm-level QFR data base.

14. The most influential paper in the literature is Fazzari, Hubbard, and Peterson (1988).

> Other examples include Devereux and Schiantarelli (1989), Hoshi, Kashyap and Scharf-stein (1991), Blundell, Bond, Devereux, and Schiantarelli (1992), Chirinko and Schaller (1993), Oliner and Rudebusch (1992), and Schaller (1993).

15. Tobin's Q is defined as the market value of the firm divided by the replacement value of its capital stock. The market value includes the stock market value of equity and the market value of debt outstanding. According to the Q theory of investment, Tobin's Q represents the shadow value of an additional dollar of investment. Thus, when Q is greater than one, the value of an additional unit of investment inside the firm is greater than its replacement cost and the firm should invest more.

16. It is worth emphasising that many papers in the literature, including Fazzari, Hubbard and Petersen (1988), either explicitly or implicitly acknowledge the identification problems posed by using Tobin's Q and are more likely to rely on comparisons across firms rather than focus on the overall cash flow coefficient when assessing the importance of financial frictions.

17. Kashyap, Lamont and Stein (1994) and Carpenter, Fazzari, and Peterson (1993) provide recent evidence on the importance of internal finance for manufacturing inventory invest-ment using reduced-form equations at the firm level.

18. Zakrajšek (1994) relies on a 'net-leverage' measure of balance sheet conditions proposed by Sharpe (1994) and recently used by Calomiris, Orphanides, and Sharpe (1994).

19. It is worth noting that the choice of sample splitting criterion is also a relevant identi-fication issue. Although we do not explicitly address the issue in this paper, evidence on the importance of financial effects in micro spending equations is robust to a wide variety of sample splitting methodologies, using both exogenously determined criteria such as size, dividend policy, and ownership structure (see Fazzari, Hubbard, and Peter-son (1988), Hoshi, Kashyap, and Scharfstein (1991), Oliner and Rudebusch (1992) and Ng and Schaller (1993)) and endogenously determined criteria derived from switching regime models (see Schiantarelli and Sembenelli (1995)).

20. The low net-leverage observations represent the firms with net-leverage less than the me-dian value over the two year estimation period. We use two rather than four classifications for ease of comparison.

21. Even though the firm-level QFR data is available from 1977:Q1–1991:Q3, we started our sample in 1979:Q1, since the 1978:Q4 data are missing, and we wanted to avoid discontinuities.

References

Attanasio, O. (1994), "The Intertemporal Allocation of Consumption: Theory and Evidence," *Carnegie-Rochester Conference Series on Public Policy*, 42:39–90.
Bernanke, B. (1993) "Credit in the Macroeconomy," *Federal Reserve Bank of New York Quar-terly Review*, 50–70.
Bernanke, B. and A.S. Blinder (1992), "The Federal Funds Rate and the Transmission of Mon-etary Policy," *American Economic Review*, 82:901–22.
Bernanke, B. and M. Gertler (1989) "Agency Costs, Net Worth, and Business Fluctuations," *American Economic Review*, 79:14–31.
Bernanke, B. and M. Gertler (1990), "Financial Fragility and Economic Performance," *Quar-terly Journal of Economics*, 105:97–114.
Bernanke, B., M. Gertler and S. Gilchrist (1995) "The Financial Accelerator and the Flight to Quality," Forthcoming, *Review of Economics and Statistics*.
Blinder, A.S. and L.J. Maccini (1991), "Taking Stock: A Critical Assessment of Recent Re-

search on Inventories," *Journal of Economic Perspectives*, 5:73–96.

Blundell, R., S. Bond, M. Devereux and F. Schiantarelli (1992), "Investment and Tobin's Q," *Journal of Econometrics*, 51:233–57.

Calomiris, C.W., C.P. Himmelberg, and P. Wachtel (1995), "Commercial Paper, Corporate Finance, and the Business Cycle: A Microeconomic Perspective," *Carnegie-Rochester Conference Series on Public Policy*, 42:203–50.

Calomiris, C.W. and R.G. Hubbard (1990), "Firm Heterogeneity, Internal Finance, and Credit Rationing," *Economic Journal*, 100:90–104.

Calomiris, C.W., A. Orphanides, and S.A. Sharpe (1994), "Leverage as a State Variable for Employment, Inventory Accumulation, and Fixed Investment," *NBER Working Paper* No. 4800.

Carpenter, R.E., S.M. Fazzari and B.C. Petersen (1994), "Inventory Investment, Internal-Finance Fluctuations, and the Business Cycle," *Brookings Papers on Economic Activity*, 2:75–138.

Chirinko, R.S. and H. Schaller (1993), "Why Does Liquidity Matter in Investment Equations," *mimeo*, Federal Reserve Bank of Kansas City.

Devereux, M. and F. Schiantarelli (1989), "Investment, Financial Factors, and Cash Flow: Evidence From U.K. Panel Data" in: R.G. Hubbard (ed.), *Asymmetric Information, Corporate Finance, and Investment*, The University of Chicago Press, Chicago: 279–306.

Fama, E.F. (1980), "Banking in the Theory of Finance," *Journal of Monetary Economics*, 6:39–57.

Fazzari, S.M., R.G. Hubbard and B.C. Petersen (1988), "Financing Constraints and Corporate Investment," *Brookings Papers on Economic Activity*, 1:141–95.

Gale, D. and M. Hellwig (1985) "Incentive-compatible Debt Contracts: The One Period Problem," *Review of Economic Studies*, 52:647–63.

Gertler, M. (1988), "Financial Structure and Aggregate Economic Activity: An Overview," *Journal of Money, Credit, and Banking*, 20:559–88.

Gertler, M. (1992), "Financial Capacity and Output Fluctuation in an Economy with Multi-Period Financial Relationship," *Review of Economic Studies*, 29:455–72.

Gertler, M. and S. Gilchrist (1991), "Monetary Policy, Business Cycles and the Behaviour of Small Manufacturing Firms," *NBER Working Paper* No. 3098.

Gertler, M. and S. Gilchrist (1993) "The Role of Credit Market Imperfections in the Monetary Transmission Mechanism: Arguments and Evidence," *Scandinavian Journal of Economics*, 95:43–64.

Gertler, M. and S. Gilchrist (1994), "Monetary Policy, Business Cycles and the behaviour of Small Manufacturing Firms," *Quarterly Journal of Economics*, 109:309–40.

Gilchrist, S. and C.P. Himmelberg (1995), "The Role of Cash Flow in Reduced-Form Investment Equations," *Journal of Monetary Economics*, 36.

Gilchrist, S. and E. Zakrajšek (1995), "Investment and the Present Value of Profits: A Panel Data Approach," *mimeo*, Boston University.

Greenwald, B.C. and J.E. Stiglitz (1993), "Financial Market Imperfections and Business Cycles," *Quarterly Journal of Economics*, 108:77–114.

Gross, D. (1994), "The Investment and Financing Decisions of Liquidity Constrained Firms," *mimeo*, Massachusetts Institute of Technology.

Jaffe, D.M. and T. Russell (1976), "Imperfect Information, Uncertainty, and Credit Rationing," *Quarterly Journal of Economics*, 90:651–66.

Jensen, M. and W. Meckling (1976), "Theory of the Firm: Managerial Behaviour, Agency Costs, and Ownership Structure," *Journal of Financial Economics*, 3:305–60.

Kashyap, A., O. Lamont and J.C. Stein (1994), "Credit Conditions and the Cyclical Behaviour of Inventories," *Quarterly Journal of Economics*, 109:565–92.

Kashyap, A., J.C. Stein and D.W. Wilcox (1993) "The Monetary Transmission Mechanism: Evidence from the Composition of External Finance," *American Economic Review*, 83:78–98.

King, S.R. (1986), "Monetary Transmission: Through Bank Loans or Bank Liabilities," *Journal of Money, Credit, and Banking*, 18:290–303.

Kiyotaki, N. and J.H. Moore (1993), "Credit cycles," *mimeo*, University of Minnesota.

Lang, W.W. and L.I. Nakamura (1995), " 'Flight to Quality' in Banking and Economic Activity," *Journal of Monetary Economics*, 36:145–64.

Leland, H. and D. Pyle (1977), "Informational Asymmetries, Financial Structure, and Financial

Intermediation," *Journal of Finance*, 32:371–87.

Long, W.F. and D.J. Ravencraft (1993), "The Quarterly Financial Report (QFR) Database," *mimeo*, Center for Economic Studies, Bureau of the Census.

Modigliani, F. and M. Miller (1958), "The Cost of Capital, Corporation Finance, and the Theory of Investment," *American Economic Review*, 38:261–97.

Morgan, D.P. (1994), "The Lending View of Monetary Policy and Loan Commitments," *mimeo*, Federal Reserve Bank of Kansas City.

Myers, S.C. and N.S. Majluf (1984), "Corporate Financing and Investment Decisions when Firms have Information that Investors Do Not Have," *Journal of Financial Economics*, 13:187–221.

Ng, S. and H. Schaller (1993), "The Risky Spread, Investment, and Monetary Policy Transmission: Evidence on the Role of Asymmetric Information," *mimeo*, Carlton University.

Oliner, S. and G. Rudebusch (1992), "The Transmission of Monetary Policy to Small and Large Firms," *mimeo*, Board of Governors of the Federal Reserve System.

Oliner, S. and G. Rudebusch (1994), "Is There a Broad Credit Channel for Monetary Policy?" *mimeo*, Board of Governors of the Federal Reserve System.

Ramey, V. (1993), "Is There a Credit Channel to Monetary Policy?," *mimeo*, University of California, San Diego.

Romer, C.D. and D.H. Romer (1990), "New Evidence on the Monetary Transmission Mechanism," *Brookings Papers on Economic Activity*, 1:149–98.

Schaller, H. (1993), "Asymmetric Information, Liquidity Constraints, and Canadian Investment," *Canadian Journal of Economics*, 26:552–74.

Schiantarelli, F. and A. Sembenelli (1995), "Form of Ownership and Constraints: Panel Data Evidence from Leverage and Investment Equations," *mimeo*, Boston College.

Sharpe, S.A. (1994), "Financial Market Imperfections, Firm Leverage, and the Cyclicality of Employment," *American Economic Review*, 84:1060–74.

Stiglitz, J.E. and A. Weiss (1981), "Credit Rationing in Markets with Imperfect Information," *American Economic Review*, 71:393–410.

Takeo, H., A. Kashyap and D. Scharfstein (1991), "Corporate Structure, Liquidity, and Investment: Evidence from Japanese Industrial Groups," *Quarterly Journal of Economics*, 106:33–60.

Townsend, R.M. (1979), "Optimal Contracts and Competitive Markets with Costly State Verification," *Journal of Economic Theory*, 21:265–93.

Williamson, S.D. (1987). "Costly Monitoring, Loan Contracts, and Equilibrium Credit Rationing," *Quarterly Journal of Economics*, 102:135–45.

Zakrajšek, E. (1994) "Retail Inventories, Internal Finance, and Aggregate Fluctuations: Evidence from Firm-Level Data," *mimeo*, New York University.

6 Investment and Debt Constraints: Evidence from Dutch Panel Data

Hans van Ees, Harry Garretsen, Leo de Haan and Elmer Sterken

6.1 Introduction

It is generally accepted that real markets are sometimes affected by imperfections and their consequences. In contrast, the functioning of financial markets is mostly assumed to be perfect and efficient. Modigliani and Miller (1958, hereafter MM) show that in the latter case the capital structure of the firm is irrelevant for the user cost of capital and does neither affect investment nor the value of the firm. Since the publication of their seminal paper a large number of studies has pointed out that the empirical relevance of the MM-proposition is small. Financial markets suffer from several (institutional) market imperfections. In the first twenty years after the MM-publication, papers showed that distortionary taxes, transaction, bankruptcy and agency costs cause the financial structure to affect investment decisions. In recent studies the focus is directed towards information asymmetries. Agents have different sets of information. Firms know more about the quality of their investment project than, for instance, a bank. It can be shown that in such a case external funds (for instance bank credit) are more expensive than internal funds, because problems like adverse selection and moral hazard may occur. This might explain the empirical phenomenon that firms preferably use internal funds to finance investment.[1]

Some firms are likely to be confronted more with problems of asymmetric information on capital markets than others. In order to identify these firms for the Dutch case we use panel data. In the literature two lines of panel data research can be discerned. The first line estimates reduced-form investment equations, including Tobin's Q and some financial variables like cash flow or liquidity, as a proxy of financial constraints. The seminal paper in this field is by Fazzari, Hubbard and Petersen (1988). Oversensitiveness of investment to these financial variables might indicate financial constraints as it reflects the preference for internal finance. Kaplan and Zingales (1997) show that these reduced-form regressions give no conclusive evidence on the sensitivity of investment to retained earnings though. A large investment-cash flow correlation could even indicate a solid financial position. The second line of research uses the structural neoclassical investment model as a starting point (see Whited, 1992 and Bond and Meghir, 1994). Investment being an intertemporal decision, this type of model starts from the Euler condition. The major advantage is that the structural parameters are estimated directly, so that both the Lucas and Kaplan-Zingales critique do not apply.

In this paper we estimate a structural investment model for the Dutch economy. We use a panel including not only publicly traded but also non-listed manufacturing firms. We compare the fit of the standard neoclassical model with the fit of a model including debt constraints. After that we try to pin down the class of firms that is confronted with financial problems.

The organisation of the paper is as follows. Section 6.2 presents the theoretical investment model, Section 6.3 its empirical implementation. In Section 6.5 some data and estimation issues are discussed. Section 6.6 presents the estimation results, after which Section 6.7 concludes.

6.2 The theoretical investment model

6.2.1 The neoclassical model

The investment model in this paper is quite standard in the finance literature and elaborates on a basic finance-hierarchy model. For the sake of simplicity, we assume that debt is the only source of external funds. This means that, in line with related research, our model abstracts from the quantitatively unimportant new share issues.[2]

The basic assumption is that arbitrage in perfect capital markets yields equality of ex post to ex ante returns to the firm's shareholders. Then the (after-tax) required return on investment can be defined as

$$R_{it} = \frac{E_t[V_{i,t+1}] - V_{it} + E_t[d_{i,t+1}]}{V_{it}}$$

where $E_t[.]$ denotes the expectation operator, d represents after-tax dividends, V is the expected present value of cash flows and subscripts i and t denote the firm and year observations in the panel data set.

The firm maximises its value, which is equal to the present value of future dividends

$$V_{i0} = E_0[\sum_{t=1}^{\infty} \Pi_{j=0}^{t-1}\beta_{ij}d_{it}]$$

where Π is the product operator and $\beta_{ij} = 1/(1+R_{ij})$ is the firm's discount factor. The firm's amount of dividends follows from the cash-flow identity equating cash inflows and outflows (all variables are in real terms unless stated otherwise)

$$d_{it} = (1-\tau)[F(K_{i,t-1}, L_{it}) - w_t L_{it} - G(I_{it}, K_{i,t-1}) - i_{t-1}B_{i,t-1}] + B_{it} - (1-\pi_t^e)B_{i,t-1} - p_{it}^I I_{it}$$

where τ is the corporate income tax rate, $F(.,.)$ represents the revenue function, K_{it} the capital stock, L_{it} labour (or more broadly defined a vector of variable production factors), w_t the wage rate (or the prices of variable production factors), $G(.,.)$ an adjustment cost function, I_{it} investment, i_t the nominal interest rate paid on corporate debt, B_{it} debt, π_t^e expected inflation and p_t^I the investment price index.[3] Apart from the cash-flow constraint two additional conditions have to be met: first, dividends of course cannot be negative, second, the standard transversality condition that the firm is not allowed to borrow an infinite amount to pay out as dividends (no-Ponzi game condition).[4]

Let λ_{it} be the series of multipliers associated with the condition of non-negative dividends. The firm decides on the capital stock K_{it} and the amount of debt B_{it}.[5] For real investment the first-order condition is

$$\beta_{it}E_t[\frac{1+\lambda_{i,t+1}}{1+\lambda_{it}}(F_K(K_{it}, L_{i,t+1}) - G_K(I_{i,t+1}, K_{it})+$$

$$(1-\delta)(G_I(I_{i,t+1}, K_{it}) + \frac{p_{i,t+1}^I}{1-\tau}))] = G_I(I_{it}, K_{i,t-1}) + \frac{p_{it}^I}{1-\tau} \quad (6.1)$$

This is the neoclassical version of the Euler investment equation, which assumes absence of capital market imperfections. The right-hand side of condition (6.1) represents the marginal installation costs G_I and the current (tax-adjusted) price of investing in new capital in the current period t. The left-hand side shows the costs of postponing investment until the next period $t+1$ ('costs of waiting'), which consists of the marginal product of the new capital foregone and the purchasing and installation costs of investing tomorrow. The opportunity cost of investing in the future is weighted by the relative shadow value of residual earnings (slack parameter λ) in this period and the next. In equilibrium both costs are equal.

The first-order condition for borrowing reads

$$(1 + \lambda_{it}) - \beta_{it}(1 + (1 - \tau)i_t - \pi^e_{t+1})E_t[(1 + \lambda_{i,t+1})] = 0 \qquad (6.2)$$

Hence, the discount factor of each firm must be equal to the inverse of the real after-tax market interest rate, identical to all firms. Equation (6.2) implies that the discount factor in Equation (6.1) reads

$$\beta_{it}E_t[\frac{1 + \lambda_{i,t+1}}{1 + \lambda_{it}}] = \frac{1}{1 + (1 - \tau)i_t - \pi^e_t} \qquad (6.3)$$

In the neoclassical model the firm equates the shadow costs of residual earnings over time so that $\lambda_{it} = \lambda_{i,t+1}$.

6.2.2   The debt-constraint augmented model

In order to allow for the impact of possible debt constraints we add the borrowing constraint, $B_{it} \leq B^*_{it}$ associated with multiplier γ_{it}. The idea is, see also Whited (1992, pp. 1426-1427), that a firm which faces this constraint will in fact behave as if it faces a higher discount rate than an unconstrained firm. This idea also implies that changes in B^*, due to for instance a perceived change in financial healthiness of the firm by banks, are reflected in corresponding changes in the discount rate.[6] In order to allow for the possibility that (changes in) the borrowing constraint are picked up by (changes in) the discount rate, the first-order conditions for borrowing (Equation (6.2) needs to be modified as follows:

$$(1 + \lambda_{it}) - \beta_{it}(1 + (1 - \tau)i_t - \pi^e_{t+1})E_t[(1 + \lambda_{i,t+1})] - \gamma_{it} = 0 \quad (6.4)$$

The additional firm-specific term γ_{it} introduces a wedge between the shadow value of residual earnings in the current and next period, as the borrowing constraint prevents the firm from equating the shadow value over time. Suppose the shadow value of dividends is higher today than it is expected to be tomorrow and the firm hits its debt capacity. Then, the present value of the firm can be increased with γ by relaxing the debt constraint.[7] The inclusion of the multiplier γ_{it} in the first-order condition for borrowing implies that the discount factor (6.3) now becomes

$$\beta_{it}E_t[\frac{1 + \lambda_{i,t+1}}{1 + \lambda_{it}}] = \frac{1 - \gamma_{it}/(1 + \lambda_{it})}{1 + (1 - \tau)i_t - \pi^e_t} \qquad (6.5)$$

Compared to an unconstrained firm, a firm facing a binding borrowing constraint has a higher value of γ_{it} and thus higher marginal opportunity cost of investing today versus delaying it until tomorrow. Again, this effectively means that if the borrowing constraint bites, a firm will in fact behave as if it has a higher discount rate than in case the borrowing constraint does not do so. An increase in the discount rate means that the return of an additional unit of investment also needs to increase in order for this investment to be still acceptable to the firm. To put it differently, future earnings are discounted more heavily if firms face a binding debt constraint in the current period (Whited, 1992). If this higher discount factor is captured by the capital markets, then debt-constrained firms will face higher rates of interest on debt. In other words, the borrowing constraint will lead to a firm-specific risk premium on corporate debt.[8] In case the firm-specific interest rates on debt i_{it} can be measured directly, and assuming that the difference between i_{it} and i_t, the interest rate in case of the neoclassical model, can reasonably be attributed to the existence of a debt constraint, the discount factor (6.5) can be rewritten as follows (see also Section 6.4)

$$\frac{1 - \gamma_{it}/(1 + \lambda_{it})}{1 + (1 - \tau)i_t - \pi_t^e} = \frac{1}{1 + (1 - \tau)i_{it} - \pi_t^e} \tag{6.6}$$

where $i_{it} = i_t$ holds for $\gamma_{it} = 0$.

6.3 Empirical implementation

6.3.1 Production, competition and adjustment costs

The empirical specification of the model requires some technical assumptions with respect to the production function, the degree of competition on the goods market and the form of the adjustment cost function. These assumptions are not important for the modelling of financial constraints, but are needed in order to keep the empirical implementation of the model tractable. In our empirical model we assume that firms have some degree of market power. If the firm faces a known demand curve $P_t = Y_t^{-1/\epsilon_D}$, then the mark-up reads

$$\mu = \frac{|\epsilon_D|}{|\epsilon_D| - 1} \tag{6.7}$$

We further assume nonconstant returns to scale. Then we can write for the marginal productivity of capital

$$F_K(K_{i,t-1}, L_{it}) = \frac{\eta Y_{it} - \mu C_{it}}{K_{i,t-1}} \tag{6.8}$$

where η denotes the scale parameter, C_{it} are variable costs. In case of perfect competiton μ approaches unity (see Equation (6.7)). It is hard to identify both parameters. We experimented with a scale parameter but obtained a rather unreliable set of estimates. It thus appeared to us, as it did to Hubbard, Kashyap and Whited (1995), that the data only allow for a joint estimation of the mark-up and scale parameters. Therefore we had to set η equal to 1, so that possible scale effects are captured by μ. Increasing returns lead to an underestimation of the mark-up parameter μ.

For the adjustment cost function we use the convex linear homogenous power function, proposed by Whited (1995):

$$G(I_t, K_t) = (\alpha_0 + \sum_{m=2}^{M} \frac{1}{m}\alpha_m(\frac{I}{K})_t^m)K_t \qquad (6.9)$$

Experiments showed that a second-order power approximation ($M = 2$) is sufficient, which basically boils down to the quadratic adjustment cost function as proposed by Summers (1981).[9]

6.4 Debt constraints

Substitution of (6.6), (6.8) and (6.9) into (6.1) yields our empirical specification of the debt-augmented Euler investment equation

$$\frac{1}{1+(1-\tau)i_{it}-\pi_t^e}[\frac{Y_{i,t+1}-\mu C_{i,t+1}}{K_{it}} + \frac{\alpha_2}{2}((\frac{I_{i,t+1}}{K_{it}})^2 + \frac{2\alpha_0}{\alpha_2}) +$$

$$\alpha_2(1-\delta)\frac{I_{i,t+1}}{K_{it}} + (1-\delta)\frac{p_{i,t+1}^I}{1-\tau}] - \alpha_2\frac{I_{it}}{K_{i,t-1}} - \frac{p_{it}^I}{1-\tau} +$$

$$f_i + s_t = e_{i,t+1} \qquad (6.10)$$

where f_i are firm effects and s_t year effects. The firm effects f_i in (6.10) are firm-specific constants which take account of the fact that average values of the variables may vary between the firms due to firm-specific reasons. The year effects s_t serve a similar purpose, since they are meant to take account of the year-to-year variation in our model variables during the sample period, the effects thus capturing the business cycle components in our variables. Note that we substituted observed values for the expectations terms under rational expectations introducing a white noise expectational error term $\epsilon_{i,t}$, which is orthogonal to any information known at period t.

The neoclassical model version, i.e. without debt constraints, is the special case of (6.10) for which $i_{it} = i_t$ (and hence γ_{it}=0). The debt-constrained firm is confronted with a higher discount rate. By dealing with borrowing constraints in this way, we circumvent the problem that the optimality conditions of the debt-constraint augmented model do not provide an analytical solution for the slack parameter λ in Equation (6.5). In her attempt to find an empirical solution Whited (1992) proposes to parameterise the slack parameter as an *ad hoc* function of financial variables. But as Chirinko (1993, p. 1903) correctly observes: '...the endogenous variables that parameterise the multiplier—such as cash flow and net worth sensitive to the firm's decisions—are not accounted for explicitly in specifying the econometric equation, thus blurring economic interpretations of the statistical tests.' The parameterisation of the slack parameter λ, by means of various proxies for the financial distress of a firm, is meant to take account of only that part of the difference between the unobserved i_{it} and i_t which might be attributed to the alleged existence of a debt constraint. The advantage of our approach of directly measuring i_{it} is that our firm-specific interest rate automatically includes any such premium.

This conclusion seems subject to one important caveat: a difference between the observed firm-specific interest rate and the 'neoclassical' interest rate i_t need not necessarily be the result of the existence of a borrowing constraint, but might also reflect firm-specific business risk factors which are unrelated to the workings of the debt market. In a similar vein one could argue that *even if* it is granted that the difference between i_{it} and i_t reflects the existence of a debt constraint, this need not have an impact on the firm's investment behaviour as long as the firm can resort to other means of external finance, notably the issuance of new shares. To start with the second point, the model assumes that the equity market is shut down for the firms in our sample, which, as has already been mentioned in section 6.2.1, is not a far-fetched assumption for the case of the Netherlands. The first point, the level of i_{it} is (partly) determined by firm-specific risk characteristics, cannot stand up to close scrunity either. From a strictly theoretical viewpoint firm-specific effects are not an issue at all, since firms are assumed to be risk-neutral in our theoretical model. If one relaxes this assumption, however, a relatively high interest rate on debt may indeed reflect high risk profiles for some firms. Empirically, this does not undermine our results, either. As Equation (6.10) shows we do allow for firm-specific (and year-specific) effects in our empirical specification of the Euler equation. The riskiness of any individual firm should be primarily related to the variability of its earnings. Hence, to the extent that this variance is constant over time, it should be picked up by the fixed effect in our estimation model (see also Whited, 1992, p. 1449). Moreover, if the risk varies over time, it should be taken care of by the fixed year effects (see Equation (6.10)). We measure the discount factor

(Equation (6.6)) using the information our panel data provide for each firm on interest payments on debt and the amount of (short-term and long-term) debt. This information enables the direct measurement of the firm-specific interest rates i_{it} by taking the ratio of interest payments to interest-bearing debt. We adjust the average interest rates for differences between debt maturity structures by taking account of the term structure of interest as follows. We define the observed firm-specific interest rate on interest-bearing debt r_i as (omitting time subscripts)

$$r_i = \kappa_i(i^s + \omega_i^s) + (1 - \kappa_i)(i^l + \omega_i^l)$$

where i^s and i^l denote the 'risk-free' short- and long-term corporate interest rate, equal to all firms, respectively, κ_i the ratio of short-term debt to total debt and ω_i^s and ω_i^l the firm-specific short- and long-term risk premiums, respectively. In order to correct the interest rate on total debt for interfirm differences in the composition of debt between short- and long-term debt, we make the simplifying assumption that the risk premiums on long-term and short-term debt are equal[10]

$$\omega_i^s = \omega_i^l = \omega_i$$

By substitution we then derive the corrected interest rate

$$i_i = i^l + \omega_i^l = r_i - \kappa_i(i^s - i^l)$$

6.5 Data and estimation

The panel data on individual firms, used for the estimation of the Euler investment models in the empirical part of this paper, are taken from the annual SFGO microdata files of the Netherlands Central Bureau of Statistics (CBS), which compiles and publishes these data only in aggregate form.[11] Beginning in 1977, balance sheet and income accounts data are collected by the CBS in annual questionnaires among by now over three-thousand non-financial firms with balance sheet totals of 10 million Dutch guilders or more. As only a minority of some 200 non-financial firms in the Netherlands is publicly traded, the SFGO microdata files do not contain data on stock market valuations, which fortunately is not a requirement to estimate the Euler model. From 1983 onwards, the questionnaires use a system of unique company-numbers for identification of firms over the years, so that only from that year a consistent panel data set can be constructed. The SFGO microdata file takes account of cross-ownership through consolidation, so that the firms are a good representation of the economic decision units. We selected firms from the manufacturing industry. In panel data studies of investment behaviour it is common practice to focus only on manufacturing firms in order to have a reasonably homogeneous group of firms as regards capital intensity and investment behaviour.

The quality and representativeness of the SFGO microdata files are high, thanks to the practical absence of non-response, as the filling in of the questionnaire forms is compulsory on the basis of the Dutch law on economic statistics and the CBS makes special efforts to realise a maximum response rate. The financial account data is quite detailed. For instance, SFGO microdata reports nine sources of changes in the book value of fixed assets, which allows for an accurate definition of cash outlays for investment in capital goods. Moreover, the level of detail allows for a rich menu of categorisations of firms by firm-specific characteristics, which is relevant for our estimation procedure.

From the SFGO microdata files we selected a balanced panel of 427 Dutch manufacturing firms. The sample period is 1983–1992. A balanced panel data set is necessary because a balanced GMM-estimation technique is used.[12] As a result of the selection of firms with continuous data there may be some selectivity bias towards healthy firms. In terms of the present research, we would expect that this survivor bias reduces the probability of finding debt constraints in the sample. Hence, any findings which support the debt-constraint hypothesis should be interpreted as being even more powerful. The Data Appendix gives details on the construction of the variables.

Table 6.1 presents some descriptive statistics for the full sample as well as for the subgroups publicly traded and private companies. The majority (80 per cent) of firms in the sample are private firms. Publicly traded and private firms differ mainly by size. The median values of the capital stock for the publicly traded firms is 4 to 5 times larger than for private firms. Further, private firms have low dividend payouts compared to publicly traded firms. The average payout ratio of public firms with stock quotations is even higher in the Netherlands, by the way, and those firms are often also much larger (see for instance Van Ees and Garretsen, 1994, and De Haan, 1995). It is an advantage that our sample also includes relatively small to medium-sized firms.

As the model is a dynamic panel model we follow Hansen's (1982) approach in estimating rational expectations equilibrium models using Generalised Method of Moments (GMM). We estimate the model in first differences to get rid of the firm-specific effects. First differencing implies that second-order lags of the model variables must be on the instrument list. While Arellano and Bond (1988) propose to use all available lags, Whited

	Whole sample	Publicly traded firms	Private firms
Number of firms	427	82	345
Investment rate			
Mean	0.27	0.30	0.27
Median	0.19	0.19	0.19
Cash flow/Capital stock			
mean	0.15	0.13	0.16
median	0.14	0.12	0.14
Capital stock (mln guilders)			
mean	131	397	68
median	19	81	15
Debt ratio			
mean	0.61	0.58	0.62
median	0.62	0.57	0.64
Payout ratio			
mean	0.07	0.14	0.06
median	0.01	0.14	0.00
Interest coverage ratio			
mean	3.8	4.3	3.6
median	4.6	6.1	4.4

Table 6.1: Summary statistics for Dutch manufacturing,1983-1992

(1992) sticks to a second-lag order only.[13] We follow Whited's procedure.

As we need both lags and leads the effective sample is 1985-1991. Taking first differences to deal with fixed firm effects yields seven difference equations, *i.e.* one for each year. We use six instruments: the second-order lags of the model variables (operating income $Y_{i,t-1}/K_{i,t-2}$, operating costs $C_{i,t-1}/K_{i,t-2}$, investment $I_{i,t-1}/K_{i,t-2}$ and the investment price P_{t-2}^I] as well as corporate tax payments and investment tax credits (see the Data Appendix for details). The structural parameters to be estimated are μ and α_2, the mark-up/scale parameter and the adjustment cost parameter, respectively, and in the augmented model parameter α_0.[14] We use the χ^2−statistic (Sargan-test) to test for possible model misspecification. The χ^2−statistic tests the overidentification restrictions, i.e. the null-hypothesis that the set of instruments is *not* correlated with the error terms. The reported p-value gives the corresponding probability of the type II-error of rejecting the null-hypothesis when the null is true. Hence, a low p-value indicates that the overidentification restrictions must be rejected.

	Neoclassical	Debt-constraint augmented
$\hat{\mu}$	1.06	1.04
$t(\hat{\mu})$	44.04	42.38
$\hat{\alpha}_2$	0.94	1.06
$t(\hat{\alpha}_2)$	4.10	4.26
$\hat{\alpha}_0$		2.84
$t(\hat{\alpha}_0)$		0.24
χ^2	39.87	36.48
Degrees of freedom	28	27
p-value	0.07	0.11
Number of firms	427	427

Sample period: 1985-1991. Standard errors are computed from robust White heteroscedastic-consistent estimates using a first-order autocorrelation scheme. All year effects (not reported) are significant at the 5%-level. The χ^2 is overidentifying test statistic. The degrees of freedom are determined as the number of instruments used minus the parameters to be estimated. The p-value indicates misspecification for low values.

Table 6.2: GMM-estimates of the Euler investment equation

6.6 Estimation results

6.6.1 Full-sample estimates

Table 6.2 presents the full-sample estimates for both the neoclassical model and the debt-constraint augmented investment model. If the augmented model outperforms the neoclassical model, we will interpret that as evidence of influences of financial constraints on investment behaviour. Our specification of the neoclassical model is obtained by assuming in (6.10) that all firms have identical interest rates on debt. In the first column of Table 6.2 we present the results for the neoclassical model. The estimated values for the parameters in the neoclassical model are significant (at the 5%-level) and have the theoretically expected signs. It is essential that α_2 is significant in order to be able to determine the optimal level of investment in an analytically meaningful way. If α_2 is not significant the Euler investment equation reduces to the statement that the average capital productivity equals the relative price of investment. Our parameter estimate $\alpha_2 = 0.94$ implies on average an adjustment cost of 12 per cent of total investment expenditure. This compares well to Whited's 10 per cent for the U.S.A.. The value of μ seems to imply that the corresponding demand elasticities are about 17, which seems rather high. Note, however, that μ captures the mark-up as well as the returns-to-scale effect. Whited (1992) finds similar full-sample results for her U.S.A. sample. Even though the p-value of 0.07 indicates that the neoclassical investment model is not misspecified at the 95%-confidence level, one would particularly like to know whether the model can be improved by allowing for debt-constraints.

The second column of Table 6.2 gives the estimates for the debt-constraint augmented model, using observed firm-specific interest rates on debt, i_{it} (Equation (6.10)). The higher p-value of 0.11 indicates a slightly better fit for the augmented model than for the neoclassical model.[15] Moreover, the estimates of the markup-cum-scale parameter and the adjustment parameter are statistically significant and still have the correct signs. Their estimated values differ slightly from their neoclassical counterparts. Hence, from these two full-sample estimates the main conclusion is that the debt-constraint augmented investment model is better equipped to describe investment behaviour of our sample of Dutch firms than the neoclassical investment model. The selectivity-bias of the sample (see section 6.5) towards healthy firms enhances this outcome, as 'survivor bias' may lead to some underestimation of the influence of debt constraints (Chirinko and Schaller, 1995).[16]

As an illustration of our direct measurement of debt constraints, Figure 6.1 shows the computed firm-specific interest rates on corporate debt for each of the 427 firms in our sample, ordered from low to high. Up to the 70th percentile the interest rate is more or less constant at a level of about 10 per cent. Given an average interest rate on government bonds of 8 per cent during this period, a risk premium of about 2 per cent is implied. However, the group of firms between the 70th and 100th percentile face much higher, sometimes extremely high, interest rates on debt, which suggests that these firms in particular may face rather severe debt-constraints. One possible objection against this interpretation is that the alleged risk premiums need not necessarily be caused by debt constraints alone, but may as well be due to firm-specific business risk characteristics. As mentioned before, the inclusion of firm-specific effects is meant to adjust the risk premiums for the risk characteristics.

6.6.2 Sub-sample estimates

Next, in order to test the robustness of the debt-constraint augmented model, we split the sample into different subsets according to certain firm-characteristics, which *a priori* reflect the probability that a particular type of firm suffers from debt constraints. In previous studies, such sample splits have been based on univariate measures like the dividend payout (for instance,

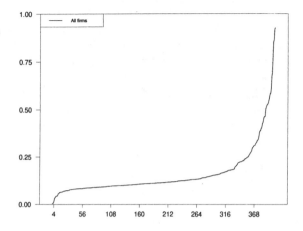

Figure 6.1: Firm-specific interest rate premium (scale of 100 %).

Fazzari *et al.*, 1988), debt and interest coverage ratios (for instance Whited, 1992), etc. The dividend ratio is traditionally claimed to represent liquidity constraints, with payout being low when constraints are binding. The debt and coverage ratios are considered to be proxies for 'financial distress.' Although the procedure may be intuitively appealing, we have some methodological objections against this univariate categorisation. Firstly, the choice of the proxies is at best arbitrary. For example, Kaplan and Zingales (1997) found no evidence that the low-payout firms in the Fazzari *et al.* panel were in fact liquidity constrained. Hence, it seems better to us not to choose one categorisation variable in advance, but instead to use the information in the dataset in a more systematic way to come to a more objective classification of firms. Secondly, the univariate categorisation procedure does not take the apparent interrelationships between different classification variables into account. For example, it should be clear that debt and interest coverage ratios are interrelated, so that it may be problematic to make two different classifications using both variables.

To overcome these drawbacks of the common practice of univariate sample-classification, we employ a multivariate procedure: factor analysis. The power of factor analysis is that it simplifies the complex and diverse relationship among different variables by uncovering the common dimensions that link them together, thus providing insight into the hidden structure of the panel data. We start with a complete set of potential proxy-variables for debt-constraints. We try the traditional proxies from the literature—payout, debt, coverage ratios and capital stock (for firm-size). We also introduce some new variables. We include the amount of bonds issued as a proxy for access onto the capital market. Moreover, we include a variable that indicates the percentage of external debt provided by banks as a proxy for bank-dependence. We also use the classifications provided in the data file between private and public firms and international and national orientation (in terms of foreign share holdership and/or foreign subsidiairies). Possibly public firms and internationals have relatively more access to the national and international capital markets, respectively, and therefore are perhaps less likely to be confronted with domestic debt constraints.

The factor analysis clusters the variables into factors based on their correlations. Variables that are highly correlated are formed into a factor with the condition that this factor is not related to the second factor, and so on. To improve the interpretation of the results from factor analysis, an orthogonal rotation is performed to obtain a simple structure. This reduces the problem of having too many variables loading on one factor or a variable showing significant loadings on more than one factor. A scree plot method, an analysis of the eigenvalues of the factors and a χ^2-test (yielding a $\chi^2 = 201.67$) on the sufficiency of three factors were employed in determining the number of factors to retain.[17]. Consequently, three factors are retained in the analysis. The resulting factor loadings are presented in Table 6.3.

Factor number	1	2	3
Factor label	size	Leverage	Payout
Variable			
Debt ratio	-0.03	0.51	0.06
Bonds issued	0.07	0.00	-0.01
Bank credit	-0.02	0.03	0.00
Payout ratio	-0.02	-0.03	1.01
Interest coverage ratio	0.02	-0.42	-0.04
Capital stock	0.92	0.07	-0.09
Public/private	-0.02	0.04	0.01
International/national	-0.01	0.08	0.01
Explained variance	1.62	1.25	1.09

Table 6.3: Factor pattern matrix

Capital stock shows the highest loading on factor 1. Therefore we label this factor 'size.' The second factor is dominated by the debt and interest coverage ratios and is therefore labeled 'leverage.' Finally, the payout ratio is the one significant loading on the third factor, which we therefore label 'payout'. As mentioned before, the payout is often used as a proxy for liquidity constraints. Fazzari *et al.* (1988) and Bond and Meghir (1994) find significant liquidity constraint effects on investment of low-payout firms. The other proxy-variables have all relatively small loadings.

In Table 6.4 we reestimate the debt-augmented Euler investment model for four subsamples. The first subset (in the first column) clusters one third of the sample firms for which the 'size'-factor has the lowest value. This subset typically includes smaller firms. We expect that smaller firms encounter more asymmetric information problems on the public capital markets, so that they are more likely to be financially constrained. The second cluster consists one third of the firms with highest values for the 'leverage'-factor, that is, firms in high debt. The third subgroup is the one-third with the lowest leverage. We include both extremes for leverage, because both high and low leverage may indicate financial problems. On the one hand, high leverage may *ex ante* indicate financial distress, but on the other hand low leverage may *ex post* indicate severe credit rationing for reasons that are not related to the debt ratio. Theoretically, there is a causality problem with the leverage criterion, as debt is an endogenous variable in our investment model. The fourth subgroup consists of one-third of the firms with lowest values for the 'payout'-factor. We expect low-payout firms to be relatively more constrained, which would imply that the augmented model performs relatively better for this group of firms.

Factor number	Factor 1	Factor 2	Factor 2	Factor 3
Label	size	Leverage	Leverage	Payout
Subsample	Smallest 33 %	Highest 33%	Lowest 33%	Lowest 33%
$\hat{\mu}$	0.62	0.72	1.34	1.34
$t(\hat{\mu})$	4.09	5.13	29.98	23.69
$\hat{\alpha}_2$	0.36	0.42	0.15	0.36
$t(\hat{\alpha}_2)$	2.14	3.01	0.60	2.28
$\hat{\alpha}_0$	0.99	-1.36	-3.40	-2.33
$t(\hat{\alpha}_0)$	0.32	-0.54	-1.62	-0.98
χ^2	40.03	31.34	38.94	28.23
Degrees of freedom	27	27	27	27
p-value	0.05	0.26	0.06	0.40
Number of firms	140	145	140	140

Sample period: 1985-1991. Standard errors are computed from robust White heteroscedastic-consistent estimates using a first-order autocorrelation scheme. All year effects (not reported) are significant at the 5%-level. The χ^2 is the overidentifying test statistic. The degrees of freedom are determined as the number of instruments used minus the parameters to be estimated. The p-value indicates misspecification for low values.

Table 6.4: GMM-estimates of debt-constraint augmented Euler investment equation: subsamples

We concentrate on the augmented-model fit for the subsamples in Table 6.4.[18] The augmented-model fit is not stronger for the small-size cluster. The p-value is 0.05, against 0.11 for the full sample (*cf.* Table 6.3). Hence, our results yield no clear-cut evidence that smaller firms are more debt constrained. Perhaps that smaller firms in our panel are not small enough, having balance sheet totals of at least 10 million guilders (see Section 6.5). Hence, the variation in size in our sample may not be large enough, especially at the lower end, to effectively discriminate between small and large firms. Whited (1992), who neither finds an independent size effect in het U.S.-panel, observes that constrained firms can be either small or large.

The augmented-model fit is relatively strong for the high-leverage group, whereas it is relatively weak for the low-leverage firms. This is consistent with the possibility that high-leverage firms are more financially constrained than low-leverage firms. On the other hand, a closer analysis of the implied interest rates on debt (not reported) reveals that the low-leverage firms pay higher, not lower, interest rates on debt as compared to both the full-sample and high-leverage group. This observation would suggest that low leverage is an *ex post* proxy for severe credit rationing rather than high leverage is an *ex ante* proxy of financial distress. An argument against this latter interpretation is, however, that the estimated value of the adjustment-cost parameter for the low-leverage group is not significantly differing from zero, leaving the investment model undetermined for this particular subsample. In view of these mixed results for the leverage groups and the earlier-mentioned theoretical objection against the leverage criterion, we are cautious making inferences on the basis of the leverage-clustering results.

The fit of the augmented model is markedly strong for the low-payout group (0.40 as against 0.11 for the full sample). This confirms our expectation that this particular group of firms is *a priori* more likely to be financially constrained. Low payout, or high retention of earnings, apparently indicates that asymmetric problems in the capital market drive a wedge between the price of external and internal funds. This result is in line with the ones found by Fazzari *et al.* (1988) for the U.S.A. and Bond and Meghir (1994) for the U.K. As is apparant from Figure 6.2, low-payout firms indeed pay higher interest rates on debt than the average firm. Hence, our results for the low-payout subsample confirm the notion that particularly financially constrained firms are confronted with higher cost of external funds.

We conclude from the estimation results for the low-payout sub-sample and the full sample that the better fit of the debt-constraint augmented model as opposed to the neoclassical model indicates that debt constraints are relevant for some firms in the Netherlands.

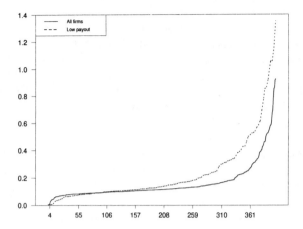

Figure 6.2: Corporate debt rate for all and low-payout firms (scale of 100 %)

6.7 Conclusion

In this paper we have examined the influence of debt restrictions on corporate investment behaviour of Dutch firms. We applied an Euler investment equation-approach on panel data of 427 Dutch manufacturing firms over the period 1983-1992. Two model specifications of the Euler investment equation have been estimated: a neoclassical version assuming perfect capital markets and a debt constraint-augmented version incorporating capital market imperfections. Our estimation results for the full sample indicate that the augmented investment model is better equipped to describe the investment behaviour of firms than the neoclassical investment model.

A problem of the Euler approach is the empirical measurement of the debt-constraint multiplier. In this paper we dealt with this problem in a novel way, by measuring debt constraints directly by means of observed firm-specific interest rates on debt, as opposed to the indirect measurement of debt constraints in related studies by means of a parameterisation of the slack or the interest rate function. Not only is our measurement more straightforward theoretically, it also yields comparatively better empirical results when applied to our sample of Dutch firms.

Next, following previous work, the robustness of the debt-constraint augmented model was tested by reestimation for subsamples of firms, which *a priori* are expected to face relatively more binding debt constraints. However, instead of the traditional univariate stratification procedure used in previous studies, we employed a multivariate factor analysis approach to define clusters of firms using the hidden structures in the data set. The factor analysis showed that the firms in the sample can be clustered in three independent ways: according to size, leverage and payout. The estimation results for the subsample splits, based on these factors, suggest that debt-constraints are especially relevant for firms with low dividend payout, while size did not appear to be an independent factor for the presence of liquidity constraints and the leverage criterion gave mixed results. Hence, according to our sample about one third of all firms are confronted with debt constraints. A low dividend payout, that is, high retention, apparently indicates that problems of asymmetric information in the capital market make external finance relatively expensive compared to internal finance. Firms with low payout ratios in fact pay higher interest rates on debt. These results confirm the theory that firms which are confronted with financial constraints face higher costs of external capital.

Data appendix

The panel is labelled SFGO (in Dutch: *Statistiek Financiën van Grote Ondernemingen (SFGO)*) and produced by the Central Bureau of Statistics. All variables are in real terms (we omit i-subscripts for individual firms). As deflator we used the gross output price, py_t, the implicit deflator of gross output per two-digit SIC-industry (source: Central Planning Bureau). For the price of capital goods we used the implicit deflator of gross investment per two-digit SIC-industry, p_t^I (source: CBS). Real value of gross investments, I_t is defined as production and purchases of capital goods (SFGO-code M03).

For the real replacement value of the capital stock, K_t, the perpetual inventory method is used in order to derive replacement values from book values. As a working hypothesis, the replacement value of the first observation year is set equal to the book value. The replacement value for subsequent years is derived from

$$K_t = \left(K_{t-1}\frac{p_t^I}{p_{t-1}^I} + I_t\right)\left(1 - \frac{x}{L}\right)$$

where

$$L = \frac{K_{t-1}^b + I_t}{D_t}$$

with x equal to 1 or 2, depending on the assumption of a single or double declining balance depreciation method, respectively, L is the depreciable life of capital goods, K_{t-1}^b the book value of fixed physical assets (SFGO-code B04) and D reported depreciation (SFGO-code M07). Experiments showed that the choice of $x = 1$ described the data in SFGO microdata most accurately. The depreciation rate δ equals $1/L$. In the average productivity of capital,

$$\frac{Y_t - \mu C_t}{K_{t-1}}$$

where Y_t is real sales (SFGO-code R01), C_t real variable costs (SFGO-code R02), K_{t-1} the real beginning-of-period stock of capital defined above and μ a parameter to be estimated.

The observed firm-specific interest rate on debt is obtained from the ratio of interest expenses (SFGO-code R12) to the amount of interest-bearing debt (SFGO-codes MINI07B through MINI10B). For the short-term risk-free interest rate i^s we took the money market rate on three-month loans to local authorities and for the long-term risk free interest rate i^l the capital market rate on medium-term central government bonds. τ is the statutory corporate tax rate. π^e is set equal to the annual consumer price inflation rate.

As instrumental variables (in addition to the lagged independent variables) we used the real value of corporate income taxes paid (SFGO-code R17) and the real value of investment tax credits (SFGO-code B61).

Notes

1. By now, a large body of literature has been developed on this subject. Examples of a theoretical nature include Myers and Majluf (1984), Stiglitz and Weiss (1981), and Bernanke and Gertler (1989, 1990), amongst others. Gertler (1988) provides a review of the different issues in this field.

2. In our sample, only 5 per cent of the median firm's capital is stock, and in 75 per cent of the firm-year observations no share issues took place.

3. Inflation erodes debt, which accounts for the inclusion of the term $\pi_t^e B_{i,t-1}$ in the real dividend equation.

4. De Haan's (1996) study of borrowing and dividend behaviour indicates that Dutch listed firms normally do not borrow to finance dividend payments.

5. The firm also takes a labour demand decision following: $F_L = w$.

6. This method of dealing with the borrowing constraint in the Euler investment equation approach implies that the debt limit "can be loosely interpreted either as a literal credit constraint or as an interest rate premium, since both interpretations lead to the same qualitative conclusions and results" (Whited, 1992, p. 1429).

7. The Euler approach uses two periods only. This implies that only a change in the debt constraint between the two periods is picked up by Equation (6.4). Generally a two-period analysis will not be sufficient to deal with the intertemporal aspects of debt constraints. In reality, future expected contraints may lead to a reduction of current investment. Likewise firms with former debt problems might have a higher interest burden today.

8. An alternative approach to account for debt constraints is followed by Bond and Meghir (1994) and Estrada and Vallés (1995). They define the firm-specific interest rate on corporate debt i_{it} as the sum of the market interest rate i_t and a risk premium. The latter is assumed to be a function $H(.)$ of the endogenous debt and capital variables:

$$i_{it} = i_t + H(B_{it}, K_{it})$$

This implies that both the Euler conditions for capital (6.1) and for debt (6.2) need to be adjusted because in this approach the firm does take the above equation, which can be thought of as the supply curve of debt, into account in the maximisation of its present value. Instead of the borrowing constraint, the firm now faces the above equation as an additional constraint:

$$\beta_{it} E_t \Big[\frac{1 + \lambda_{i,t+1}}{1 + \lambda_{it}} (F_K(K_{it}, L_{i,t+1}) - G_K(I_{i,t+1}, K_{it})$$
$$-H_K(B_{it}, K_{it}) B_{it} + (1 - \delta)(G_I(I_{i,t+1}, K_{it}) + \frac{p_{i,t+1}^I}{1 - \tau})) \Big]$$

$$= G_I(I_{it}, K_{i,t-1}) + \frac{p_{it}^I}{1-\tau}$$

$$(1 + \lambda_{it}) - \beta_{it}(1 + (1-\tau)i_t - (1-\tau)H_B(B_{it}, K_{it}) - \pi_{t+1}^e)E_t[(1 + \lambda_{i,t+1})]$$
$$= 0$$

We estimated this model version without success (see also Section 6.5).

9. We experimented with alternative assumptions on the relative magnitude of the risk premiums on short-term and long-term debt, but did not find qualitatively different results from the ones presented hereafter.

10. Hamermesh and Pfann (1996) discuss alternative adjustment cost functions: asymmetric convex, piecewise linear or lumpy specifications. For labour demand it is known from micro studies that the alternative specifications dominate the symmetric convex adjustment cost function. For investment less evidence is known. Considering the complexity of the model dynamics we refrain from testing alternative cost functions.

11. Published as *Statistiek Financiën van Grote Ondernemingen*.

12. We used TSP 4.3 for GMM-estimation. An alternative is the DPD-algorithm for Gauss, written by Arellano and Bond (1988), but this can be applied to linearised equations only whereas we estimate the Euler equation in the original non-linear form.

13. Barran and Peeters (1996) test for the inclusion of additional lags or differenced instruments and conclude that instruments in levels lagged once perform the best in an undifferenced model. As we estimate the model in first differences this boils down to the same.

14. In the empirical specification of the neoclassical equation the parameter α_0 will be subsumed by the fixed time effects.

15. A significance test is impossible in this case, as the interest rate variables in the two models are different (*i.e.* the models cannot be nested).

16. In order to compare our version of the debt-constraint augmented model with a model version that parameterises the slack λ we also estimated Equation (6.10) with $1 - \frac{1 + \lambda_{i,t+1}}{1 + \lambda_{it}} = \delta_0 + \delta_1 \frac{CF}{K}$ where CF represents cash flow. This parameterisation is used by Hubbard, Kashyap and Whited (1995). An increase in cash flow is assumed to imply a relaxation of the debt-constraint ($\delta_1 < 0$). Moreover we estimated the model proposed by Bond and Meghir (1994) and Estrada and Vallés (1995) with the parameterisation of the interest rate function: $i_{it} = i_t + b\frac{B_{it}}{K_{it}}v$. In both cases we were not able to let the model parameters converge.

17. The null hypothesis of no common factors was clearly rejected: $\chi^2 = 6265.83$

18. A closer look at the parameter estimates for the subsample estimations learns that the adjustment parameter α_2 has lower values compared to the full-sample estimates (*cf.* Table 6.4), indicating slower adjustment towards desired levels of capital. Further, the value of μ is smaller than one for the small-size and low-leverage clusters, suggesting that the assumption of constant returns to scale is too rigid and leads to underestimation of the mark-up parameter for these particular subsamples.

References

Arellano, M. and S.R. Bond (1988), "Dynamic Panel Data Estimation Using DPD- a Guide for Users," *Working Paper Series* 88/15, The Institute for Fiscal Studies.

Bernanke, B. and M. Gertler (1989), "Agency Costs, Net Worth and Business Fluctuations,"

American Economic Review, 79:14–31.
Bernanke, B. and M. Gertler (1990), "Financial Fragility and Economic Performance," *Quarterly Journal of Economics*, 105:87–114.
Bond, S.R. and C. Meghir (1994), "Dynamic Investment Models and the Firm's Financial Policy," *Review of Economic Studies*, 61:197–222.
Chirinko, R.S. (1993), "Business Fixed Investment Spending: Modeling Strategies, Empirical Results, and Policy Implications," *Journal of Economic Literature*, 31:1875–1911.
Chirinko, R.S. and H. Schaller (1995), "Why Does Liquidity Matter in Investment Equations?," *Journal of Money, Credit and Banking*, 27:527–48.
Fazzari, S.M., R.G. Hubbard, and B.C. Petersen (1988), "Financing Constraints and Corporate Investment," *Brookings Papers on Economic Activity*, 141–95.
Gertler, M (1988), "Financial Structure and Aggregate Economic Activity," *Journal of Money, Credit, and Banking*, 20:559–88.
Haan, L. de (1995), "Dividendbeleid van Nederlandse beursondernemingen," (Dividend Policy of Listed Dutch Firms) *Maandblad voor Accountancy en Bedrijfseconomie*, 69:774–85.
Haan, L.de (1996), "The Empirical Relationship between Investment, Dividend and Financing Decisions of Dutch Firms," *DNB-Staff Report 2*, De Nederlandsche Bank.
Hamermesh, D.S. and G.A. Pfann (1996), "Adjustment Costs in Factor Demand," *Journal of Economic Literature*, 34:1264–92.
Hansen L.P (1982), "Large Sample Properties of Generalised Methods of Moments Estimators," *Econometrica*, 50:1029–54.
Hubbard, R.G., A.K. Kashyap and T.M. Whited (1995), "International Finance and Firm Investment," *Journal of Money, Credit and Banking*, 27:683–701.
Kaplan, S.N. and L. Zingales (1997), "Do Investment-Cash Flow Sensitivities Provide Useful Measures of Financing Constraints," *Quarterly Journal of Economics*, 112:169–215.
Modigliani, F. and M.H. Miller (1958), "The Cost of Capital, Corporate Finance, and the Theory of Investment," *American Economic Review*, 48:178–97.
Myers, S.C. and S. Majluf (1984), "Corporate Financing Decisions when Firms Have Investment Information that Investors Do Not," *Journal of Financial Economics*, 13:187–220.
Stiglitz, J.E. and A. Weiss, (1981), "Credit Rationing in Markets with Imperfect Information," *American Economic Review*, 71:393–410.
Summers, L.H. (1981), "Taxation and Corporate Investment: A Q-Theory Approach," *Brookings Papers on Economic Activity*, 67–127.
van Ees, H. and H. Garretsen (1994), "Liquidity and Business Investment: Evidence from Dutch Panel Data" *Journal of Macroeconomics*, 16:613–27.
Whited, T.M. (1992), "Debt, Liquidity Constraints, and Corporate Investment: Evidence from Panel Data," *Journal of Finance* 47:1425–60.
Whited, T.M. (1995), "Why and to What Extent Do Euler Equations Fail?," Working paper, University of Delaware, 1995.

PART IV
Labour
Market
Imperfections

7 Sticky Consumption and Rigid Wages

Tore Ellingsen and Steinar Holden

7.1 Introduction

A common explanation for the rise in European unemployment in the 1970s and early 80s has been based on the increase in import prices due to the oil and commodity price shocks.[1] If workers resist the negative impact on the real wage, an import price shock will cause a rise in unemployment. In their influential book on unemployment, Layard, Nickell and Jackman (1991, p. 31) explain the existence and consequences of real wage resistance in the following way:

> Workers value not only the level of their real consumption wage, but also how it compares with what they expected it to be. ... When external shocks like import price shocks, tax increases, or falls in productivity growth reduce the feasible growth of real consumption wages, this generates more wage pressure, which (in equilibrium) requires more unemployment to offset it.

The empirical study by Alogoskoufis and Manning (1988) provides a related view, the authors claiming that 'in Europe wage aspirations are extremely persistent...'

But to many economists, the idea that past expectations matter for the wage that workers demand is based on irrational behaviour. In his comprehensive review article of Layard *et al.* (1991), Phelps (1992, p. 1484), articulates these worries:

> I am more reluctant to accept the argument that individual workers put up wage resistance when...it is irrational for them to do so. In part it is because I am skeptical that whole populations behave irrationally for long periods. In part it is because, in company with most other economists, I would prefer to model behaviour as rational and to turn to irrational mech-

anisms as a last resort...

In this paper we propose a rational explanation for why past expectations affect the current wage.[2] The idea is that consumption patterns are costly to change. For example, one cannot costlessly sell a large house and move into a smaller. Besides the pure transaction costs (frequently up to 5 per cent-10 per cent of the sale price, OECD, 1994, p. 67), come other monetary and non-monetary costs associated with the move. For small discrepancies between anticipated and actual income, the transaction costs will prevent adjustment of the consumption of durable goods. Thus, the full adjustment will be on consumption of non-durable goods. However, for large expectational errors in the income, it is preferable to incur the transaction costs and adjust consumption of durable goods also.

Transactions costs may have several effects on wage setting. Workers that unexpectedly lose their job will face a large loss in their income, and the transactions costs will increase the costs of adjusting consumption, and thus also the costs of becoming unemployed. Greater costs of becoming unemployed tend to reduce unions' wage claims.

The wage moderating effect may be further strengthened if economic growth (and thus income growth) is above expectations. Consumption of durable goods will be 'low,' due to 'pessimistic' expectations in the past, while consumption of non-durables will be 'high' due to high income. Thus, consumption of non-durables will be suboptimally high compared to the consumption of durables. A further wage rise will lead to even higher consumption of non-durables, with little gain in utility due to the suboptimal consumption profile. The small gain from a wage rise will also induce moderation in the wage setting. The same effect could also be caused by rationing of credit, which prevents purchases of durables, thus causing a suboptimally low consumption of durable goods. Both these factors may contribute to an explanation of wage moderation in many European countries in the 1950s and 1960s.

However, if the room for wage growth is below expectations, as in Europe in the 1970s and early 1980s, the transaction costs may induce wage aggressiveness. With transaction costs, a modest but unexpected reduction in the real wage (or a lower increase than earlier expected) would typically lead not to a sale of consumer durables or other major changes in a worker's lifestyle, but to a number of possibly quite painful adjustments on the margins that are left: food, clothes, holidays. We argue that it could well be rational for a utilitarian union to fight for the expected real wage when faced with an adverse shock, so as to preserve the consumption pattern of a majority of its members, while letting a minority go unemployed and make major alterations in the way that they live: For all the members, the lottery is preferable to the sure disappointment.[3]

The rest of the paper is organised as follows. In Section 7.2, we set out the basic model, which is analysed in Section 7.3. Section 7.4 contains some final remarks. All proofs are in the Appendix.

7.2 The model

We consider a trade union comprising L members. The union sets the wage W, while one (or more) employer decides how many workers to employ at this level of pay; the standard specification of the monopoly union model. (The same qualitative results could be derived in a model where the wage is set in a bargain between union and employer.)

The labour demand function is $N(\alpha, W) = \alpha - \alpha_1 W$, where α is a stochastic parameter which indicates the product demand facing the firm ($N = 0$ for $W > \alpha/\alpha_1$). More specifically, we assume that $\alpha = \alpha^* + u$, where α^* is non-stochastic, and u is uniformly distributed over the interval $[u^L, u^U], u^U > 0 > u^L$. The upper bound, u^U is assumed to be 'large,' to avoid boundary solutions. (Linearity of the labour demand simplifies the analysis, but is not important for the qualitative results.) $L - N$ of the union members are unemployed.

Workers consume non-negative quantities of two kinds of goods; durable and non-durable. They all have the same utility of consumption, $V(C, D) = C^\beta D^{1-\beta}, \beta \in (0, 1)$, where C is the quantity of non-durables and D is the quantity of durables. The main reason for choosing the Cobb-Douglas specification is tractability, but the main results do not hinge on this particular specification. As will become apparent below, the chosen utility function entails that workers are risk neutral for income levels where the consumption profile is optimal. This simplifies the interpretation of the model, because it implies that any risk aversion or love of risk is the consequence of an nonoptimal consumption profile. Consumption good prices are taken to be exogenous and fixed, and we normalise units so that both prices are equal to 1.[4]

The model has four stages. At stage one, each union member makes a durable goods investment (buys a house, say), financed by borrowing at a zero rate of interest. At this point in time, the workers do not know the realisation of the labour demand parameter α, and hence their income. Each worker chooses D to maximise expected utility, and the choice is denoted D^*. At stage two, the labour demand parameter α is revealed. Knowing the labour demand, the union then sets the wage in order to maximise the sum of the members' utilities. Given the wage chosen by the union, the firm at stage three chooses how many workers to retain, and these are randomly selected among the union members. Finally, at stage four, workers decide their consumption baskets, subject to the budget constraint $C + D = Y$, where Y is income, equal to W for employed workers and unemployment benefits $B > 0$ for unemployed workers. The workers may decide to sell their durable good and reinvest, or they may stick with the original level. Crucially, there is a fixed transaction cost of $T > 0$ associated with altering the volume of durable goods consumption from the volume chosen at stage one (for simplicity, T is assumed to be independent of the sise of D^*). We shall assume that the transaction costs are relatively small compared to the uncertainty in the income, so that if the wage turns out to be very small (large), the workers will indeed sell their durable good and buy a larger (smaller) quantity. This simplifies the analysis by preventing tedious boundary solutions. Moreover, we assume that the transaction cost is incurred even if the initial quantity of D^* is zero. This assumption prevents the workers waiting to buy the durable good in order to avoid the transaction costs. We want to capture that, in the real world, there is ample time for shocks to disposable income after durable consumption goods are acquired.[5]

Before turning to the analysis, let us briefly discuss some assumptions and properties of the model. The assumption that the labour demand parameter α is unknown when durables are purchased, while known when wages and employment levels are set, reflects that wages and employment levels usually are set much more frequently than for example a family purchases a house. There is clearly much more information about the economic situation the next year (which is the duration of most wage contracts) than about the next 10-20 years, which is the planning horizon of many house purchases.

The assumption that the union maximises the sum of the members' utilities is not crucial; the results are very robust to the choice of the specification of union preferences. As will become apparent below, the main results hold as long as the wage is determined by weighing the marginal utility of a wage rise for the employed against the costs of becoming unemployed, a feature common to all union models we know of.[6]

The simple structure of the model with one consumption period and thus no saving involves strong limitations on the applicability of the model. Obviously, it is not well suited for an analysis of the purchases of durables and non-durables over the business cycle. However, we do believe that the model is suited for an analysis of larger and less frequent changes in the economy, for example huge oil price rises or the slowdown of productivity growth in the 1970s, where household dissaving may be a temporary solution, but not a permanent one.

The model is designed so that transaction costs is the only possible source of wage stickiness. This is done to sharpen the focus on the novel aspect of the paper and is not meant to imply that other explanations of wage stickiness discussed in the literature are unimportant.

7.3 Analysis

To solve the model, we start at the last stage. A worker that keeps his/her durable good, may buy $Y - D^*$ units of the non-durable good, and thus obtains utility $v^K(Y, D^*) = (Y-D^*)^\beta (D^*)^{1-\beta}$. A worker that sells his/her durable good, incurs the transaction costs T, and reinvests, obtains utility

$$v^S(Y, D^*) = \max\left\{ C^\beta D^{1-\beta} | C + D \leq Y - T, C \geq 0, D \geq 0 \right\} \quad (7.1)$$

and straightforward calculations show that the optimal quantities are $C = \beta(Y - T)$ and $D = (1 - \beta)(Y - T)$, which yields utility $v^S(Y) = \beta^\beta (1 - \beta)^{1-\beta}(Y - T)$. Given an optimal reinvestment decision, the utility of the worker is

$$v(Y, D^*) = \max\{v^K(Y, D^*), v^S(Y)\} \quad (7.2)$$

We assume that the worker will keep the durable good if $v^K = v^S$. The worker's choice depends on Y and D^*. Consider first variation in income, Y, for a given quantity of the durable good, D^*. There exist two critical values for the income, which we denote $Y^R(D^*)$ and $Y^I(D^*)$, where $Y^R(D^*) < Y^I(D^*)$, for which the worker is indifferent between selling and keeping the durable good, that is, where

$$
\begin{aligned}
v^K(Y^R(D^*), D^*) &= v^S(Y^R(D^*)) \\
v^K(Y^I(D^*), D^*) &= v^S(Y^I(D^*))
\end{aligned}
\quad (7.3)
$$

The fact that there exists two and only two values of Y for which $v^K = v^S$ is a consequence of the following observations:

1. v^K is strictly concave in Y, while v^S is linear in Y.
2. v^K is greater than v^S for $Y = D^*/(1 - \beta)$.
3. v^S is greater than v^K for Y 'close to' D^* and for 'a very large' Y.

This is all very intuitive: for low income levels, the worker will sell the durable good and buy less (Reduce), for high income levels the worker will sell the durable good and buy more (Increase), while for medium income levels the worker keep his/her durable good. Figure 7.1 shows a crucial part of the analysis: for income levels where the workers choose to sell their durable

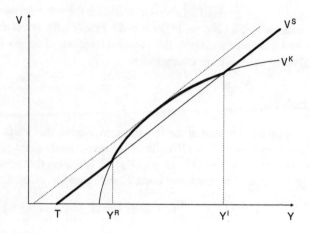

Figure 7.1: The actual utility is given by the best of the two alternatives: selling the durable good $\left(v^S\right)$ or keeping it $\left(v^K\right)$

good, utility is linear in income. However, in the interval where the workers keep their durable good, utility is a strictly concave function of income. For low income levels within this interval, the consumption of the non-durable good is nonoptimally low. Thus the marginal utility of income is high, as the consumption of the non-durable good may be increased. Likewise, for high income levels in this interval, the marginal utility of income is low, as the already nonoptimally high consumption of the non-durable good will be increased.

Then consider variation in the quantity of the durable good D^*, for a given income. Again, there exist two critical values, which we denote $d^R(Y)$ and $d^I(Y)$, where $d^R(Y) > d^I(Y)$, for which the worker is indifferent between selling and keeping the durable good, *i.e.* where

$$v^K(Y, d^R(Y)) = v^S(Y) \tag{7.4}$$
$$v^K(Y, dI(Y)) = v^S(Y)$$

The fact that there exists two and only two values of D^* for which $v^K = v^S$ follows from the following observations:

1. v^K is strictly concave in D^*, while v^S is independent of D^*.
2. v^K is greater than v^S for $D^* = (1 - \beta)Y$.
3. v^S is greater than v^K for D^* 'close to' 0 or D^* 'close to' Y.

Again, all is very intuitive: a small quantity of the durable good will be sold to buy more, a large quantity will be sold to buy less, a medium quantity will be kept. From our assumption in Section 7.2 that the workers will sell their durable good if the wage turns out to be very small, it also follows that $B < Y^R(D^*)$ (as the wage can never be smaller than B), so that the unemployed will sell their durable good.

We now turn to stage three, the wage determination. The optimisation problem of the union is to set W to maximise the sum of the members' utility[7]

$$U(W, \alpha, D^*) = v(W, D^*)(\alpha - \alpha_1 W) + v^S(B)[L - (\alpha - \alpha_1 W)] \quad (7.5)$$

The maximisation problem is complicated by the fact that the workers' utility function shifts depending on whether it is optimal to keep or sell the durable good. For expository reasons, we shall first analyse the maximisation problem of the union where the workers' decision of whether to sell the durable good is taken as given. Let

$$U^K(W, \alpha, D^*) = v^K(W, D^*)(\alpha - \alpha_1 W) + v^S(B)[L - (\alpha - \alpha_1 W)]$$

and

$$U^S(W, \alpha) = v^S(W)(\alpha - \alpha_1 W) + v^S(B)[L - (\alpha - \alpha_1 W)] \quad (7.6)$$

denote the sum of the members' utility, conditional on the employed workers keeping (K) or selling (S) the durable good. U^S is clearly independent of D^* because v^S is independent of D^*. The respective optimal wages are

$$W^K = W^K(\alpha, D^*) = \arg\max_W U^K(W, \alpha, D^*)$$
$$W^S = W^S(\alpha) = \arg\max_W U^S(W, \alpha, D^*) \quad (7.7)$$

Maximisation of U^K yields the first order condition[8]

$$\frac{\partial U^K}{\partial W} \equiv \phi^K(W, \alpha, D^*) = 0 \quad (7.8)$$

where

$$\phi^K(W, \alpha, D^*) = v_1^K(W, D^*)(\alpha - \alpha_1 W) - \left(v^K(W, D^*) - v^S(B)\right)\alpha_1 \quad (7.9)$$

A similar condition can be derived for W^S. At the optimal wage, the utility gain for the employed workers of a marginal wage rise balances the loss in utility for the workers that lose their job due to the wage rise. The corresponding indirect union utility functions are $V^K(\alpha, D^*) \equiv U^K(W^K, \alpha, D^*)$ and $V^S(\alpha) \equiv U^S(W^S, \alpha)$.

Note that W^S is independent of the size of the transaction costs, which reflects that utility is linear in income. W^S can thus be given the alternative interpretation as the wage that would be chosen if there were no transaction costs.

Let $\alpha^O(D^*)$ denote the value of the labour demand parameter that makes the initial quantity of the durable good optimal *ex post*, defined as an implicit function of D^* by $D^* \equiv (1 - \beta)W^K(\alpha^O(D^*), D^*)$. Correspondingly, let $D^O(\alpha) \equiv (1 - \beta)W^K(\alpha, D^O(\alpha))$ denote the quantity of the durable good that would be optimal ex post. (Although W^K depends on D^*, it is shown in the appendix that $D^O(\alpha)$ is unique.)

We are now in a position to describe how W^S and W^K depend on α and D^*.

Lemma 7.1 *The optimal wages W^K and W^S are continuous, differentiable and strictly increasing in α, and*

$$\frac{dW^S(\alpha)}{d\alpha} > \frac{\partial W^K(\alpha, D^*)}{\partial \alpha} > 0.$$

Furthermore, $W^S(\alpha^O) > W^K(\alpha^O, D^*)$.

Thus, the optimal wage is increasing in the labour demand parameter α. Intuitively, a rise in α leads to a higher level of employment, which increases the utility gain of a wage rise (as more workers obtain the wage rise). Hence the union chooses a higher wage. Note also that the transaction costs make the wage 'more sticky,' as W^K is less responsive to variation in labour demand than W^S is.

The intuition for the second result in Lemma 7.1 is the following. When labour demand takes the value that makes the initial quantity of the durable good optimal *ex post*, the marginal utility of income does not depend on whether there are transaction costs. However the utility loss from becoming unemployed is greater with transaction costs, which makes the union choose a lower wage, that is, $W^S(\alpha^O) > W^K(\alpha^O, D^*)$.

The next lemma shows the effect of D^* on W^K.

Lemma 7.2 *W^K is strictly increasing in D^* for $D^* \geq D^O$, while the effect of D^* is indeterminate for $D^* < D^O$.*

(The intuition behind this result will be explained after Proposition 7.1 below).

We now turn to the question of which of the two wages, W^S or W^K, the union will choose. This is closely related to the similar question for the individual workers. It is clear that for α close to $\alpha_1 B$, the union must choose a wage close to B (if $W \geq \alpha/\alpha_1$, then $N = 0$), and since $B < Y^R(D^*)$, the workers prefer to reduce their quantity of the durable good. Correspondingly, if α is very large, there will be full employment and a large W, and the workers will prefer to increase their quantity of the durable good. For medium values of α (in particular, for the α that gives the wage that makes D^* optimal *ex post*), the workers prefer to keep their initial quantity of the durable good.

In fact, there are two critical values for the labour demand parameter, which we denote $\alpha^I(D^*)$ and $\alpha^R(D^*)$, where $\alpha^R(D^*) < \alpha^I(D^*)$, for which

$$V^K(\alpha^I(D^*), D^*) = V^S(\alpha^I(D^*)) \qquad (7.10)$$
$$V^K(\alpha^R(D^*), D^*) = V^S(\alpha^R(D^*)).$$

Furthermore, $V^K \geq V^S$ for $\alpha \in [\alpha^R(D^*), \alpha^I(D^*)]$, while $V^K < V^S$ for $\alpha < \alpha^R(D^*)$ and $\alpha > \alpha^I(D^*)$. The interpretation is that when labour demand is close to its expected level, the union will set a wage which makes all employed workers keep their initial quantity of the durable good, while for large surprises, the wage will be such that all workers sell their initial quantity. It is quite clear that there can only be two critical values of α: W^S and W^K are both increasing monotonically in α, so if Keep (the durable good) is better than Reduce for α', Keep is also better than Reduce for $\alpha' + \varepsilon$, $\varepsilon > 0$. Likewise, if Increase is better than Keep for α'' Increase is also better than Keep for $\alpha'' + \varepsilon$.

The effect of D^* is similar: for very low or very high values of D^*, the union chooses a wage that makes the workers change their quantity of the durable good, while for 'medium' values of D^*, the union chooses a wage that makes the workers keep their initial quantity. Thus, there are two critical values for the quantity of durable good, denoted $D^I(\alpha)$ and $D^R(\alpha)$, where $D^R(\alpha) > D^I(\alpha)$, for which

$$V^K(\alpha, D^I(\alpha)) = V^S(\alpha)$$
$$V^K(\alpha, D^R(\alpha)) = V^S(\alpha). \qquad (7.11)$$

We have $V^K \geq V^S$ for $D^* \in [D^I(\alpha), D^R(\alpha)]$, while $V^K < V^S$ for $D^* < D^I(\alpha)$ and $D^* > D^R(\alpha)$. Thus, for intermediate values of the durable, the union will set a wage which makes all employed workers keep their initial quantity of the durable good, while for extreme levels, the wage will be such that all workers sell their initial quantity. Again there can only be two such critical values: V^S is independent of D^* while V^K is strictly increasing in D^* up till the value where D^* is optimal ex post, $D^O(\alpha)$, and decreasing in D^* thereafter.

We are now ready to state the overall solution to the union's problem.

Proposition 7.1   *(i) The union's preferred wage, W^*, is given by*

$$W^* = W^K(\alpha, D^*) \text{ for } \alpha \in [\alpha^R(D^*), \alpha^I(D^*)],$$
$$W^* = W^S(\alpha) \text{ otherwise.}$$

W^* is a increasing monotonically in α, and is continuous in α except for two discrete positive jumps at α^R and α^I. (ii) Alternatively, we may focus on how W^* depends on D^*. Then

$$W^* = W^K(\alpha, D^*) \text{ for } D^* \in [D^I(\alpha), D^R(\alpha)],$$
$$W^* = W^S(\alpha) \text{ otherwise.}$$

W^* is independent of D^* for $D^* < D^I$ and $D^* > D^R$, strictly increasing in D^* for $D^* \in [D^O, D^R]$. The effect of D^* is indeterminate for $D^* \in [D^I, D^O)$. Moreover, W^* is continuous in D^* except for two discrete negative jumps at D^I and D^R.

The results are illustrated in Figures 7.2 and 7.3, respectively. The relationship between D^* and W^* is the basis for our main result, as it shows how past expectations (that determine D^*) influence the current wage. Let us first discuss a situation where labour demand turns out to be somewhat higher than expected, so that the quantity of the durable good is somewhat lower than the quantity that would be *ex post* optimal (*cf.* the analysis of stage one below). More precisely, assume that $D^* \in (D^I, D^O)$, so that although the quantity of the durable good is *ex post* too low, it is still advantageous for the union to choose a wage which makes the workers keep their durable good. In this situation there are two reasons for why the union will choose a lower wage than what it would have done if the quantity of the durable good were so large or small that the workers would sell their durable good (*cf.* $W^K < W^S$ in this interval in Figure 7.3). First, when D^* is nonoptimally low, consumption of the non-durable good is nonoptimally high, and the marginal utility of income is low. This reduces the union's desire for a wage rise. Secondly, because the employed workers keep their durable good, their utility is higher than it would have been if they sold the durable. The utility loss of becoming unemployed is thus greater when the durable good is kept, which makes the costs of a wage rise larger.

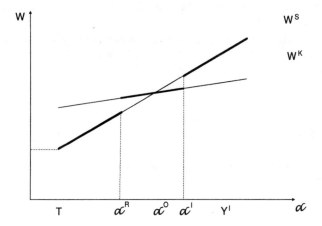

Figure 7.2: The chosen wage is increasing in labour demand, with discrete positive jumps when the workers shift between keep and sell the durable good. Transaction costs make the wage more sticky, as W^K is less steep than W^S

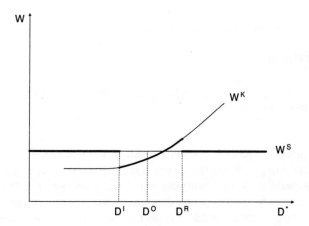

Figure 7.3: A small quantity of durables $(D^* < D^O)$ causes wage moderation (as $W^K < W^S$), while a large quantity of durables (D^* slightly lower than D^R) causes wage aggresiveness

If D^* increases within this interval, this has two opposing effects on the optimal wage. The consumption of non-durable good is reduced, and the marginal utility of income rises. This raises the gain for the employed workers of a wage rise. However, as the consumption profile becomes less suboptimal, the utility of the employed workers increases, and the associated costs of a job loss increase too. The effect of an increase in D^* within the interval $[D^I, D^O)$ is thus indeterminate.

When D^* increases within the interval $[D^O, D^R]$, both effects cause the union to push up the wage. As above, a rise in D^* reduces consumption of the non-durable good and thus increases the marginal utility of income. However, now a rise in D^* makes the consumption profile more suboptimal. This reduces the utility of the employed workers, and thus reduces the associated costs of a job loss.

We then investigate the effect of α and D^* on employment N. Consider α first. An increase in α has a direct positive impact on N via the labour demand function, but it has also a negative indirect effect via the increase in W^*. In the Appendix, we show that the direct effect dominates, so that N is increasing in α except at the jumps α^I and α^R. At the jumps, however, a marginal increase of α leads to a non-marginal increase in W, so the indirect negative effect dominates.

Employment is affected by D^* only indirectly through the wage. Thus, D^* has the opposite effect on N as it has on W.

We now turn to how wage rigidity is affected by the parameter in the utility function, β.

Proposition 7.2 $dW^K/d\alpha$ *is increasing in* β *at the point* $D^* = D^0 \equiv (1 - \beta)W^K$.

Thus, a small β entails more wage rigidity if the initial quantity of the durable good is at its *ex post* optimal quantity. The intuition is that a small β means that a large share of the income is spent on durables, and a small share on non-durables. Thus when income is lower than expected, the relative reduction in consumption of non-durables is large and causes a large loss in utility. Thus, the workers show greater resistance against a low wage the smaller the share in income of non-durable consumption is.

Somewhat more speculatively, a small β may also be associated with a large public sector, which presumably is associated with large taxes, and thus consumption on non-durables being a small share of gross income. To the extent that a large public sector involves a smaller degree of freedom in adjusting consumption, a large public sector will also lead to greater wage rigidity.

We finally turn to stage one, the determination of the initial quantity of the durable good, D^*. When purchasing the durable, the workers are assumed to know the functioning of the economy, as analysed above, but they do not know the realisation of labour demand α. Thus, they do not know which wage level that will be chosen, the level of employment, or whether they will become employed or unemployed. Worker i chooses D_i to maximise the expected utility, that is,

$$D_i^* = \arg\max_{D_i} \mathrm{E}[v(W^*(\alpha, D^*), D_i)p + v(B)(1-p)], \qquad (7.12)$$

where $p = (\alpha - \alpha_1 W^*(\alpha, D^*))/L$ is the probability that worker i keeps his/her job and $\mathrm{E}[]$ is the expectations operator. In this maximisation, worker i treats the average quantity of the durable good as a exogenous (because there are many workers). In equilibrium, all workers will of course choose the same quantity, so $D_i^* = D^*$. The first order condition is

$$\mathrm{E}[v_2(W^*(\alpha, D^*), D_i^*)p(D^*)] = 0. \qquad (7.13)$$

Now, $v_{12}^K > 0$ and $v_{12}^S = 0$ for all W and D^*. Thus, an increase in α^*, which moves the distribution of α upwards, and consequently also raises the wage distribution upwards, will increase the left hand side of (7.13). Hence, it will cause the worker to choose a higher quantity of the durable good, a higher D_i^*. Combined with the result of Proposition 7.1, this shows how past expectations, represented by α^*, affect wages through the investment in consumer durables.

7.4 Final remarks

This paper shows how consumption patterns may depend on expected wages. In particular, if there is a moderate adverse shock to labour demand, the wage chosen by a utilitarian union will be higher the more optimistic the workers were before the shock. Optimistic workers will have chosen a consumption pattern that fits a high income (that is, a large quantity of the durable good), and due to adjustment costs a moderate loss of income hurts quite a lot. This will induce the union to oppose the negative impact on the real wage of an adverse shock.[9]

Note that the consumption pattern depends on the expected future wage, not the past wage. Hence what matters is the wage compared to expectations, not the change in the wage *per se*. The resulting real wage aggressiveness may thus result from an unexpected reduction in real wage growth, and not necessarily a reduction in real wage levels.

We have focussed on the role of expectations in determining the consumption of durables. But investment in durables is also affected by credit market conditions. In many countries credit has been rationed due to interest rate ceilings and other government interventions. To the extent that such regulation has prevented purchases of durables (the purchasing boom following financial deregulations indicated that investments have been suppressed), the consumption of durables have been lower than optimal. As we have seen, a low D^* leads to wage restraint (Figure 7.3). Financial deregulation may not only have led to a boom and over-optimistic expectations; perhaps it has also removed a source of wage moderation.

In our model, the cost of changing the consumption pattern is monetary and hence works through the budget constraint. This is reasonable for durable goods like houses or cars. However, we may also think of the investment in durable good as a choice of hobbies, where to live, children's education, and so on. A lower income may cause one to choose different and cheaper alternatives in all these choices. There need not be monetary transaction costs associated with choosing a different alternative, yet it may be perceived as very costly by the individual who has to make the change.

In the model, there is an extreme dichotomy in the sense that all employed workers keep their durable good, while all unemployed workers sell. A more realistic model would allow for individual differences regarding preferences, initial wealth, income of spouse, *etc.* Moreover, one could also allow for different types of durable goods. In this more realistic model, a loss of income could be met by a reduction of some but not all durable goods, and different alternatives would be chosen by different individuals. An adverse shock would still lead to wage aggressiveness, as long as the following crucial feature prevails: A 'small' income loss (lower wage than expected) is relatively more painful, due to the stickiness of the consumption pattern, while a 'large' income loss (becoming unemployed) is relatively less painful, due to the additional freedom in choosing all types of goods freely (but incurring the costs associated with changing the quantity of durable goods). As long as this feature holds, the investment in durable goods makes workers risk seeking, and the negative impact on the wage associated with an adverse shock is resisted.[10]

Note also that once we depart from the assumption that all union members are treated identically, we no longer need the rather extreme assumption that all unemployed workers must sell their durable good.[11] Consider our model with the modification that layoffs are determined in a certain known order (seniority, say). Workers with a high risk of becoming unemployed might want to choose a low D^*, so as to avoid the need to sell if they become unemployed. Workers with low risk of becoming unemployed would choose a high D^*, and thus face the decision problem of our model. In particular, if labour demand turns out to become lower than expected, the union would have to weigh the costs of a low wage and nonoptimal consumption profile against the costs of making some low risk workers redundant, inducing them to sell their durable good. Thus, an adverse shock may lead to wage aggressiveness, in spite of the fact that many unemployed (the high risk workers) do not sell their durable good.[12]

Appendix

Proof of uniqueness of D^O: The proof follows from the fact that the inequality $\partial((1 - \beta)W^K(\alpha, D^*))/\partial D^* < 1$ holds for all D^*. To see this, assume to the contrary that an increase in D^* leads to a larger increase in W^K. It follows that C will also increase and thus v_1^K is reduced while $v^K(W^K, D^*) - v(B)$ would increase. As can be seen from the first-order conditions (7.8) and (7.9), this would lead to a decrease in W^K: a contradiction. ∎

Proof of Lemma 7.1: The first order condition (7.8) defines the optimal wage as a function of α and D^*. Implicit differentiation of (7.8) with respect to α gives us

$$\frac{\partial \phi^K}{\partial W} \frac{\partial W^K}{\partial \alpha} + \frac{\partial \phi^K}{\partial \alpha} = 0. \qquad (7.14)$$

Solving for $\partial W^K / \partial \alpha$, we obtain

$$\frac{\partial W^K}{\partial \alpha} = -\frac{\partial \phi^K / \partial \alpha}{\partial \phi^K / \partial W} > 0 \qquad (7.15)$$

where the inequality follows from the second-order condition

$$\frac{\partial \phi^K(W, \alpha, D^*)}{\partial W} = v_{11}^K(W, D^*)(\alpha - \alpha_1 W) - 2v_1^K(W, D^*) < 0. \qquad (7.16)$$

and

$$\frac{\partial \phi^K}{\partial \alpha} = v_1^K(\alpha, D^*) > 0 \qquad (7.17)$$

The same procedure can be used to show that W^S is also a strictly increasing function of α. To show that W^S rises more rapidly in α than W^K does, observe that from (7.15)-(7.17) we obtain

$$\frac{\partial W^K}{\partial \alpha} = \frac{v_1^K}{-v_{11}^K N + 2v_1^K \alpha_1} = \frac{1}{-\frac{v_{11}^K}{v_1^K} N + 2\alpha_1} > 0 \qquad (7.18)$$

and correspondingly for $\partial W^S / \partial \alpha$. However, observe that $v_{11}^K < v_{11}^S = 0$, which ensures that $\partial W^S / \partial \alpha > \partial W^K / \partial \alpha$.

To see that $W^S(\alpha^O) > W^K(\alpha^O, D^*)$, observe that

$$\phi^S\left(W, \alpha^O, D^*\right) > \phi^K\left(W, \alpha^O, D^*\right) = 0, \qquad (7.19)$$

for $W = W^K(\alpha^O, D^*)$, as $v_1^S = v_1^K$ while $v^K(W, D^*) > v^S(W)$. ∎

Proof of Lemma 7.2: The effect of D^* on W^K can be found exactly as the effect of α, by implicit differentiation of the first-order condition. We have

$$\frac{\partial W^K}{\partial D^*} = -\frac{\partial \phi^K / \partial D^*}{\partial \phi^K / \partial W} \tag{7.20}$$

The denominator is negative by the second-order condition, while

$$\begin{aligned}
\frac{\partial \phi^K}{\partial W} &= v_{12}^K N - v_2^K \alpha_1 \\
&= \beta(1-\beta)\left((W - D^*)^{\beta-1} + (W - D^*)^{\beta-2}\right) \cdot \\
&\quad (D^*)^{1-\beta}(\alpha - \alpha_1 W) \\
&\quad + \beta(W - D^*)^{\beta-1}(D^*)^{1-\beta}\alpha_1 \\
&\quad - (1-\beta)(W - D^*)^{\beta}(D^*)^{-\beta}\alpha_1.
\end{aligned} \tag{7.21}$$

The first two terms are strictly positive, while the last term is negative. However, the sum of the last two terms $(-v_2^K \alpha_1)$ is non-negative if $\beta D^* \geq (W - D^*)(1 - \beta)$, or equivalently if $D^* \geq D^O = (1 - \beta)W$. ∎

Proof of Proposition 7.1: (i) Note that $W^* = W^S$ if $U^S > U^K$, while $W^* = W^K$ if $U^K \geq U^S$. From the existence of the critical values α^I and α^R, it follows that W^S applies for $\alpha > \alpha^I$ and $\alpha < \alpha^R$, while W^K applies for $\alpha \in [\alpha^R, \alpha^I]$. When $U^S > U^K$, we must have $W^* > Y^I$ or $W^* < Y^R$, so that the individual workers find it optimal to sell (if the individual workers preferred not to sell, $U^S > U^K$ could not hold). Likewise, when $U^S \leq U^K$, we must have $W^* \in [Y^I, Y^I]$, so that the individual workers find it optimal to keep. The features of W^S and W^K are shown in Lemma 1, and what remains of (i) is to prove the existence of the discrete jumps. Assume the opposite, that $W^S(\alpha^I(D^*)) = W^K(\alpha^I(D^*), D^*) = Y^I$. From $W^S = W^K$ and $U^S = U^K$, it is clear that $v^S = v^K$. But as can be seen from Figure 7.1, $v_1^K < v_1^S$, so we cannot have $\partial U^K / \partial W = \partial U^S / \partial W = 0$, a contradiction. By the same type of argument it can be shown that $W^S(\alpha^R(D^*)) = W^K(\alpha^R(D^*), D^*) = Y^R$ leads to a contradiction.

Then consider (ii). The values of D^* for which W^S, respectively W^K, applies are based on the critical values defined in (7.11) above, and can be shown with the same argument as in (i). We then prove that there holds $W^*(\alpha, D^*) = W^K(\alpha, D^*) < W^S$ for $D^* \in [D^I, DO]$ (which also proves the existence of the negative jump). Assume the opposite, that $W^K(\alpha, D^*) = W^S = Y^I$. As in (i) above, it is clear from $W^S = W^K$ and $U^S = U^K$, that $v^S = v^K$. But as can be seen from Figure 7.1, $v_1^K < v_1^S$, so we cannot have $\partial U^K / \partial W = \partial U^S / \partial W = 0$, a contradiction. By the same type of argument it can be shown that $W^K(\alpha, D^R(\alpha)) = W^S$ leads to a contradiction. ∎

Proof of Proposition 7.2: In the point where $D^* = D^0 \equiv (1 - \beta)W^K$, we have

$$\frac{-v_{11}^K}{v_1^K} = \frac{1-\beta}{\beta W} \tag{7.22}$$

which is decreasing in β. Inspection of (7.18) shows that $dW^K/d\alpha$ is increasing in β. ∎

Proof that N is decreasing in α except at the jumps α^R and α^I: Assume the opposite, that an increase in α leads to so big an increase in W^* that N is reduced. Except at the jumps, W^* is given by the first-order condition for local maximum (either W^S or W^K). But if W^*

increases and N is reduces, $v_1 N$ is reduced while $v(W, D^*)$ rises, so the first-order condition cannot hold; a contradiction. Thus, an increase in α leads to an increase in N. ■

Acknowledgments

This paper is part of the research project 'Unemployment, institutions and policy' at SNF-Oslo. The research is partially financed by the Swedish Council for Research in the Humanities and Social Sciences and the Research Council of Norway. We wish to thank Raymond Gradus, Karl Ove Moene, Jon Strand and participants at workshops in Copenhagen and Groningen for useful comments on an earlier draft.

Notes

1. See Bean (1994) for a discussion of this and other explanations.
2. For the record; we do not share the view that one should necessarily avoid theories based on behaviour which is not fully rational. Both the 'personal construct' theory and the theory of 'cognitive dissonance' popular in social psychology may alternatively explain the sort of path dependent preferences that we are concerned with here (see, for example, Earl, 1990, for an introduction), and further investigations along these lines may be worth an effort.
3. The idea that transaction costs make people more risk seeking has been explored in more detail by Flemming (1969), DeMeza and Dickinson (1984), and the fact that indivisibilities affect risk aversion has been recognised even earlier, by Ng (1965) and Eden (1979).
4. With exogenous prices, the model is clearly partial equilibrium. This seems unproblematic for sector-specific shocks to labour demand. However, for economy-wide labour demand shocks, the relative price may well be affected, although it is not obvious in which way.
5. Alternatively we could have added another period of consumption to the model, so as to provide a more explicit rationale for purchasing the durable at stage 1.
6. In fact, there is not even need for a union; the same argument could be made in an implicit contract framework.
7. This specification implicitly assumes that v is the workers' von Neumann-Morgenstern utility, and that were it not for the adjustment costs, workers would be risk neutral (v^S is linear in Y). Our analysis would go through even if the union maximised the expectation of a concave transformation of v.
8. We assume an interior solution; we intend to address the possibility of a corner solution with full employment in a companion paper.
9. If there is a very large negative shock, even the employed workers must sell their durable good, in which case adjustment costs do not matter for the wage.
10. In more recent work, we study a similar model with two types of workers, with low and high transaction costs, and confront the model with data (Ellingsen and Holden, 1996).
11. To our knowledge, empirical evidence on the effects of becoming unemployed is relatively scarce. Colbjørnsen (1994) reports that of a sample of about 700 long-term unemployed (that is, unemployed for at least the last six months) in Norway, about 15 per

cent have sold their car or house. For 16 per cent of the households, the spouse of the unemployed had to get a job.

12. This story suggests a reason for why layoffs often are not random in reality. With differential job security, workers can adapt their consumption pattern to their degree of job security. Senior workers can safely buy a house and start with expensive hobbies, while workers with lower seniority choose a less sticky consumption pattern so as to reduce the adverse effects of unemployment.

References

Alogoskoufis, G.S. and A. Manning (1988), "On the Persistence of Unemployment," *Economic Policy*, 7: 428-69.

Bean, C.R. (1994), "European Unemployment: A Survey," *Journal of Economic Literature*, 32: 573-619.

Colbjørnsen, T. (1994), "Fra elendighetsbeskrivelse til mestringsstudier, in Perspektiv på arbeidsledigheten," *SNF's Årbok*, Fagbokforlaget, Bergen.

De Meza, D. and P.T. Dickinson (1984), "Risk Preferences and Transaction Costs," *Journal of Economic Behavior and Organisation*, 5: 223-36.

Earl, P.E. (1990), "Economics and Psychology: A Survey," *Economic Journal*, 100: 718-55.

Eden, B. (1979), "On Aversion to Positive Risks and Preference for Negative Risks," *Economics Letters*, 4: 125-29.

Ellingsen, T. and S. Holden (1996), "Indebtness and Unemployment. A Durable Relationship." Mimeo, Stockholm School of Economics.

Flemming, J. S. (1969): "The Utility of Wealth and the Utility of Windfalls," *Review of Economic Studies*, 36: 55-66.

Layard, R., S. Nickell and R. Jackman (1991): *Unemployment: Macroeconomic Performance and the Labor Market*, Oxford University Press, Oxford.

Ng, Y.K. (1965), "Why Do People Buy Lottery Tickets? Choices Involving Risk and the Indivisibility of Expenditure," *Journal of Political Economy*, 73: 530-35.

OECD (1994), "Evidence and explanation. Part II The Adjustment Potential of the Labour Market.", *Jobs Study*.

Phelps, E.S. (1992), "A Review of Unemployment," *Journal of Economic Literature*, 30: 1476-90.

8 Structural Aspects of the Labour Markets of Five OECD Countries

Geert Ridder, Niels de Visser and
Gerard van den Berg

2.1 Introduction

In the past decades, labour economists have accumulated evidence that is at odds with the hypothesis that the labour market is a standard competitive market. Wage regressions show that employer size, that is the number of employees of the firm or establishment, has a positive effect on the wage (Brown and Medoff, 1989), and that there are persistent differences between the wages in different industries (Krueger and Summers, 1988). These effects remain, if an extensive list of controls for productive differences between workers is included in the regression. Moreover, these results have been replicated for many countries.

In the same period another literature has emerged that stresses the importance of labour market flows (Mortensen, 1986, Blanchard and Diamond, 1989), for instance flows to and from unemployment and job-to-job transitions. The size of these flows is assumed to be affected by the behaviour of employers and employees, who make their decisions with incomplete knowledge of the opportunities in the market. The discovery of these opportunities is modelled as the outcome of a random process, that is random from the point of view of the individual employer or employee. The resulting delays are referred to as search frictions. There are various types of search models, that differ in the search technology of the agents, allowance for aggregate supply and demand effects, and the nature of uncertainty. The standard job search model assumes that (un)employed individuals search randomly among firms, that they take aggregate supply and demand conditions as given, and that they are uncertain on both the location of employment opportunities and on the terms of these opportunities, in particular the wage. The job search model has inspired empirical research on unemployment and job spells. This research focuses on variations in search frictions and the role of choice in transitions between labour market positions (Devine and Kiefer (1991) give a survey).

More recently, attempts have been made to integrate the two strands of research in labour economics. The impetus came from difficulties that arose in obtaining variation in the terms of employment as an equilibrium outcome (Diamond, 1971). The standard job search model is a model of labour supply, and the distribution that describes the uncertainty on the terms of employment is exogenous to this model. Hence, research started to make the determination of the terms of employment endogenous to the model. A number of such models are now available (Albrecht and Axell, 1984, Mortensen, 1990, Burdett and Mortensen, 1996). We shall refer to these models as equilibrium search models. Equilibrium search models are consistent with the observed anomalies in wage determination. In these models a firm can have a larger workforce by offering wages that are higher than those of other firms. Moreover, search frictions prevent the equalization of wages and profits among industries, and inefficient firms can survive by paying low wages. In explaining the anomalies equilibrium search models do not invoke special behavioural assumptions that are difficult to test directly, as required by, for instance, efficiency wage models.

Some of the theoretical models have been used in empirical studies (Eckstein and Wolpin, 1990, Van den Berg and Ridder, 1993). A partial survey can be found in Ridder and Van den Berg (1996). This research is facilitated by the availability of panel data on labour market histories and the relatively modest computational effort that is needed to solve these theoretical models. Moreover, if we maintain the hypothesis that firms maximize their long-run profit rate, the parameters of the model can be estimated from observed labour market histories. Data on firms are not needed, although they would allow us to test and relax some of the assumptions on employer behaviour.

The resulting models have policy implications that sometimes differ from those derived from the standard competitive model. We consider the effect of changes in the level of unemployment benefits and in the level of the minimum wage. The standard job search model predicts that an increase in unemployment benefits raises the reservation wage of the unemployed and as a consequence lengthens unemployment spells and raises the level of unemployment. This argument ignores the fact that employers may change their wage offers in reaction to a change in the reservation wage. If employers set wages, they will make their wage offer equal to some reservation wage. Hence, if firms make positive profits, they may react to an increase in the benefit level by increasing their wage offers leaving unemployment unaffected. Recent proposals to lower the benefit level, or equivalently to lower taxes on wages but not on unemployment benefits, in order to decrease reservation wages, will lower the wage offers, but not the level of unemployment. These results are not robust over all possible models and parameter values, but it seems unwise to ignore the effect of changes in the level of benefits on wage offers.

Because in equilibrium search models frictions confer some monopsony power on employers, the effect of a change in the minimum wage may differ markedly from that in the standard competitive model. In the simplest model a moderate increase of the minimum wage raises the average wage offer, but has no effect on unemployment. In a model where individuals differ in the value that they attach to unemployment income, a increase in the minimum wage may reduce unemployment, because the higher average wage offer makes more individuals willing to work. In a model where jobs have different levels of productivity the minimum wage may destroy jobs, because some activities may become unprofitable. Equilibrium search models are sufficiently rich to allow for all possibilities, and the question which situation applies can be resolved by empirical research. It is hardly surprising that Card and Krueger (1995) in their controversial study of the effect of the minimum wage on employment mention equilibrium search models as a possible explanation for their results.

The simplest equilibrium search models depend on a few parameters that determine the joint distribution of unemployment spells, job spells, and wages. In this study we use aggregate data to estimate these key parameters for five OECD countries: (West-)Germany, the Netherlands, France, the United Kingdom and the U.S.A. We show that in the simple model only information on the marginal distribution of wages and the marginal distributions of unemployment and job spells is needed to estimate the structural parameters. Thus, the methodological contribution of this paper is the demonstration that the model can be calibrated from readily available aggregate data, and that panel data on individuals are not necessary. Our estimation method provides a direct link between types of information and parameters. For example, we shall show that data on job durations allow us to estimate an index of the search frictions, without the need to estimate the other parameters simultaneously. The parameters, the arrival rate of wage offers, the rate of job destruction, the average productivity of jobs, and the variation of job productivities are of interest in their own right. We shall also use the parameter estimates to obtain estimates of structural unemployment due to wage floors, of the average level of monopsony power in the economy, and to make a decomposition of wage variation into variation due to productive differences between jobs and variation due to search frictions

The estimation results are reported in Section 8.5. In Section 8.2 we introduce the equilibrium search model that we use to obtain these results. The estimation procedure is described in Section 8.3, and Section 8.4 discusses the data. Section 8.6 contains some conclusions and questions for further research.

8.2 The Burdett-Mortensen equilibrium search model

As noted, there are several models for search markets (see Ridder and Van den Berg, 1996, for a review). Our starting point is the equilibrium search model of Burdett and Mortensen (Burdett and Mortensen, 1996, Mortensen, 1990). This model has a dispersed wage offer distribution as an equilibrium outcome, even if all workers and firms are identical. Moreover, it allows for job-to-job transitions, which can not occur in some other equilibrium search models. The model gives explicit solutions for the wage offer distribution and the distribution of wages paid in a cross-section of employees, and it specifies all relevant transition intensities up to a vector of parameters. For our purposes, it is important that the equilibrium solution is such that the parameters of the joint distribution of wages and unemployment and job spells can be identified from the implied marginal distributions of wages and job/unemployment durations. This allows us to use aggregate data on wages and unemployment/job durations that are available for a number of countries, to estimate the parameters of the equilibrium search model.

First, we introduce the Burdett-Mortensen model with identical workers and firms. Next, we extend the basic model by allowing for differences in productivity between workers and firms.

8.2.1 The basic model: homogeneous workers and firms

We consider a labour market consisting of a continuum of workers and firms. Firms set wages and unemployed and employed workers search among firms. The unemployed are looking for an acceptable job, the employed for a better job. Jobs do not last forever, but terminate at an exogenous rate. Firms compete for employees, and set their wage taking account of the wages offered by other firms and the acceptance strategies of the (un)employed. Workers use the resulting wage offer distribution to determine their acceptance strategies. In such a labour market, there are flows of workers who change jobs, who find a job from unemployment, and who become unemployed. In a steady state the flows to and from the stocks of individuals in a particular labour market position are equal. We assume that the labour market is in this steady state. The model does not consider how this steady state is reached.

We use the following notation:

λ_0	=	arrival rate of job offers while unemployed
λ_1	=	arrival rate of job offers while employed
δ	=	rate at which jobs terminate
w	=	wage rate
p	=	marginal value product of employee
b	=	value of leisure (which, among other things, depends on unemployment benefits)
m	=	number (measure) of workers
u	=	number (measure) of the unemployed
$F(w)$	=	distribution function of wage offer distribution
$G(w)$	=	distribution function of earnings distribution
r	=	reservation wage of unemployed job seekers

The distribution functions F and G have the usual properties: they are right-continuous. The left-hand limit of F at w is denoted by $F(w_-)$. Initially, we allow for discontinuities in F, that is, there may be wages with $F(w) - F(w_-) > 0$. This is important, because we must entertain the possibility that the wage offer distribution is degenerate. The wage offer distribution is the distribution of the wage offers made to employed and unemployed workers. The earnings distribution is the distribution of wages paid to a cross-section of employees at a particular moment. To derive the equilibrium of the model we must consider the behaviour of the suppliers of labour, that is, the unemployed and employed individuals, and of the employers. This behaviour, and the constraints imposed by the lags in the arrival of information, determine the flows between labour market positions.

First, we consider the workers. The unemployed obtain wage offers from $F(w)$ at an exogenous rate λ_0. The optimal acceptance strategy maximizes the expected wealth of the unemployed. It is characterized by a reservation wage r (Mortensen and Neumann, 1988)

$$r = b + (\lambda_0 - \lambda_1) \int_r^\infty \frac{1 - F(w)}{\delta + \lambda_1(1 - F(w))} dw \qquad (8.1)$$

This reservation wage takes account of search on the accepted job. As a result, it depends on the difference between the arrival rates while unemployed and employed. In particular, the reservation wage is equal to the value of leisure if the arrival rates are equal. The unemployed may accept offers below b if $\lambda_1 > \lambda_0$. Here and in the sequel, we assume that future income is not discounted. A comparison of Equation (8.1) with the usual expression for the reservation wage in the infinite horizon case, shows that wage offers are implicitly discounted at a rate $\delta + \lambda_1(1 - F(w))$, which is the job-leaving rate as we shall see shortly.

The acceptance strategy of the employed workers is simple. They accept any wage offer, that exceeds their current wage. We assume that job-to-job transitions are costless.

Next, we consider the flows of workers, that result from these acceptance strategies. The flow from unemployment to employment is $\lambda_0(1 - F(r_-))u$, the product of the offer arrival rate, the acceptance probability, and the measure of unemployed workers. The flow from employment to unemployment is $\delta(m - u)$. In a steady state these flows are equal and the resulting measure of unemployed workers is

$$u = \frac{m}{\delta + \lambda_0(1 - F(r_-))} \tag{8.2}$$

Let the distribution of wages paid to a cross-section of employees have distribution function G. The wages paid to a cross-section of employees are on average higher than the wages offered, because of the flow of employees to higher paying jobs. Consider the stock of employees with a wage less or equal to w, which has measure $G(w)(m - u)$. In the steady-state the flows into and from this stock are equal, and this equality gives a relation between the wage offer and earnings distributions. The flow into this group consists of the unemployed that accept a wage less than or equal to w, and this flow is equal to $\lambda_0(F(w) - F(r_-))u$ if $w \geq r$ and is 0 otherwise. The flow out of this group consists of those who become unemployed, $\delta G(w)(m - u)$ and those who receive a job offer that exceeds w, $\lambda_1(1 - F(w))G(w)(m - u)$. In a steady state the inflow and outflow are equal, and we can express G as a function of F

$$G(w) = \frac{F(w) - F(r_-)}{1 - F(r_-)} \frac{\delta}{\delta + \lambda_1(1 - F(w))} \tag{8.3}$$

where we have substituted for u from Equation (8.2). This equation holds if $w \geq r$, and $G(w) = 0$ otherwise. Note that if jobs last forever, that is, $\delta = 0$, the steady-state unemployment rate is 0, and transitions to higher paying jobs would continue until all workers have a wage equal to p. In the sequel we only consider the case that $\delta > 0$.

From the two wage distributions we derive the supply of labour to an employer that offers wage w. There are $(G(w) - G(w - h))(m - u)$ employees that earn a wage in the interval $(w - h, w]$ and there are $F(w) - F(w - h)$ employers that offer a wage in that interval. Because firms that offer the same wage have the same steady-state employment level, the supply of labour to a firm that offers w is obtained by dividing the number of employees by the number of firms and letting h approach 0. This supply is denoted by $l(w \mid r, F)$ where we explicitly indicate its dependence on the acceptance strategy of the unemployed and the wages offered by other firms that compete for the same workers

$$
\begin{aligned}
l(w \mid r, F) &= \lim_{h \to 0} \frac{(G(w) - G(w - h))(m - u)}{F(w) - F(w - h)} \\
&= \frac{\frac{m\delta\lambda_0(\delta + \lambda_1(1 - F(r_-)))}{\delta + \lambda_0(1 - F(r_-))}}{(\delta + \lambda_1(1 - F(w)))(\delta + \lambda_1(1 - F(w_-)))} \quad \text{, for } w \geq r \\
&= 0 \text{ for } w < r
\end{aligned}
$$
(8.4)

Note that it is allowed that a positive measure of employers offers wage w. It is easily seen that l increases in w. Due to search frictions and competition for workers employers face an upward sloping supply curve for labour with a finite wage elasticity. If F is differentiable at w, this elasticity is proportional to the fraction job leavers, that leave for a higher paying job and the measure of firms that pay a comparable wage. The supply function is discontinuous at points of discontinuity of F.

Finally, we consider optimal wage setting by the employer. We assume that the marginal value product p does not depend on the number of employees, that is, we assume that the production function is linear in employment. In that case the profit flow of the firm that pays wage w is $(p - w)l(w \mid r, F)$. The wage offer of the firm maximizes this profit flow

$$
w = \underset{s}{\operatorname{argmax}}[(p - s)l(s \mid r, F)]
$$
(8.5)

We make the implicit assumption, that the firm is only interested in the steady state profit flow. Hence, in setting its wage the firm does not try to smooth its level of employment in response to short run random fluctuations in the level of employment. Because all workers and all firms are identical, each worker is equally productive at each firm. This completes our description of the search market.

Next, we characterize equilibrium in this search market. Because firms that offer wages that are strictly smaller than r have no employees and 0 profits, while a firm that offers r has strictly positive profits, we have $F(r_-) = 0$, that is, there are no wage offers below r. Because firms, that offer a wage

equal to p have 0 profits, and again a firm that offers r has strictly positive profits, wage offers are bounded above by p. The fact that the profit
per employee $p - w$ is continuous in w, puts restrictions on the equilibrium
wage offer distribution. For let w be offered by a positive measure of firms,
that is, $F(w) - F(w_-) > 0$. Then $l(w_+) - l(w_-) > 0$, that is, there
is a positive measure of workers employed at wage w. If one of the firms
that offer w increases its wage offer by a small amount, it will eventually
attract all the workers employed at firms with wage offer w. Because the
profit per employee is continuous in w, the firm increases its profit rate by
$[l(w_+) - l(w_-)]w > 0$. Hence, competition for employees eliminates the
discontinuities in the wage offer distribution. An equilibrium wage offer distribution has no mass points, and in particular, it can not be degenerate. We
have already noted, that we also need $\delta > 0$ to preclude that the wage offer
distribution is degenerate at p.

The wage offers also are a connected set. For firms that offer a wage
at the upper bound of a gap in the set of wage offers, can lower their wage
to the lower bound of the gap without losing any employees, because l is
constant, if F does not change with w. In doing so they increase their profits.
Hence, profit maximization eliminates the gaps in the set of wage offers. As
a consequence F is strictly increasing for all wage offers. Finally, we derive
an expression for F. In equilibrium, firms have no incentive to change their
wage offer. This implies, that all wage offers must give the same profit flow
π. We already know, that the lowest wage offer is equal to r. Firms that offer
r only attract unemployed workers. Their profits are equal to

$$\pi = (p - r)l(r \mid r, F) = \frac{m\delta\lambda_0}{(\delta + \lambda_0)(\delta + \lambda_1)} \frac{p - r}{(\delta + \lambda_1(1 - F(w)))^2} \quad (8.6)$$

Hence, this equation expresses the common profit rate as a function of
the arrival rates, p and r. All equilibrium wage offers yield the same profit
rate π

$$\frac{m\delta\lambda_0(\delta + \lambda_1)}{\delta + \lambda_0} \frac{1}{(\delta + \lambda_1(1 - F(w)))^2} \quad (8.7)$$

Substituting for π from Equation (8.6) we can solve for F

$$F(w) = \frac{\delta + \lambda_1}{\lambda_1}\left(1 - \sqrt{\frac{p - w}{p - r}}\right) \quad (8.8)$$

This expression holds for all equilibrium wage offers. The lowest wage offer is r. By setting F equal to 1 we obtain the highest offer $\overline{w}$

$$\overline{w} = \left(\frac{\delta}{\delta + \lambda_1}\right)^2 r + \left(1 - \left(\frac{\delta}{\delta + \lambda_1}\right)^2\right) p \qquad (8.9)$$

Of course, $F(w)$ is 0 for $w < r$ and 1 for $w > \overline{w}$. Note that F is differentiable. The density function is

$$f(w) = \frac{\delta + \lambda_1}{2\lambda_1\sqrt{p-r}}\frac{1}{\sqrt{p-w}} \quad \text{,for } r < w < \overline{w}$$

$$\qquad (8.10)$$

$$= 0 \quad \text{otherwise}$$

We substitute the equilibrium wage offer distribution in Equations (8.1), (8.2), (8.3), and (8.4) to obtain the equilibrium reservation wage, unemployment rate, earnings distribution and employment.

$$r = \frac{(\delta + \lambda_1)^2 b + (\lambda_0 - \lambda_1)\lambda_1 p}{(\delta + \lambda_1)^2 + (\lambda_0 - \lambda_1)\lambda_1} \qquad (8.11)$$

$$u = \frac{\delta}{\delta + \lambda_0} \qquad (8.12)$$

$$G(w) = \frac{\delta}{\lambda_1}\left(1 - \sqrt{\frac{p-r}{p-w}}\right) \quad \text{, for } r < w < \overline{w} \qquad (8.13)$$

$$g(w) = \frac{\delta\sqrt{p-r}}{2\lambda_1}\frac{1}{(p-w)^{\frac{3}{2}}} \quad \text{, for } r < w < \overline{w} \qquad (8.14)$$

$$l(w \mid r, F) = \frac{m\delta\lambda_0}{(\delta + \lambda_0)(\delta + \lambda_1)}\frac{p-r}{p-w} \quad \text{, for } r < w < \overline{w} \qquad (8.15)$$

The model has dispersed equilibrium wage offer and earnings distributions. Because all workers and firms are identical, this implies that the law of one price does not hold in equilibrium. However, we obtain the competitive equilibrium, in which all wages are equal to p, and the monopsonistic equilibrium, in which all wages are equal to b, as limits of the equilibrium solution. If λ_0 approaches ∞, that is, if the unemployed find jobs instantaneously, then the wage offer and earnings distributions degenerate in p. If λ_1 approaches 0, that is, if the employed do not receive alternative job offers, then the distributions degenerate at b. For $\delta > 0$ the maximum offer $\overline{w}$ is strictly smaller than p, but for $\lambda_1 > 0$ it is also strictly larger than b. Hence, the equilibrium offers are those of firms that have a finitely elastic labour supply. This is confirmed by the wage elasticity of $l(w \mid r, F)$, which is equal to $(p - w)/w$, as it is for a monopsonistic firm.

The basic equilibrium search model is a highly stylized model with strong implications for the distribution of unemployment and job spells. Are these predictions consistent with empirical evidence? Of course, not much should be expected from a model that assumes that all workers and firms are identical. In equilibrium the lowest wage offer is equal to the reservation wage of the unemployed. Hence, all job offers are acceptable to the unemployed, and the re-employment hazard is equal to the offer arrival rate. This is consistent with the empirical evidence in, for instance, Devine and Kiefer (1991) and and Van den Berg (1990). Although job search models originally were introduced as a potential explanation for the existence of unemployment, most empirical studies find that rejection of job offers is rare. In the basic model equilibrium unemployment is due to lags in the arrival of job offers. The homogeneous model does not allow for structural unemployment. The rate at which job spells end, decreases with the wage. This is consistent with empirical evidence (Lindeboom and Theeuwes, 1991). In equilibrium there is a positive association between firm size and wage. Hence, the model is consistent with the employer size wage effect.

The wage offer and earnings distributions have an increasing density. In Figure 8.1 these densities are drawn. Observed distributions of wages do not resemble this earnings distribution. In particular, they do not have an increasing density. As shown in Ridder and Van den Berg (1996), allowing for heterogeneity in p improves the fit to observed wages dramatically, and we use such an extension of the basic model to obtain our estimates.

There are empirical results that the model can not describe. In labour economics there has been a lively debate on the positive relation between wages and labour market experience. Although the debate is still active, the available evidence suggests, that wage growth is due to both wage growth on the job and wage increases that are associated with transitions from lower to higher paying jobs (Abraham and Farber, 1987, Altonji and Shakatko, 1987, Topel, 1991, Wolpin, 1994). The present model only allows for the second type of wage growth. Attempts have been made to construct an equilibrium search model in which firms offer a wage path, but thus far the resulting models are unappealing from an empirical viewpoint, because they do not allow for direct job-to-job transitions, and as a consequence have counterfactual implications for the relation between wages and firm size (Coles and Burdett, 1992).

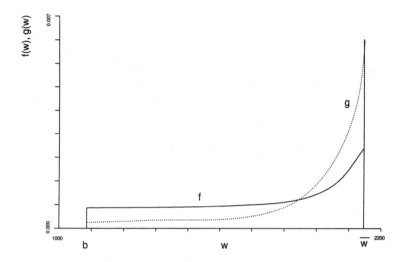

Figure 8.1: Earnings and wage offer density; $\lambda_0 = \lambda_1 = .047$, $\delta = .025$, $b = 1192$, $p = 2208$

8.2.2 The minimum wage and heterogeneity in p

We consider two extensions of the basic model. First, we allow for a minimum wage w_L. Next, we introduce heterogeneity in productivity. If the minimum wage is lower than the reservation wage of the unemployed, then it does not affect the equilibrium solution of the model. If it exceeds this reservation wage, than the lowest wage offer is equal to the minimum wage. The maximum offer is as in Equation (8.9) with the reservation wage r replaced by the minimum wage w_L. With a binding minimum wage the equilibrium is independent of the level of unemployment income b. Hence, the equilibrium depends either on b or on w_L but not on both.

As long as the minimum wage is lower than p, the level of unemployment is independent of the level of the minimum wage. An increase in the minimum wage lowers the profits of the employers and raises the income of the workers. If the minimum wage exceeds the productivity p, firms will close, and all workers become permanently unemployed.

Although we could let all parameters vary in the population, we choose to have heterogeneity in p. As argued in Ridder and Van den Berg (1996), heterogeneity in p is essential to obtain an acceptable fit to the observed wage distribution. The fit to the duration data is also improved. We can distinguish between within-market and between-market heterogeneity in p. In the first case, we consider a single or a few markets, in which firms with different levels of productivity coexist. This alters the equilibrium solution. Here, we consider the second case, in which we have a large number, in the sequel we assume a continuum, of markets, each with its productivity level p. In each market, the equilibrium is as in the basic model. With between-market heterogeneity it does not matter whether we associate the productivity with the worker or with the firm. We shall not relate the productivity to characteristics of workers and/or firms. Our aggregate data do not allow us to make distinctions. Instead, we assume that p has some distribution with p.d.f. h and c.d.f. H.

Although between-market heterogeneity in p does not alter the equilibrium solution, it enriches the model by adding the possibility of structural unemployment. If $p < max(w_L, b)$, then the firms in the corresponding market close down, and the workers become unemployed. If the measure of the affected workers is $h(p)mdp$, then the unemployment rate is equal to

$$\frac{u}{m} = \frac{\delta}{\delta + \lambda_0}(1 - H(\underline{w})) + H(\underline{w}) \qquad (8.16)$$

The first term on the right-hand side of this equation reflects frictional unemployment and the second-term structural unemployment. A further distinction could be made between voluntary and involuntary (structural) unemployment, but because the data will not allow us to make this distinction, this is of little importance.

8.3 Inference

The equilibrium search model with between-market heterogeneity in p specifies the joint distribution of wages and unemployment and job spells. Panel data, in which individuals are followed during some period, contain the required information. Ridder and Van den Berg (1996) discuss the estimation of the model with panel data. Here we use aggregate data to estimate the parameters of the model. The advantage of aggregate data is that they are available for a larger number of countries and for more years. However, aggregate data on the joint distribution of wages and spells are not available. The data that are available refer to the marginal distributions of wages and unemployment and job spells.

Fortunately, all parameters can be identified from the marginal distributions. The basic model implies that the marginal distribution of unemployment spells t_0 is exponential with parameter λ_0. Hence, the average length of an unemployment spell is $\frac{1}{\lambda_0}$. To obtain the marginal distribution of job spells t_1, we note that upon substitution of Equation (8.8) in the job-leaving rate we obtain

$$\delta + \lambda_1(1 - F(w)) = (\delta + \lambda_1)\sqrt{\frac{p - w}{p - r}} \tag{8.17}$$

If we integrate with respect to the density of earnings of Equation (8.14), we obtain the marginal density of t_1

$$k(t_1) = \frac{\delta(\lambda_1 + \delta)}{\lambda_1} \int_\delta^{\lambda_1 + \delta} z \exp(-zt_1) \frac{1}{z^2} \mathrm{d}z \tag{8.18}$$

This is a mixture of exponentials with a fully specified mixing distribution with bounded support. Note that this distribution does not depend on p. Hence, we obtain the same marginal distribution of job spells, irrespective of the assumed distribution of p. The average job spell is

$$E(t_1) = \frac{\lambda_1 + 2\delta}{2\delta(\lambda_1 + \delta)} \tag{8.19}$$

In theory, we can recover λ_0, λ_1 and δ from the marginal distributions of t_0 and t_1. Because for some countries we only have the average spell lengths, we can only identify two parameters. For that reason we assume in the sequel that

$$\lambda_0 = \lambda_1 = \lambda \tag{8.20}$$

In words, we assume that the offer arrival rate is the same when employed or unemployed. This implies that the reservation wage r is equal to unemployment income b (see Equation (8.1)). In an empirical study with individual panel data we found that the two arrival rates do not differ by much (Koning, Ridder, and Van den Berg, 1995).

The marginal distribution of wages in a cross-section of employees is obtained by integration of the density in Equation (8.14) with respect to the density of p truncated at $max(b, w_L)$. The mean and variance of this distribution are

$$E(w) = \underline{w} + \frac{\lambda}{\lambda + \delta}(\mu_T - \underline{w}) \qquad (8.21)$$

and

$$Var(w) = \left(\frac{\lambda_0}{\lambda + \delta}\right)^2 \left(1 + \frac{\delta}{3(\delta + \lambda)}\right)\sigma_T^2 + \frac{\delta\lambda^2}{3(\delta + \lambda)^3}(\mu_T - \underline{w})^2 \quad (8.22)$$

with μ_T and σ_T^2 the mean and variance of the distribution of p truncated at $\underline{w} = max(b, w_L)$.

Data on unemployment spells allow us to estimate λ. After substitution in the density of t_1, we estimate δ with data on job spells. Substitution of the estimates in Equations (8.21) and (8.22) gives two equations in two unknowns, which after substitution of the observed mean wage and wage variance, can be solved for μ_T and σ_T^2. Finally, we can estimate the structural unemployment rate $H(\underline{w})$ by solving Equation (8.16) after substitution of the observed unemployment rate and the estimates of λ_0 and δ.

8.4 The aggregate data

We have aggregate data for five OECD countries: The Netherlands (NL), Western-Germany (D), France (F), United Kingdom (U.K.) and the U.S.A. The aggregate data are not reported in a uniform format, but fortunately our estimation procedure is flexible in that respect. Here we give a short description of our data.

Job spells. Data on job spells categorized in 6 intervals for NL, D, F and U.K. can be found in the OECD Employment Outlook of June 1993. These data have been obtained either from special panel surveys (NL, D) or from the yearly labour Force Survey that is conducted in all countries of the European Union (F, U.K.). These data are for 1990 (NL, D) or 1991 (F, U.K.). The Employment Outlook also contains similar data for the U.S.A., obtained from the Current Population Survey. These data are for 1991.

Unemployment spells. The distribution of unemployment spells categorized in 7 intervals have been obtained from the labour Force Survey (NL, D, F, U.K.). These data refer to 1990 (NL, D) or 1991 (F,U.K.). For the U.S.A. we have the average spell length as reported by the Bureau of labour Statistics, which obtains these from the Current Population Survey.

Wage data. Categorized data on before-tax monthly wages of full-time em-

	NL	D	F	U.K.	U.S.A.
Average	3825	4074	8286	1241	1416
Stand. dev.	1602	1635	3720	585	
Min. wage	2041	2000	2588	-	663

Table 8.1: Average, standard deviation of monthly wage and minimum wage in local currency 1990/1991

ployees (NL from Sociaal-economische maandstatistiek, U.K. from Annual Abstract of Statistics). Mean and standard deviation of before-tax monthly wages of full-time employees (D from Löhne und Gehälter, Statistisches Bundesamt). Categorized data on after-tax monthly wage (F from Annuaire Statistique de la France). For the U.S.A. we have only average monthly before-tax wages from the Current Population Survey.

Unemployment rate. Average rate during the year from the OECD Quarterly labour Force Statistics (all countries).

If the wage data are categorized we compute the mean and variance by fitting a lognormal distribution. For France we have after-tax wages. If the tax is approximately proportional, then it can be shown that the tax rate is applied to the moments of p as well. The resulting estimates are in Table 8.1. Note that the U.K. does not have a minimum wage.

8.5 Results

First, we report our estimates of λ and δ in Table 8.2. For NL, D, F and U.K., the estimate of λ is obtained by maximum likelihood. We assume that individuals with an unemployment spell longer than 23 months are structurally unemployed, that is, these spells are not used in the estimation of λ. This is rather arbitrary, and an alternative procedure is to estimate a mixture model that allows for differences in λ among the (un)employed. This is left to future research. We do not report standard errors, because for some countries we only have the relative frequency of the duration intervals. For the U.S.A. the estimate is obtained from the reported average spell length. The estimates of δ are obtained by maximum likelihood. The likelihood takes account of the length bias in the stock sample.

The US has the largest offer arrival rate and also the largest job destruction rate. The next largest arrival rate is that of the U.K.. However, the job destruction rate is larger in The Netherlands. It is almost as large as that in the U.S.A. The offer arrival rate and job destruction rates are the smallest in France and Western-Germany.

	NL	D	F	U.K.	U.S.A.
λ	.162	.147	.143	.195	.316
δ	.00591	.00360	.00376	.00534	.00616

Table 8.2: Offer arrival rate and job destruction rate per month

	NL	D	F	U.K.	U.S.A.
Unempl. rate	.075	.049	.094	.087	.066
Frictional	.034	.023	.024	.025	.018
Structural	.041	.026	.070	.062	.048

Table 8.3: Unemployment rate: frictional and structural

From the estimates we obtain a decomposition of the observed unemployment rate into a frictional and structural component. The results are in Table 8.3. Note that structural unemployment is due to a wage floor, which is equal to $max(b, w_L)$. For the computation of the structural component of the unemployment rate, it does not matter which is larger. The frictional rate is highest in the Netherlands and lowest in the US. The structural rate is highest in France and the U.K., and relatively small in Western-Germany.

Finally, we estimate the mean and standard deviation of the productivity distribution in active markets. The results are in Table 8.4. We use these estimates to compute an average monopsony index

$$\frac{\mu_T - E(w)}{\mu_T} \tag{8.23}$$

and a decomposition of the wage variance into a component due to heterogeneity in p and a component due to search frictions. These quantities do not depend on the currency. We assume that the lowest wage is equal to the minimum wage, except for the U.K. The lowest monthly wage is set equal to 400 for the U.K.

The results show that the search frictions do not give a substantial monopsony power to the employers. They are only able to set the average wage about 1.5 per cent lower than the competitive wage. This is a direct consequence of the relative size of arrival rate and the job destruction rate. The ratio of λ and δ is the expected number of job offers during an employment spell, which is an index of the search frictions in the market. The larger this index, the smaller the frictions. The monopsony index decreases in the wage floor. We also report the index for a wage floor equal to 0. The the role of search frictions in explaining wages is limited. The fraction explained by

	NL	D	F	U.K.	U.S.A.
μ_T	3890	4125	8436	1264	1430
σ_T	1539	1658	3763	593	
Monopsony index	.017	.012	.018	.018	.0098
Frac. var. due to p	.94	.98	.97	.97	
Monopsony index, $\underline{w}=0$	.035	.024	.026	.027	.019
Frac. var. due to p, $\underline{w}=0$	.79	.90	.95	.61	

Table 8.4: Mean and standard deviation of productivity in active markets (national currency), average monopsony index, and decomposition of wage variation

search frictions increases if the wage floor decreases. Again we report the upper bound. In particular, in the Netherlands and in the U.K. wage floors keep the market equilibrium close to the competitive equilibrium.

8.6 Conclusion

This paper is a first attempt to use aggregate data to estimate the key parameters of a simple equilibrium search model. The estimates suggest, that the equilibrium in the five labour markets under consideration is not far from the competitive outcome, at least for the employed. Wage floors play a role in keeping the equilibrium close to the competitive outcome. However, these wage floors also lead to structural unemployment.

The model is simple. In particular, the assumed equality of the offer arrival rate in unemployment and employment may give an underestimate of the job destruction rate, and hence an underestimate of the level of frictional unemployment, and the monopsony index. Data on employment spells, in addition to data on job spells, would allow us to investigate this.

References

Abraham, K. G. and H. S. Farber (1987), "Job Duration, Seniority, and Earnings," *American Economic Review*, 77:278-97.

Albrecht, J. W., and B. Axell (1984), "An Equilibrium Model of Search Unemployment," *Journal of Political Economy*, 92:824-40.

Altonji, J., and R. Shakatko (1987), "Do Wages Rise with Job Seniority?," *Review of Economic Studies*, 54:437-59.

Berg, G. J. van den (1990), "Search Behavior, Transitions to Non-Participation, and the Duration of Unemployment," *Economic Journal*, 100:842- 65.

Berg, G. J. van den and G. Ridder (1993), "An Empirical Equilibrium Search Model of the Labour Market," *Research Memorandum* 1993-39, Faculty of Economics and Econometrics, Vrije Universiteit Amsterdam.

Blanchard, O. and P. Diamond (1989), "The Beveridge Curve," *Brookings Papers on Economic*

Activity, 1:1-74.

Brown, C. and J. Medoff (1989), "The Employer Size Wage Effect," *Journal of Political Economy* 87:1027-59.

Burdett, K. (1990a), "Empirical Wage Distributions: A Framework for Labour Market Policy Analysis," in: J. Hartog, G. Ridder and J. Theeuwes, (eds.), *Panel Data and Labour Market Studies*, North-Holland, Amsterdam: 297-312

Burdett, K. and D. T. Mortensen (1996), "Equilibrium Wage Differentials and Employer Size," *International Economic Review*, forthcoming.

Card, D. E. and A. B. Krueger (1995), *Myth and Measurement: The New Economics of the Minimum Wage*. Princeton University Press, Princeton.

Coles, M. E. and K. Burdett (1992), Steady-State Price Dynamics. *Discussion Paper*, Department of Economics, University of Essex.

Devine, T. J. and N. M. Kiefer (1991), *Empirical Labour Economics*, Oxford University Press, New York.

Diamond, P. A. (1971), "A Model of Price Adjustment," *Journal of Economic Theory*, 3:156-68.

Eckstein, Z. and K. I. Wolpin (1992), "Estimating a Market Equilibrium Search Model from Panel Data on Individuals," *Econometrica*, 58:783- 808.

Koning, P., G. Ridder and G. J. van den Berg (1995), "Structural and Frictional Unemployment in an Equilibrium Search Model with Heterogeneous Agents," *Journal of Applied Econometrics*, 10:S113-51.

Krueger, A. B. and L. H. Summers (1988), "Efficiency Wages and the Inter- Industry Wage Structure," *Econometrica*, 56:259-93.

Lindeboom, M. and J. Theeuwes. (1991) "Job Duration in The Netherlands," *Oxford Bulletin of Economics and Statistics*, 53:243-264.

Mortensen, D. T. (1986), "Job Search and Labour Market Analysis," In O. Ashenfelter and R. Layard (eds.), *Handbook of labour Economics*, North-Holland, Amsterdam: 849-920.

Mortensen, D. T., (1990) "Equilibrium Wage Distributions: A Synthesis," in J. Hartog, G. Ridder and J. Theeuwes (eds.), *Panel Data and Labour Market Studies*, North-Holland, Amsterdam: 849-920.

Mortensen, D. T. and G. R. Neumann (1988), "Estimating Structural Models of Unemployment and Job Duration," in: W.A. Barnett, E.R. Berndt and H. White (eds.), *Dynamic Econometric Modeling*, Proceedings of the Third International Symposium in Economic Theory and Econometrics, Cambridge University Press, Cambridge.

Ridder, G. (1984), "The duration of single-spell duration data," in: G. R. Neumann and N. Westergard- Nielsen (eds.), *Studies in Labour Market Analysis*, Springer, Berlin.

Ridder, G. and G. J. van den Berg (1996), "Empirical Equilibrium Search Models," Forthcoming in *Proceedings of the Seventh World Meeting of the Econometric Society*.

Topel, R. H. (1991), "Specific Capital, Mobility, and Wages: Wages Rise with Job Seniority," *Journal of Political Economy*, 99:145-76.

PART V
Short-run
Rigidities

9 Short-run Rigidities and Long-run Equilibrium in Large-scale Macroeconometric Models

Keith B. Church, Peter R. Mitchell and Kenneth F. Wallis

9.1 Introduction

A widespread response to the persistence of high unemployment in many OECD countries has been an increased emphasis on structural features of the economy, especially in respect of the functioning—or malfunctioning—of labour markets, and a corresponding reduction of emphasis on demand management policies. Indeed, to stimulate demand would be seen as an inappropriate response to 'structural' unemployment, causing inflation to rise instead. This is consistent with a neo-classical view of the world, at least in respect of its long-run equilibrium properties, in which the level of real activity is independent of the steady-state inflation rate. In the short run, however, a more Keynesian view might be appropriate, due to real and nominal rigidities in the determination of wages and prices which result in a relatively slow process of dynamic adjustment to equilibrium.

The large-scale macroeconometric models that are regularly used for forecasting and policy analysis in national and international government agencies, central banks, and independent research institutes commonly reflect the same view of the world. The supply side has received increased emphasis in recent years, correcting the previous over-emphasis on effective demand. The supply side of a model determines its long-run properties, and the model of Layard and Nickell (1986) has been especially influential (see also Nickell, 1988, and Layard *et al.*, 1991). Goods and labour markets are imperfectly competitive, and price setting and wage bargaining equations determine domestic prices and wages. Adjustment costs and contractual arrangements imply that markets do not clear instantaneously, although the specification

221

of the price and wage equations ensures that the model is inflation-neutral in the long run. The potential persistence in output and employment disequilibria may nevertheless leave considerable scope for fiscal policy whose effects, although temporary, strictly speaking, may still be well worth having.

This article explores these issues in the context of two leading large-scale quarterly models of the UK economy. Although this is a particular national-economy context, the questions raised and the methods used to answer them are general. Section 9.2 presents a simple analytical framework which illustrates the determination of the non-accelerating inflation rate of unemployment (NAIRU) in the models and acts as a benchmark against which they can be compared. Section 9.3 describes some simulation experiments designed to evaluate the impact of short-run rigidities and the convergence to long-run equilibrium in the large-scale empirical models. Section 9.4 contains concluding comments.

The research reported in this article forms part of the general research programme of the ESRC Macroeconomic Modelling Bureau, a unique third-party model comparison project; Wallis (1993) contrasts this with the more common style of model comparison conference. The models studied are those of the London Business School (LBS) and Her Majesty's Treasury (HMT). The version of the LBS model is that deposited at the Bureau in early 1995 and analyzed in the Bureau's most recent model comparison exercise (Church *et al.*, 1995); the version of the Treasury model is a new version publicly released in September 1995 (see Chan *et al.*, 1995).

9.2 A framework for analysis

9.2.1 Long-run equilibrium

We first examine the nature of the long-run equilibrium for the variables of interest, and then consider the short-run adjustment process. For the first purpose our core supply-side framework, following Joyce and Wren-Lewis (1991), Turner (1991), Wren-Lewis (1992) and Turner *et al.* (1993), consists of the long-run versions of the price, wage and employment equations, assumed to exhibit inflation neutrality. A constant returns to scale Cobb-Douglas technology is assumed, with labour elasticity α, and firms' factor demands are assumed to be consistent with this in the long run.

Prices are determined as a mark-up on costs, with the mark-up a negative function of the demand elasticity. Whether this is pro-cyclical or counter-cyclical is uncertain theoretically, and empirical evidence is also ambiguous (Layard *et al.*, 1991, pp.340-1). Nevertheless in the HMT model a commonly-used indicator of the pressure of demand, namely a measure of capacity utilization in manufacturing, is found to have a significant positive effect. Thus, neglecting dynamics and with lower-case letters denoting variables in logarithms, we may write

$$p = cost + \beta_1 cu \tag{9.1}$$

The capacity utilization term (cu) would drop out in an equilibrium state in which output coincides with 'normal' output.

The wage equation is generated by a collective bargaining model, as discussed by Layard and Nickell (1985). It is assumed that firms and unions bargain over the real wage, and that the setting of employment is part of the firm's 'right to manage,' although Creedy and McDonald (1991) demonstrate the similarity of wage behaviour in a range of different bargaining models. The result is an equation relating real wages to profits, unemployment and various 'wage pressure' variables, then profits are often substituted out and the equation rearranged to give

$$(w - p) = \gamma_0 + pr - \gamma_1 U + z^w \tag{9.2}$$

Here wages, w, include employers' taxes, pr is average labour productivity, U is unemployment and z^w represents wage pressure variables, possibly including the 'wedge', that is, the difference between employers' real wage costs and employees' real consumption wage. In the HMT model this appears with a coefficient of 0.5.

The long-run demand for labour satisfies the relevant marginal productivity condition, which in the Cobb-Douglas case can be written

$$w - p = pr + \log \alpha \tag{9.3}$$

Equating the right-hand sides of (9.2) and (9.3) then gives the equilibrium rate of unemployment—the NAIRU—as

$$U = (\gamma_0 - a + z^w)/\gamma_1 \tag{9.4}$$

where $a = \log \alpha$. This is consistent with the view of Blanchard (1988) 'that models should be constructed in such a way that large productivity shocks affect unemployment for a while but not forever.'

The empirical models of the UK economy under consideration are more complicated than this simple theoretical model, the different equations typically being disaggregated across the various sectors of the economy, for example. The key long-run features of the HMT model nevertheless conform closely to the above specification, with Cobb-Douglas technology being assumed. The LBS model instead adopts a translog flexible functional form, although the estimates imply long-run constancy of factor shares; in this more general setting the NAIRU also depends on the cost of capital.

In the LBS model the price system is not completely statically homogeneous, since rises in commodity prices are not passed on completely to the output price. A time trend is included in the output price equation, however, which may offset this effect empirically, although not in any policy simulation. The capacity utilization term is replaced by the ratio of output in manufacturing to GDP. A wage bargaining model is again employed, although profits are retained on the right-hand side of the wage equation, as discussed immediately above Equation (9.2). A further time trend represents additional productivity effects, but no other wage pressure variables appear.

9.2.2 Short-run rigidities

The process of dynamic adjustment to a new equilibrium following a shock depends on the degree of real and nominal rigidities in the model. This in turn is a function of the dynamic specification and coefficient estimates in the empirical wage and price equations. These are commonly of an 'error correction' form, describing a process of short-run adjustment in response to deviations from the long-run relationships presented above.

The extent of real rigidity is measured by the responsiveness of real wages and the mark-up of prices over costs to changes in demand conditions. With respect to prices it thus depends on the size of the coefficient β_1 in Equation (9.1), the smaller this coefficient the greater the degree of real rigidity in price determination. Likewise, in wage determination the coefficient γ_1 in 9.2) measures the sensitivity of wages to labour market conditions, a small value again indicating a high degree of real rigidity in wage determination.

Nominal rigidity implies a relatively slow response of nominal wages and prices to changes in their determinants, and to each other, and is thus reflected directly in the dynamic specification of the wage and price equations. The error correction formulation implies that these are expressed in terms of the first differences of the (log) levels of wages and prices, that is, the wage and price inflation rates, together with lagged values of these and other variables expressed in similar form, except for the levels variables that appear in the error correction term itself. For example, a wage equation of this form might be written

$$\phi(L)\,\Delta w_t = \theta(L)\,\Delta p_t + \varphi(L)\,\Delta^2 p_t + ec_{t-1} + \ldots \qquad (9.5)$$

where $\phi(L)$, $\theta(L)$ and $\varphi(L)$ are polynomials in the lag operator L, and the error correction term, ec, is proportional to the residual in Equation (9.2) above. Dynamic homogeneity implies that the long-run equilibrium is independent of the steady-state inflation rate, and is satisfied in this example if $\phi(1) = \theta(1)$. This is sometimes stated as the requirement that the 'mean lag' between p and w is zero (Currie, 1981), although for this to hold mathematically some of the distributed lag coefficients must be negative, hence the relevant quantity can no longer be interpreted as a statistical mean; this also implies that the dynamic adjustment path overshoots during its approach to equilibrium. The inclusion of terms in the change in an inflation rate does not affect dynamic homogeneity but may also capture nominal rigidities in the adjustment of prices to a change in wages, or *vice versa*.

The empirical models contain further disaggregation, as noted above, and relatively few conclusions about their dynamic implications can be drawn from an examination of their structures; simulation methods are needed instead, as in the next section. Two observations are relevant, however. The first is that the price system in the LBS model is not dynamically homogeneous. The second is that the earnings equation in the HMT model is statically but not dynamically homogeneous with respect to productivity. Thus productivity growth has a long-run effect on the NAIRU, consistent with evidence presented by Manning (1992) that the slowdown in productivity growth is an important explanation of the increase in unemployment in many OECD countries. Turner *et al.* (1993) report a similar finding for Germany, but not for Japan or the United States; Turner and Rauffet (1994) extend the 'European' result to France and the United Kingdom.

9.3   Simulation experiments

9.3.1   Experimental design

In this section we report the results of simulation experiments designed to illustrate the issues discussed above. We consider first a demand-side shock and then a supply-side shock. Macroeconomic responses to a shock are estimated by comparing the results of two solutions of a model, one a base run and the other a perturbed run in which a relevant variable, treated as exogenous, is assigned values that deviate from its base-run values. We conduct a range of variant simulations to assess partial responses, by repeating the perturbation while holding relevant endogenous variables fixed at their base-run values or making other changes to the specification of the models.

The base run corresponds to a published forecast, extended into the future for simulation purposes; in the case of the HMT model this is the forecast of the Ernst and Young ITEM (Independent Treasury Economic Model) Club, kindly supplied by the Club, since the Treasury does not publish its forecast assumptions. The simulation horizon is 25 years in the LBS model and 10 years in the HMT model.

Cross-model comparison exercises require the specification of standardized conditions under which they are carried out, in particular with respect to the stance of fiscal and monetary policy, in order that perceived differences in the results represent genuine differences in the models rather than differences in side conditions or other adjustments that might be made. The standardized policy environment used by Church *et al.* (1995) reflects the broad objectives of current U.K. macroeconomic policy—sound public finances and low inflation—by using feedback rules for income tax and interest rates. Thus monetary policy in the LBS model is represented by a nominal interest rate rule in which the interest rate responds to deviations of inflation from some target level. Such a rule is not part of the HMT model, but it can be operated in a comparable way by using optimal control to set a path for interest rates which keeps inflation near to its target. A common practice is to set the target values in a perturbed run at the base-run values, but the disadvantage in the present context is that returning the inflation rate to base might hide some of the characteristics of the equilibrium of the model. For example, after a positive permanent demand shock we might expect the model to settle down at a new, higher rate of inflation rather than back at base. In this exercise we specify a policy regime in which nominal interest rates are moved period-by-period to ensure that real interest rates remain at baseline values, corresponding to the default monetary policy setting in an earlier version of the LBS model. The method of achieving this is referred to as a 'Type II fix'. To focus on the domestic economy implications of this exercise, we also hold the real exchange rate at its base-run values in all experiments except the first.

The LBS model also includes a fiscal closure rule which ensures that the public sector remains solvent, by adjusting tax rates in response to deviations from target of the public sector borrowing requirement. This acts relatively slowly, however (Church *et al.*, 1995), so that the question of fiscal solvency is not a critical issue in the simulation horizon of the present exercise, and the rule is suppressed for comparability.

9.3.2 Demand shock

The demand shock is a permanent increase in central government procurement expenditure of £2 billion per annum in 1990 prices, approximately 0.4 per cent of GDP. Figure 9.1 shows the impact of the government expenditure increase on the unemployment and inflation rates. The models have similar responses in that they both give real effects which persist for 10 years before the unemployment rate returns to its baseline values. There is also agreement that in the long run there is no Keynesian type benefit from a permanent increase in government expenditure. However, the path to this equilibrium varies substantially between the models with the temporary improvement in unemployment in the HMT model being achieved at a much lower cost in terms of higher inflation. In accordance with a common finding in our model comparison exercises, it is the behaviour of the exchange rate that dominates the differences in simulation results. In the HMT model the real exchange rate falls below base initially but appreciates sharply above base in the middle years of the simulation, putting downward pressure on the inflation rate. The real exchange rate has almost returned to base by the end of the tenth year. By contrast the real exchange rate in the LBS model is close to its baseline values for the first ten years and then falls gradually.

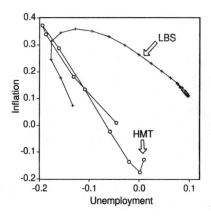

Figure 9.1: Inflation-unemployment responses (percentage-points difference from base)

The importance of different exchange rate behaviour is shown in Figure 9.2, which presents the response of the inflation and unemployment rates to the same shock but with both real interest rates and the real exchange rate held at baseline values. The inflation response of the HMT model is

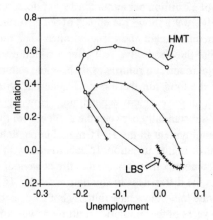

Figure 9.2: Inflation-unemployment responses (percentage-points difference from base)

now much closer to that of the LBS model because the large appreciation of the real exchange rate in the middle years of the previous simulation is eliminated. The results from the LBS model are little changed as the real exchange rate was close to base in the first experiment. There is slightly less inflation in the final years because the real exchange rate no longer depreciates gradually. The unemployment responses are close to each other, with the peak reduction in the unemployment rate of 0.18 percentage points in the LBS model compared to 0.21 in the HMT case. The dynamic response of the LBS model is quicker, reaching this peak in the third year, whereas it takes one year longer in the HMT model. From the third year onward the HMT model also features a larger inflation response. The difference between these responses increases over the course of the simulation with a difference of 0.32 percentage points at the end of the tenth year.

The responses of some other key macroeconomic indicators are given in Table 9.1. The level of real GDP remains above base until the eighth year of the simulation in both models and the initial impact on output is similar in both models. The reason for the large initial deviation in unemployment responses is the difference in the initial reaction of prices. Labour demand in the LBS model is determined by the ratio of total costs to labour costs, weighted by the share of labour costs. In the first quarter, total costs increase by 0.33 per cent while average earnings increase by just 0.07 per cent, lead-

	Year	LBS		HMT	
GDP[a]	1	0.31		0.30	
	3	0.28		0.38	
	5	0.20		0.28	
	7	0.07		0.09	
	n	0.02		−0.21	
Unemployment[b]	1	−38		−12	
	3	−52		−50	
	5	−46		−51	
	7	−25		−31	
	n	−1		5	
Price level[a,c]	1	0.07	(0.07)	0.00	(0.00)
	3	0.50	(0.26)	0.49	(0.35)
	5	1.23	(0.39)	1.56	(0.58)
	7	2.04	(0.40)	2.80	(0.63)
	n	2.26	(0.03)	3.49	(0.50)
Average earnings[a]	1	0.17		0.09	
	3	0.70		0.65	
	5	1.47		1.70	
	7	2.29		2.90	
	n	2.62		4.40	
PSBR[e]	1	1654		1741	
	3	1604		1057	
	5	2054		1677	
	7	2497		2129	
	n	2174		1600	
Current Account[e]	1	−945		−1408	
	3	−1122		−2017	
	5	−593		−1951	
	7	336		−1267	
	n	1956		775	

[a] Percentage difference; [b] Thousands; [c] Percentage point difference in inflation from base in brackets [d] Effective exchange rate; [e] Absolute difference, £m, 1990 prices.

Table 9.1: Government expenditure simulation: increase of £2bn (1990 prices) in procurement spending

ing to an initial increase in employment of 39 000 people. Employment in the HMT model is also determined by relative labour costs, together with output in manufacturing with a unit long-run elasticity and trends representing exogenous technical progress. Although the initial impact of the increase in demand creates upward pressure on employment this is offset by an increase in relative wage costs. The central cost variable in the HMT model falls by 0.15 per cent in the first quarter, and it is not until the fifth quarter that it moves above base, increasing gradually during the rest of the simulation. This reaction can be traced to the behaviour of the unit labour cost component of total costs which is driving this initial downward movement, because with output initially reacting faster than both earnings and employment, unit labour costs fall.

Further examination of the employment equations shows that the reaction
of employment to movements in relative factor prices is sluggish and less
pronounced in the HMT model in comparison with the LBS. The single-
equation properties reveal that an increase of 1 per cent in relative prices has
no contemporaneous effect in the HMT model, with only a quarter of the
adjustment to the long run occurring by the fifth period. Eighty-four per cent
of the adjustment has been achieved after five years, the long-run effect of the
increase being a 0.051 per cent increase in manufacturing employment. In
contrast there are no dynamics in the LBS equation and a 1 per cent increase
in relative costs leads to an instantaneous increase in total employment of
0.72 per cent.

The HMT model produces far more inflation than the LBS model, despite
profiles for the unemployment rate which are comparable, especially from
the fourth year onwards. The differences in these responses can be explained
at least in part by differences in nominal and real inertia in the models.

Nominal rigidities occur when nominal variables exhibit a sluggish response to changes in the factors that influence them. In simulation, wages
and prices should move together in the long run if static homogeneity holds
throughout the model, but an empirical finding embodied in the LBS model
is that rises in commodity prices (fuels and non fuels) do not appear to be
passed on completely to the output price. This manifests itself in the equation for the price of imports of final manufactures which has an elasticity
of 0.68 with respect to the price of world exports of manufactures. Under
the assumption of a fixed real exchange rate this leads to a more subdued
increase in prices than would otherwise be the case, as the decline in the
exchange rate and consequent rise in the price level is not as large as in a
situation where full neutrality exists. This is illustrated in Figure 9.3 where
in a variant simulation the price of imports of manufactures is forced to rise
period-by-period by the same amount as the world price of export goods.
The outcome, given by the line LBS*, shows little difference in the first
three years but from this point until the sixteenth year, and once more in the
final three years of the simulation, inflation is higher.

Turning to real rigidities we examine the sensitivity of the results to the
treatment of demand conditions in the price and wage equations. Corresponding to the expression for the NAIRU in Equation (9.4), capacity utilization has no effect on the long-run level of unemployment in the present
exercise provided that it returns to its base-run level in the perturbed run.
Clearly the nature of the adjustment that the model makes to this equilibrium depends on both the magnitude of the utilization response and how
protracted that response is. In the HMT model the measure of capacity uti-

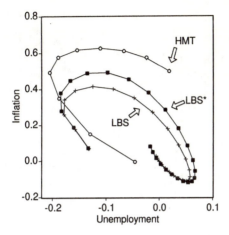

Figure 9.3: Inflation-unemployment responses (percentage-points difference from base)

lization is based on the measure of spare capacity from the CBI Industrial Trends Survey. It is determined by the movements in the ratio of output in manufacturing to the capital stock, the latter being proxied by the cumulation of investment over seven years. The variable that fulfils the same role in the key price equation in the LBS model is the ratio of manufacturing output to GDP. However, this is simply equated to its own lagged value, implying that there can be no feedback from this source to prices during a simulation. Clearly this difference in behaviour is a key determinant of the different dynamic responses of the two models. Following the government expenditure increase already described, capacity utilization in the HMT model increases by 0.62 percentage points in the first quarter, rising to 0.87 percentage points above base two years later. From this point it gradually returns to base due to a combination of the increase in manufacturing output first being halted and then reversed back towards base, and an increase in the cumulative investment variable. However although capacity utilization is within 0.12 percentage points of its baseline values at the end of the seventh year, from this point it increases again rapidly because the investment variable falls faster than output, with the gap between their responses giving an increase in utilization of 0.89 percentage points by the end of the simulation.

With utilization movements playing a key role in the determination of

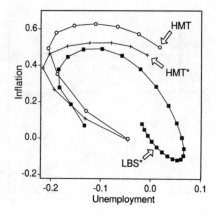

Figure 9.4: Inflation-unemployment responses (percentage-points difference from base)

price responses in one model it is important to try and quantify this difference. Figure 9.4 compares the LBS model with static homogeneity imposed to the fully operational HMT model and to a variant HMT* where capacity utilization has been exogenized. The result of removing utilization effects from the price equation takes the HMT* response nearer to that of the LBS model. The lack of an endogenous utilization term in the LBS model results in an underestimate of the inflationary pressures that normally follow a demand shock.

The final source of real rigidity we consider is the effect of unemployment on wage determination. The long-run coefficient on the (log) unemployment rate is -0.058 for the LBS and -0.092 for the HMT model, and to assess their impact we simply exchange coefficients between the models. This coefficient dictates how much the demand side influences wage setting: a small coefficient (in absolute terms) gives a more protracted movement towards equilibrium with more substantial real effects; the more negative the coefficient then the more decreases in unemployment push up inflation, leading to quicker crowding out of gains following the initial demand shock. We note that each coefficient lies within the other model's 95 per cent interval estimate.

The above analysis is confirmed by the results for the LBS model shown in Figure 9.5. Clearly the dynamic adjustment of the model is sensitive to

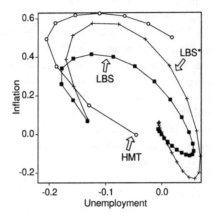

Figure 9.5: Inflation-unemployment responses (percentage-points difference from base)

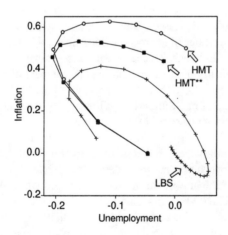

Figure 9.6: Inflation-unemployment responses (percentage-points difference from base)

the larger coefficient on unemployment from the HMT model. The impact response is unchanged, but with slightly less rigidity. The following years are characterized by higher inflation and a lower unemployment response. In the original case the unemployment rate returns to base in the ninth year of the simulation, as opposed to the seventh year in the variant case.

The difference that the change to the HMT model makes, as shown in Figure 9.6, is less dramatic than that seen in the LBS model, but the increase in real rigidity does deliver less inflation as expected. The difference between the two HMT unemployment responses, although less pronounced than in the LBS case, seems to be growing at the end of the simulation. In year six the increase in real rigidity means that there are 2 000 fewer unemployed than in the base HMT case, rising to 6 000 in year eight and 11 000 fewer in the tenth and final year of the simulation. In contrast to this there is only a slight tendency for the inflation rates from the two variants to converge during the course of simulation, whereas in both LBS variants the inflation rate is back to base by the end of the tenth year.

9.3.3 Supply shock

Although the supply side of macroeconometric models has been much developed in recent years, as noted above, only a small range of the issues addressed in the policy debate has featured in empirical models. Many of the issues concern questions of microeconomic reform, whose short-to-medium term macroeconomic consequences are hard to evaluate. Model-based exercises have typically focused on measures that impact directly on costs, a popular simulation in the light of the experience of the 1970s being a change in the world price of oil. Similarly, with respect to wage costs, changes to wage-push factors represented by the variable z^w in Equation (9.2) can be readily simulated although, again, wedge effects or unemployment benefit effects are sometimes hard to pin down empirically. Relatively less attention has been given to technical progress, despite the debate about the causes and consequences of the slowdown in productivity growth in the 1970s. The OECD's INTERLINK model seems to have been the only model used to address this question (Englander and Mittelstädt, 1988, Torres and Martin, 1990, Giorno *et al.*, 1995) until a recent comparative study of Australian models (Hargreaves, 1994), and we attempt to fill this gap in a U.K. context in what follows.

In the framework of Section 9.2, with a constant returns to scale Cobb-Douglas technology, technical change is typically assumed to be Harrod-neutral or labour-augmenting, the model's steady state then conforming to the stylized fact of a constant investment/output ratio. In an empirical model with an explicit production function, technical change can then be measured as an index of labour efficiency, calculated as the residual output growth unexplained by growth in the two primary factor inputs. This is done by Turner *et al.* (1993), for example, who go on to show that the difference between this measure and the labour productivity variable that appears in the wage equation plays no part in the determination of the NAIRU. In a model without an explicit production function but with factor demand equations consistent with an underlying production function, exogenous technical change appears in those equations, typically as time trends. The LBS and HMT models are of this form; the LBS model imposes cross-equation restrictions that ensure consistency with the underlying technology, whereas the HMT model does not.

The supply shock we consider is a 'productivity slowdown' associated with a one per cent increase in employment, brought about by a change in exogenous technical progress. The change is implicitly a step change in the exogenous technical progress trend, and not a change in its slope, thus it has no effect on the NAIRU in the HMT model. The shock is applied to the employment equation and implies a reduction in labour efficiency of one per cent in a single-equation context, although in a full-model exercise feedbacks from the rest of the model to employment, notably from output and relative factor prices, also influence the labour productivity outcome. The responses of key macroeconomic indicators are given in Table 9.2.

The inflation and unemployment responses to the shock are shown in Figure 9.7. As with the demand shock the response is characterized by substantial real gains in the short term which do not persist in the long run. Some of the difference in the employment response between the LBS and HMT models is explained by their different approaches to the aggregation of employment. In the HMT model non-manufacturing and manufacturing employment are separately modelled, each with a demand equation including output in the respective sector (with a unit long-run elasticity), relative factor prices and time trends representing exogenous technical progress as explanatory variables; other components of total employment (central government, local authorities, the offshore oil and gas industry) are treated as exogenous and do not change in this simulation. In contrast the LBS model uses a central labour demand variable with smaller coverage than the manufacturing and non-manufacturing employment variables of the HMT model, and treats different components as exogenous. Consequently a one per cent increase in labour demand in the LBS model delivers a first quarter increase in the number of people employed of 184 thousand compared to 219 thousand in the HMT model. However this difference is dominated by the differences in the linkages between employment and unemployment in a full

	Year	LBS		HMT	
GDP[a]	1	0.00		0.31	
	3	0.00		−0.18	
	5	−0.00		−0.78	
	7	−1.00		−1.16	
	n	−1.00		−1.34	
Unemployment[b]	1	−188		−143	
	3	−188		−139	
	5	−142		−64	
	7	−82		−4	
	n	−33		30	
Price level[a,c]	1	0.06	(0.07)	0.42	(0.43)
	3	1.63	(1.02)	2.40	(1.05)
	5	4.27	(1.39)	4.31	(0.88)
	7	6.96	(1.27)	5.61	(0.56)
	n	11.46	(0.33)	6.48	(0.14)
Average earnings[a]	1	0		0	
	3	2		2	
	5	5		3	
	7	8		4	
	n	13		5	
PSBR[d]	1	−2596		−1772	
	3	−2340		−373	
	5	−705		1371	
	7	35		1728	
	n	9572		125	
Current Account[d]	1	−262		−1393	
	3	−1233		−385	
	5	1057		1031	
	7	4208		1506	
	n	17108		2093	

[a] Percentage difference; [b] Thousands; [c] Percentage point difference in inflation from base in brackets; [d] Absolute difference £m1990 prices.; n denotes final year of simulation, 25 for LBS and 10 for HMT

Table 9.2: Productivity shock: increase in level of employment of 1 per cent

simulation.

There are two obvious routes by which the negative shock to productivity has an effect on inflation. First, in the wage equation downward pressure on wages occurs either because productivity falls as in the HMT model or because the increase in employment gives a decrease in the profit rate, as in the LBS case. This is counteracted by the reduction in the unemployment rate which works in the opposite direction. Second, the shock also feeds directly into prices through unit labour costs, because initially employment increases sharply without any corresponding rise in output or fall in wages.

In contrast to the demand shock, it is the LBS model that features the larger increase in inflation. However the initial response is fairly sluggish

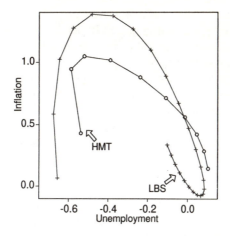

Figure 9.7: Inflation-Unemployment responses (percentage-points difference from base)

and it takes four years for the increase in the inflation rate to exceed that seen in the HMT model. The tendency for higher inflation in the medium term in the LBS model can be attributed to the lack of pressure from profits onto wages. Although profits do fall a diagnostic check reveals that it makes little difference to our results if profits are removed from the wage equation. The absence of this transmission mechanism ensures that inflation is greater than in the HMT model, where productivity falls do subdue the increase in wages. Despite the absence of this effect the LBS model still has the more subdued price response in the first three years of the simulation. Manufacturing prices in the HMT model increase by 0.64 per cent in the first year, driven by increases in unit labour cost in manufacturing and non-manufacturing of 1.26 per cent and 1.35 per cent, respectively. Although whole economy unit labour costs increase by 1.03 per cent in the LBS model, the rise in the price of gross output, the key output price variable, is only 0.19 per cent. Unlike in the HMT model, unit labour costs do not feed into total costs and then into prices. Instead, average wage costs faced by employers rise in line with earnings and enter the cost function, but their reaction is comparatively sluggish. By the end of the third year of the simulation the inflation rate in the LBS model has risen by more than that in the HMT model, because of the absence of a productivity effect in the wage equation and a much larger decrease in the unemployment rate. The real rigidity in the wage equation indicates that reductions in unemployment in the HMT model give rise to more inflationary pressure than in the LBS model, but here the larger magnitude of the unemployment fall in the LBS model outweighs the effect of the lower wage responsiveness.

The HMT model's smaller reduction in the unemployment rate is attributable to the powerful discouraged worker effect in its modelling of labour force participation and its use of a different measure of unemployment. The number of extra people employed in the HMT model in the first year is 244 thousand compared to 188 thousand in the LBS model. However, many of these people do not come from the measured unemployed, because this number is reduced by only 144 thousand in the HMT model, whereas the full increase in employment in the LBS model is reflected in the numbers unemployed. The HMT model has recently switched from modelling the unemployment benefit claimant count measure of unemployment, still retained by the LBS model, to the ILO measure of unemployment, based on the Labour Force Survey. This measure relates to those who when surveyed are available to start work in the next fortnight and have looked for work in the last four weeks. Its inclusion in the model is justified by its more accurate description of those people who have an impact on wage setting because, unlike the claimant count measure, it includes those not receiving benefit but looking for work, while excluding benefit recipients not actually participating in the labour force. The HMT single-equation properties show that a one unit increase in employment in manufacturing and non-manufacturing gives respectively a 0.95 and 0.52 unit decrease in unemployment in the long run.

The interaction of wages and prices is also an important factor when comparing the response of unemployment in the models. The length of time that the improvement in employment persists is dictated by the speed with which prices and wages react and reestablish equilibrium following the disturbance. In the LBS model the real wage remains comparatively close to its baseline values with wages and prices rising together, unlike in the demand shock. There is a slight increase in the real wage during the first years of the simulation, with a maximum increase of 0.41 per cent at the start of the fifth year, after which real wages slowly return to base. This slightly higher real wage is a partial explanation for the slow return of unemployment to base in the LBS model. The HMT model shows the opposite response. Unit labour costs rise quickly pushing up prices, but earnings lag behind. Real wages are 1.79 per cent lower by the end of the simulation, but this favourable situation in terms of the price of labour is outweighed by output effects.

The combination of the different treatment of unemployment and the interaction of wages and prices explains the response of employment. The first year in the LBS model is characterized by wage and total costs rising in line with each other, ensuring that the initial increase in labour demand and hence employment is maintained. In the first three years of the simulation more employment is created in the HMT model, but a greater reduction in unemployment in the LBS model. After this point added employment in the HMT model falls below that in the LBS, as output falls. This dominates the fall in real wages described above, returning employment to base in the eighth year of the simulation, while the rise in the LBS model persists until the eleventh year.

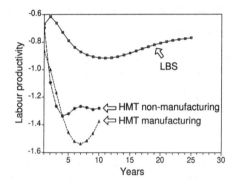

Figure 9.8: Labour productivity over time

The distinction between the exogenous technical progress shock and the measured labour productivity outcome is noted above, and Figure 9.8 shows the response of labour productivity. In the HMT model the manufacturing and non-manufacturing sectors are distinguished, employment in both sectors starting the simulation one per cent higher. The initial effect of this extra labour, despite its reduced efficiency, is an output increase in the first year of 0.17 per cent in manufacturing and 0.43 per cent in non-manufacturing. This pushes up employment slightly in subsequent periods. The direct impact of unit labour costs on prices results in a gap between prices and wages, making labour more attractive than capital. However this influence is outweighed by the output effect. As real wages fall the positive short-run output effects also disappear. Output in non-manufacturing returns to base in the eighth quarter and in manufacturing one quarter later. Both continue to decline throughout the simulation, finishing 1.39 per cent and 1.77 per cent below base respectively, by which time they appear to have settled down. This effect together with the offsetting relative factor price movements gives employment responses that are 0.11 per cent and 0.44 per cent below base for non-manufacturing and manufacturing at the end of ten years, productivity falling by 1.24 per cent and 1.26 per cent in the respective sectors.

The behaviour of whole-economy productivity in the LBS model is similar to the responses already described in the HMT case. Both output and employment rise in the short run, but after four years output returns to base. After 25 years the responses appear to have settled at a point where total economy output defined as GDP at factor cost declines by 0.65 per cent while employment, after being slightly below base for nearly ten years in the middle of the simulation, finishes 0.12 per cent higher, giving a reduction in productivity of 0.77 per cent. Curiously, productivity in manufacturing (not shown) declines continuously throughout the simulation, to a reduction of 4.17 per cent after 25 years. This is calculated postrecursively from statistical relations describing manufacturing's share in total output and employment, which are clearly unreliable in this simulation.

9.4   Discussion

This article shows how a broadly neoclassical view of macroeconomic equilibrium coexists with a new Keynesian view of short-to-medium term adjustment in two large-scale econometric models of the U.K. economy. This coexistence is becoming the leading paradigm in single-country and multi-country models of the OECD economies, although the extent of real and nominal rigidities in wage and price setting varies among them. Layard *et al.* (1991), for example, find that Japan shows very little real or nominal wage rigidity, whereas both are substantial in Germany, France and the U.K., with the United States somewhat intermediate between these extremes; Turner *et al.* (1993) report similar findings for the United States, Germany and Japan based on exercises carried out under comparable assumptions to those of section 9.3 above.

The diagnostic simulations show how, within a common general framework, different choices by different modellers may lead to differences in the estimated responses, which can nevertheless be traced back to specific characteristics of the models. Of course both models are representations of the same economy, but its macroeconomic time series data cannot always discriminate between competing specifications, and until econometric testing has determined this question, our sensitivity analyses provide an explanation for the differences that emerge.

References

Blanchard, O.J. (1988), "Comment on Nickell (1988)," in: R.C. Bryant *et al.* (eds.), *Empirical Macroeconomics for Interdependent Economies*, Brookings Institution, Washington, DC: 218-9.

Chan, A., D. Savage and R. Whittaker, (1995), "The New Treasury Model," *Government Economic Service Working Paper* No.128, *Treasury Working Paper No.70*, H.M. Treasury, London.

Church, K.B., P.R. Mitchell, P.N. Smith and K.F. Wallis, (1995), "Comparative Properties of Models of the UK Economy," *National Institute Economic Review*, 153:59-72.

Creedy, J. and I.M. McDonald, (1991), "Models of Trade Union Behaviour: a Synthesis," *Economic Record*, 67:346-59.

Currie, D. (1981), "Some Long Run Features of Dynamic Time Series Models," *Economic Journal*, 91:704-715.

Englander, A.S. and A. Mittelstädt, (1988), "Total Factor Productivity: Macroeconomic and Structural Aspects of the Slowdown," *OECD Economic Studies*, No.10:7-56.

Giorno, C., P. Richardson and W. Suyker, (1995), "Technical Progress, Factor Productivity and Macroeconomic Performance in the Medium Term," *OECD Economic Studies*, No.25:153-177.

Hargreaves, C.P. (ed.) (1994), "A Comparison of Economy-Wide Models of Australia: Responses to a Rise in Labour Productivity," *Commission Paper* No.2, Economic Planning Advisory Commission, Commonwealth of Australia.

Joyce, M. and S. Wren-Lewis (1991), "The Role of the Real Exchange Rate and Capacity Utilisation in Convergence to the NAIRU," *Economic Journal*, 101:497-507.

Layard, P.R.G. and S.J. Nickell (1985), "The Causes of British Unemployment," *National Institute Economic Review*, 111: 62-85.

Layard, P.R.G. and S.J. Nickell (1986), "Unemployment in Britain," *Economica*, 53 (supplement): S121-69.

Layard, P.R.G., S.J. Nickell, and R. Jackman (1991), *Unemployment: Macroeconomic Performance and the Labour Market*, Oxford University Press, Oxford.

Manning, A. (1992), "Productivity Growth, Wage Setting and the Equilibrium Rate of Unemployment," *Centre for Economic Performance Discussion Paper* 63, London School of Economics.

Nickell, S.J. (1988), "The Supply Side and Macroeconomic Modeling," in: R.C. Bryant *et al.* (ed.), *Empirical Macroeconomics for Interdependent Economies*, Brookings Institution, Washington, DC: 202-218.

Torres, R. and J.P. Martin (1990), "Measuring Potential Output in the Seven Major OECD Economies," *OECD Economic Studies*, 14:127-149.

Turner, D.S. (1991), "The Determinants of the NAIRU Response in Simulations on the Treasury Model," *Oxford Bulletin of Economics and Statistics*, 53:225-242.

Turner, D.S. and S. Rauffet (1994), "The Effect of the Wedge and Productivity on the NAIRU in Five Major OECD Economies," *Discussion Paper* 38, ESRC Macroeconomic Modelling Bureau, University of Warwick.

Turner, D.S., P. Richardson and S. Rauffet (1993), "The Role of Real and Nominal Rigidities in Macroeconomic Adjustment: A Comparative Study of the G3 Economies," *OECD Economic Studies*, No.21:90-137.

Wallis, K.F. (1993), "Comparing Macroeconometric Models: A Review Article," *Economica*, 60:225-237.

Wren-Lewis, S. (1992), "Between the Medium and Long Run: Vintages and the NAIRU," in: C.P. Hargreaves (ed.), *Macroeconomic Modelling of the Long Run*, Edward Elgar, Aldershot: 323-38.

10 High and Low-Skilled Labour in a Macroeconometric Model of the Netherlands

D.P. Broer, D.A.G. Draper, A. Houweling,
F.H. Huizinga and P.A. de Jongh

10.1 Introduction

This paper presents a preliminary version of a new macroeconometric model developed at the Netherlands Bureau for Economic Policy Analysis. It is the result of a research effort with two major objectives. The first one is to build a model that is suitable for short, medium and long-term analyses. The rationale behind this goal is that many current policies and policy proposals are aimed at improving the structure of the economy. Thus, these policies aim to change the equilibrium of the economy. However, the new equilibrium obviously is not reached overnight, and these policies may in effect have very different short and medium-term effects. In particular, the short/ medium and long-term welfare effects of these policies may be of opposite sign, following the saying of 'no pain, no gain,' and it is this trade-off between 'current pain' and 'future gain' that often makes it unclear whether a policy should be adopted, or at least makes the proposal controversial.

A proper economic analysis of such a policy, or of a certain exogenous shock of a similar nature, obviously requires a discussion of both the equilibrium or long-run effects and the process of transition towards the new equilibrium. Also, obviously, these two elements of the analysis should be internally consistent. For this reason, it is advisable that both elements be carried out with the same model. Such a model, therefore, must be able to describe the short and medium-term as well as the long-term effects of that policy or exogenous shock.

In terms of the macro models currently in use at the Bureau, the new model will bridge a gap between FKSEC and MIMIC (see CPB, 1992, and

Gelauff and Graafland, 1994). FKSEC is an econometric model for short and medium-term analyses. It has been estimated almost exclusively in first differences. Moreover, no Malgrange-type adjustment on the coefficients has been imposed to ensure that, at least for first differences, the model has reasonable long-term properties. The horizon for reasonable forecasts or simulations with FKSEC is therefore limited. MIMIC is an applied general equilibrium model meant for long-term analyses. It is calibrated to have reasonable properties, but the empirical foundation would be improved if modern estimation techniques were used. Moreover, MIMIC contains virtually no dynamics. There is, therefore, a clear gap in the Bureau's model instruments, and the current project seeks to bridge that gap.

To achieve this goal, the model will be formulated in levels and estimated in error correction form. We have not yet completed this process of reestimating the entire model, and so some of the equations have been taken directly from FKSEC or are reestimated, but only in first difference form. The most important of these are the consumption equation and the equations for the so-called Cumulative Production Structure. However, we have paid a lot of attention to the modelling of the supply side which has been disaggregated into the exposed and sheltered sectors. For both sectors symmetric generalized McFadden production functions have been estimated.

The second objective in this research effort is to provide a richer description of the labour market. Unemployment of low-skilled workers, and the effects of the welfare system, technological change and international trade on unemployment of low-skilled workers, have become increasingly important policy issues. For this reason, the labour market in the model is disaggregated by skill level. High and low-skilled labour are treated separately in the production functions and in wage formation. Furthermore, we introduced vacancies and a matching function into the model, which allows the model to track labour flows and unemployment durations by skill level. Such a disaggregation already exists in the MIMIC model mentioned above, and we use a similar approach. What we add is empiry and dynamics.

Because little data on a disaggregation of the labour market by skill level existed, we undertook a separate project, together with the Statistics Netherlands, to construct the time series data required to estimate the model. This resulted in a unique and consistent dataset consisting of time series from 1969 to 1993 for employment, jobs, unemployment and vacancies, hours worked, wages and wagecosts, all by skill level times economic activity. This dataset is described in a separate publication, CBS (1996).

Seen from an international perspective, our project fits in a general trend towards greater theoretical and econometric consistency in econometric model building that occurred in response to both theoretical and empirical challenges. These were posed to model builders by structural shifts in the economy and by the emergence of the new classical macroeconomics. Lucas (1976) argued that econometric models were not suited for policy analysis, as the model parameters are not invariant across policy regimes. Sims (1982) identified three main points on which the then current models were at fault, *viz.* the theoretical foundation of the individual equations of the model, the econometric identification of the parameters, and the treatment of the expectations of forward-looking agents.

A response to this and similar criticism came forth along several lines, both theoretical and econometric (see also Hall, 1995). On the theoretic side, one part of Sims's critique was that the long run of macroeconometric models was essentially unrestricted, as the models used to be constructed on an equation-by-equation basis, generally neglecting cross-equation restrictions. Since then, more effort has been put into a consistent modelling of individual blocks of the model. A notable example of this approach is the production sector, in which both factor demands and cost/price equations are usually derived from an underlying common production function or dual cost function. *E.g.* both the Freia model (CPB, 1983) and the NIESR model (National Institute, 1994) use a vintage production structure to derive demand equations for capital and labour that are consistent in the long term. The LBS model (Allen, 1994) uses a translog dual cost function to derive long-term demands for capital, labour, energy, and materials. The FRB/MPS model (Brayton and Mauskopf, 1985) uses a Cobb-Douglas production function in capital, labour, and energy to specify long-term factor demands. As stated above, our model employs a generalized quadratic cost function to specify long-term demand equations for capital and labour.

The trend towards better theoretical foundations of the model equations also helps to answer the second problem raised by Sims, the structural identification of the parameters of the model in the absence of *a priori* restrictions on the lag structure of the model. Current econometric practice entails estimation of the model's equations by estimation of a restricted VAR for the equations of separate blocks of the model, using theory to identify cointegrating relationships among the variables. Short-run dynamics are largely data-determined, but the use of an ECM guarantees long-term behaviour that is consistent with economic theory. This approach is now standard in the newer versions of the NIESR model, the LBS model, the FRB/MPS model, and others. In our model, the production block and the wage equations are examples of the application of this method.[1]

The rational expectations revolution started by Lucas and Sargent has had a large impact on macroeconometric model building, particularly in the U.K. (see, for instance, Hall, 1995). However, the general picture arising from these efforts is that the empirical issue is far from resolved. Rational expectations models typically predict a substantial amount of overshooting in agents's behaviour, that is not generally observed. As a result, rational expectations models do not always perform well in practice. Current research focuses on learning behaviour as a possible solution, but it seems fair to say that the choice of learning rule is still arbitrary. In our model, expectations are not considered explicitly. This low key approach to expectations is partly related to the fact that, because of the virtually fixed exchange rate regime that exists within most of the EC, the model does not contain a financial sector. The exchange rate is exogenous and the long-term interest rate is tied directly to the German one.[2]

Our modelling of the labour market builds on the recent theoretical and empirical literature attempting to explain the increasing unemployment and unemployment durations in Western Europe by some form of wage bargaining (see, for instance, Layard, Nickell, and Jackman, 1991, and Bean, 1994). This work has also been incorporated into macroeconometric models, for example in the NIESR model and the LBS model. Our model uses a similar setup. The disaggregation of the labour market into high-quality, high-skilled jobs and low-skilled jobs is a novel feature of our model. These jobs are distinguished in the production function, with respect to wage formation, and in the matchings function. As far as we know, matching functions also have not yet been incorporated into large macroeconometic models.

The focus of the paper is on the more innovative feature within the model, namely the disaggregation of the labour market by skill level. First, in the next section, we give an overview of the model and indicate how vacancies and the matching function are introduced. Then, in Sections 3 and 4, we describe in more detail how the division between high and low-skilled labour enters the model, by describing labour demand, production and wage formation. Section 5 discusses the CPS matrix, which connects the demand and supply side of the model and is rather unusual in macroeconomic models. Section 6 presents some simulation exercises which indicate how these features operate within the larger model. Section 7 summarizes and offers some conclusions.

10.2 A bird's-eye view

Figure 10.1 gives an overview of the model excluding the public sector and highlights the levels of disaggregation in the different model blocks. We outline the structure of the model starting at the box labelled production and then follow the arrows.

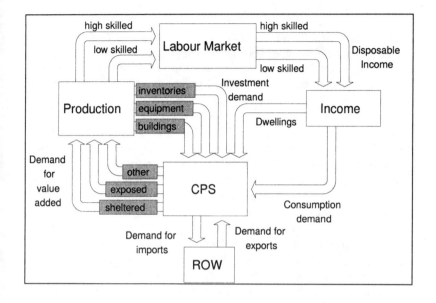

Figure 10.1: Overview of the model

The supply side is divided into six sectors, namely the exposed, sheltered, mining and quarrying, residential, non-market and government sectors. The exposed and sheltered sectors are the main sectors of interest. Value added for these sectors is modelled by symmetric generalized McFadden production functions with equipment, buildings and low and high-skilled labour as inputs. We chose this type of production function because it is globally concave and fully flexible in the sense of allowing a full range of substitution elasticities between inputs. The other sectors do not contain production functions, and employment in these sectors is exogenous in the model. Employment in the mining and quarrying sectors contribute only little to total employment, while their production structures are very different from that of the others. Employment in the non-market and government sectors are exogenous because they are strongly influenced by government policy.

The production of value added leads to demand for production factors. We assume that the markets for factor inputs operate at the macro level, so we aggregate the demand for low-skilled labour from all sectors to a macro demand for low-skilled labour, and do the same for demand for high-skilled labour, investment demand for equipment and investment demand for buildings. The demand for high and low-skilled labour enters the labour market block, and the demand for investment enters the so-called cumulated production structure (CPS) matrix. This CPS will be described below.

The labour market confronts labour supply and demand by skill level. Labour supply by skill level is exogenous in the current version of the model. In the next version, it will be determined by demographic factors and schooling, the unemployment rate (proxying the discouraged worker effect), and the net real wage. These equations are simple reduced forms of the extensive labour supply model in MIMIC, so that the two models may be able to work together in simulations in which labour supply is the central issue. Wage levels for high and low-skilled workers are estimated jointly using a wage curve approach. In addition, the aggregate wage growth is split into a growth in negotiated contractual wages and the wage drift. This latter split is important because many institutional income components (such as welfare payments) as well as the minimum wage are indexed only to contractual wages. A novel feature is the introduction of vacancies into the model by means of a matching block. This block describes vacancy levels, vacancy and unemployment durations and gross labour market flows. Unemployment durations feed back into wage formation. A more detailed outline of the matching block is given in Figure 10.2 below.

We assume that for the purposes of determining consumption demand, income levels of high and low-skilled labour may be aggregated. This aggregate wage income, together with aggregate non-wage income, transfers, wealth and the interest rate then determines consumption demand. Aggregate income, together with the interest rate and exogenous government policy, also determines investment in residential buildings. As noted above, investment demand for equipment and for buildings are the aggregates of the investment demands generated in the different sectors of production. Inventory demand is linked directly to total production and the interest rate.

Next to consumption and investment, the third main endogenous component of final demand is exports, in the figure indicated as the demand from the rest of the world (ROW). Exports of goods (excluding energy) depend on world trade, the difference between Dutch and competing export prices and a sectoral supply factor, namely the ratio of investment to gross value added in the exposed sector relative to a weighted average of the same term for nine competing countries. Exports of services are assumed to follow exports of goods, while the exports of energy are exogenous.

To close the circle of the demand for goods and services, we need to translate the components of final demand into demand for the different sectors of production. This is done by the Cumulated Production Structure matrix or CPS. The CPS is the reduced form of an input-output matrix in which the matrix of intermediary deliveries has been substituted away. It provides, therefore, a direct link between the components of final demand on the one side and imports and sectoral value added on the other. Import demand goes to the rest of the world, and demand for value added to the different sectors of production.

The arrows in Figure 10.1 indicate the demands for goods and services. Prices are linked in a similar way, although in reverse order. The price of value added is determined by unit costs of labour and capital and by demand components, such as the price of foreign competitors and the capacity utilization rate. The CPS links value added prices and the price of intermediary and final imports with final demand prices. The price of investment goods, together with the interest rate, determines the cost of capital. The value added and consumer prices influence wages.

As mentioned above, the model does not contain a financial sector. The exchange rate is exogenous, and the long-term interest rate is tied directly to the German one.

10.2.1 The public sector

In addition to the private sector, the model contains a rather extensively modelled public sector. The receipts and outlays of the public sector include many potential instruments of economic policy. Most of the equations in the public sector block reflect policy rules rather than behavioural relationships. Indeed, while it may be argued that some aspects of policy are endogenous, we purposely made no attempt to model government policy, so as to be able to analyse the effects of government policy as is and as proposed. That is, all forecasts and simulations are made conditional on the current policy stand. The public sector block is the same as in FKSEC and is described in CPB, 1992.

10.2.2 High and low-skilled labour

A novel feature of the model is the disaggregation by skill level. As mentioned in the introduction, the main reason for this disaggregation is to be able to analyze policies specifically targeted to low-skilled labour. Therefore, we divided the labour market into two components, low-skilled and the rest. The term high skilled is thus a slight misnomer, indicating not low skilled, or medium and high-skilled together.

It is somewhat arbitrary to decide which skill levels are included in the category low-skilled. The group of low-skilled workers should ideally be restricted to those with weak labour market positions. On the other hand, for the division between low and high-skilled to be meaningful in a fairly general macro model, the group also cannot be too small. For instance, to get sensible substitution elasticities between low and high-skilled workers in a production function, both groups have to be of reasonable and not too unequal size.

For this reason, we also made no further subdivisions which in principle could be useful, such as one based on orientation (technical versus adminis-trative) or one based on age. Age is potentially an important variable because of the steady increase in education levels that has occurred since World War II. It is unlikely that skill levels have matched this rise in education, so that low education attainment for a middle-aged person may be less indicative of low skills than for a young person. A different approach to the one we take here is to set up a more detailed model or submodel. We assume that both approaches are useful and, in fact, complementary.

The division we ended up choosing defined a worker as low-skilled if the highest completed education level belonged to levels one, two or three of the Standaard Onderwijs Indeling (CBS, 1993). This corresponds roughly to nine years of schooling, starting from elementary school. This division matches that of other studies in this area, such as Van Ours and Ridder (1995).

10.2.3 Overview of the labour market

As indicated in Figure 10.1, the distinction between high and low-skilled labour enters the model in the production structure and in the labour market. These model blocks determine employment, unemployment and wages for both skill levels. As mentioned above, labour supply is exogenous in this version of the model. In addition, the model contains a vacancy-matching block which determines the number of vacancies, the vacancy and unem-ployment durations, and gross and net flows into and out of (un)employment. High and low-skilled labour are conceptually treated symmetrically in the model, so the sets of linkages for both types of labour are the same. Figure 10.2 gives an overview of the sets of causal links for each type of labour.

In Figure 10.2, the outer block linking labour demand, employment, un-employment and wages is standard. The novel part is in the middle of the figure, where variables related to labour flows are modelled. We model all labour flows as occurring between employment and unemployment. Em-ployment to unemployment flows are called quits, and unemployment to employment flows are called matches. Employment to employment flows are conceptually thought of as flows from employment to unemployment to employment with quits and matches in between, but with an unemploy-ment duration of zero. Within the model, an increase in such flows shows up as an increase in the quit rate and a reduction in the average vacancy and

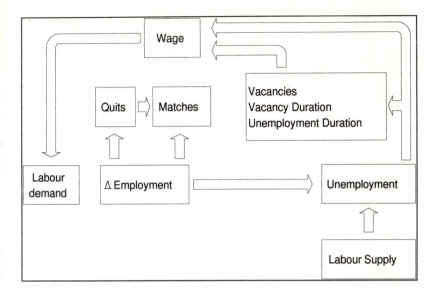

Figure 10.2: The labour market

unemployment durations of the specific type of labour.

The quits are linked to the change in employment. By definition, the change in employment is equal to the matches minus the quits. So, in terms of the arrows in Figure 10.2, given the change in employment and the quits, the matches follow. The matching function links the flow of matches to the stocks of unfilled vacancies and unemployed job seekers, and the replacement rate (as a proxy for search intensity). Given matches and the number of unemployed, vacancies follow, as well as the average vacancy fill rate (= matches/stock of vacancies) and the average unemployment duration (=matches/stock of unemployed). The unemployment duration influences wage formation, and in this way the matching block feeds back into the outer block of Figure 10.2.

10.3 Labour demand and the production function[3]

Figures 10.3 and 10.4 show the development of employment and wages of high and low-skilled labour over time. As the figures indicate, the educational level of Dutch workers has increased dramatically in recent decades. Demand for low-skilled workers halved, while demand for high-skilled work-

ers tripled between 1969 and 1993. As a consequence, unemployment hit
low-skilled workers much harder. In the same period wage differentials decreased considerably. In 1969 a high-skilled worker earned 60 per cent more
than his low-skilled colleague did. This earnings differential decreased to a
mere 10 per cent in 1993.

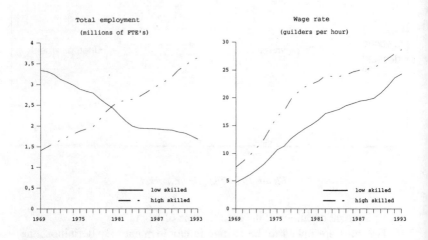

Figure 10.3: Development of employment and wages by skill level

Given these changes in wages and employment, substitution on the demand side seems a likely explanation for the structural upgrading of the
workforce. Employers may have replaced increasingly expensive, low-skilled workers by relatively inexpensive, high-skilled workers. Another plausible explanation for the shift in demand is labour-saving technological change
as indicated by American research (for example, Berman, Bound, Griliches,
1994). In the U.S.A., and in contrast to the Netherlands, wage differentials
rose sharply over the 1980s, while demand for educated labour continued
to grow compared to the demand for unskilled labour. Substitution due to
changes in relative wage costs cannot explain these patterns.

Two other possible explanations seem less plausible. First, changes in
the growth pattern by industry have only a limited effect on the composition
of employment. The increased demand for educated labour is due to shifts
within rather than between industries. Second, one could argue that the rapid

growth in the educational level of the workforce could lead skilled workers to occupy unskilled jobs. Van Ours and Ridder (1995), however, find empirical evidence suggesting that job competition between low- and high-skilled workers is unimportant. Indeed, job competition occurs only between workers with an academic and a higher vocational education. Both categories are high-skilled.

Empirical studies on the demand of different types of labour in the Netherlands are relatively scarce. Examples are Broer and Jansen (1989), Hebbink (1991) and Huigen *et al.* (1993). Lack of appropriate data is the main reason for this limited research. Well known is the missing data on capital, but also information on employment and wages by educational level were, until recently, available only for a few years. Recently, Statistics Netherlands and CPB released new labour market data by educational level for the period 1969-1993 (CBS, 1996). We used this data in the estimation.

10.3.1 The production function

One of the most vexing problems facing applied economists in estimating factor demand relations is to find functional forms that are flexible, use only a few parameters and satisfy the theoretical restrictions implied by economic theory. For our purposes we require the estimated relations to satisfy the global concavity restriction. A (nested) CES function is globally concave but restrictive with respect to the substitution possibilities. For example, a nested CES function that aggregates low- and high-skilled labour into one production factor excludes the possibility of complementarity between high-skilled labour and capital. Moreover, this particular structure implies identical substitution between capital and both types of labour. A widely used form, the translog cost function, is flexible, but global concavity is not guaranteed. Diewert and Wales (1987) show that imposing this restriction may lead to biased estimates. We apply a functional form suggested by Diewert and Wales (1987, 1995), the so-called *Symmetric Generalized McFadden* (SGM) cost function. Global concavity in this function may be imposed without destroying its flexibility.

We distinguish three production factors: low-skilled labour x_l, high-skilled labour x_h and capital x_k. Technology can be represented by a production function $y = f(\mathbf{x}, t)$, where $\mathbf{x}$ is the vector of factor inputs ($\mathbf{x} = [x_l, x_h, x_k]^T$), y is output and t is the time trend representing the state of technology. Given the vector of factor prices $\mathbf{p}$ ($\mathbf{p} = [p_l, p_h, p_k]^T$), costs are defined by $C = \mathbf{x}^T \mathbf{p}$. We assume cost minimization and constant returns to scale. Given these assumptions and an appropriate functional form for the cost function, factor demand relations can be derived. Diewert and

Wales (1987) propose the following functional form for the input-output ratios:

$$\frac{x^*}{y} = g\left(\mathbf{p}\right) + \alpha + \beta t + \gamma t^2 \tag{10.1}$$

$$g\left(\mathbf{p}\right) = \frac{\Xi \mathbf{p}}{\theta^T \mathbf{p}} - (1/2)\,\theta \frac{\mathbf{p}^T \Xi \mathbf{p}}{\left(\theta^T \mathbf{p}\right)^2} \tag{10.2}$$

where Ξ denotes a matrix and α, β, and γ are vectors of parameters. All prices are normalized to unity and the time trend equals zero in the base year 1990. The matrix Ξ is symmetric and $\Xi \iota = 0$, where ι is the unity vector. The asterisk denotes a long-run variable.

Each factor-demand relation in the system given in (10.1) and (10.2) consists of three parts. The first term $g(\mathbf{p})$ represents the effect of prices; the second term α denotes the input-output ratio in the base year, while the third and fourth term $\beta t + \gamma t^2$ stand for the influence of technology. The input-output ratios together with factor prices yield total cost per unit output:

$$c = \frac{\mathbf{p}^T x}{y} = (1/2)\frac{\mathbf{p}^T \Xi \mathbf{p}}{\theta^T \mathbf{p}} + \mathbf{p}^T \alpha + \mathbf{p}^T \beta t + \mathbf{p}^T \gamma t^2 \tag{10.3}$$

Diewert and Wales labelled this function the *Symmetric Generalized McFadden* (SGM) cost function. They show that it is flexible and satisfies the conditions implied by economic theory. The function is linear homogeneous in factor prices. Global concavity in the prices implies that the matrix is negative semi- definite. If necessary this restriction can be imposed without destroying the flexibility of the functional form.

10.3.2 Short run

The short run is modelled as an error correction model in which the changes in and the deviations from optimal demand enter, in addition to the change in output and, for investment, the profit rate.

10.3.3 Estimation results

We focus on the long-run parameters. We estimated the model both for the exposed (manufacturing, transport) and the sheltered sectors (trade, market services and construction). The disturbance terms are assumed to feature a multivariate normal distribution. We used Non-Linear Least Squares. For both sectors global concavity was enforced.

	Exposed sector		Sheltered sector	
Price elasticities				
ε_{ll}	−0.23	(2.3)	−0.60	(2.3)
ε_{lh}	0.18	(1.8)	0.49	(1.9)
ε_{lk}	0.04	(3.8)	0.11	(3.8)
ε_{hl}	0.22	(1.8)	0.39	(2.0)
ε_{hh}	−0.21	(1.7)	−0.39	(2.1)
ε_{hk}	−0.01	(0.4)	0.00	(0.1)
ε_{kl}	0.06	(3.8)	0.15	(3.9)
ε_{kh}	−0.01	(0.5)	0.00	(0.1)
ε_{kk}	−0.05	(6.0)	−0.15	(7.9)
ε_{aa}	−0.02		−0.05	
Elasticities of technological change				
ε_{lt}	−0.064	(57.3)	−0.053	(20.6)
ε_{ht}	0.001	(0.3)	0.021	(8.6)
ε_{kt}	−0.004	(10.8)	0.015	(16.3)

notes: 1. Absolute t-values in parentheses; 2. Elasticities are evaluated in the midpoint of the sample; 3. Price elasticities are defined as $\varepsilon_{ij} = \partial \ln x_i^* / \partial p_j$; 4. Elasticities of technological change are defined as $\varepsilon_{it} = \partial \ln x_i^* / \partial t$

Table 10.1: Price elasticities and elasticities of technological change

Table 10.1 contains the price elasticities for low- and high-skilled labour, for capital, and for labour as a whole (ε_{aa}) and the semi-elasticities for each factor input. The elasticities are evaluated in the midpoint of the sample. Price elasticities are considerably below unity. Price effects on employment are only modest. Own price elasticities decrease with skill, indicating that the (direct) negative effect of an increase in wages for low- rather exceeds that of high- skilled labour. The cross-price elasticities indicate that the two types of labour are substitutes in both sectors. However, substitution in the sheltered sector is considerably higher than in the exposed sector. High-skilled labour is complementary with capital in both sectors. Substitution between low-skilled labour and capital is only small, especially in the exposed sector. The price elasticity for aggregate labour is small, indicating hardly any substitution between labour as a whole and capital; the cross-price elasticity between labour and capital (minus the own- labour demand elasticity) ranges from 0.02 in the exposed sector to 0.05 in the sheltered sector.

Using these elasticities one can compute how a decrease in wage costs affects employment. A 10 per cent decrease in overall wages, without any change in other prices and output, increases low-skilled employment by 0.8 per cent (corresponding to 16 000 labour years). High-skilled employment is not affected, due to complementarity with capital. A 10 per cent decrease

in wage costs for low-skilled labour *only* raises low-skilled employment by
4.2 per cent (82 000 labour years) and reduces the demand for high- skilled
labour by 3.2 per cent (54 000 labour years). These elasticities assume a
level of constant output. When taking output effects into account, one may
find much larger effects on employment.

Technological progress is asymmetric. It is strongly labour saving for
low-skilled labour. Due to exogenous technological change the demand for
low-skilled labour is reduced by 6.4 per cent a year in the exposed sector
and by 5.3 per cent in sheltered sector. In the sheltered sector, technological
progress is labour-using for high-skilled labour. One might argue that it is
difficult to distinguish technological progress from substitution in response
to changes in relative factor prices. However, estimating the model under
the constraint of no technical change, we do not find substantially higher
estimates for the substitution elasticities.

	Annual growth rate long-run employment	Decomposition		
		Output	Techn. progress	Substi- tution
Exposed Sector				
Low-skilled	−3.9	2.4	−5.5	−0.8
High-skilled	2.7	2.4	−0.1	0.4
Sheltered Sector				
Low-skilled	−3.0	2.1	−4.1	−1.0
High-skilled	4.3	2.1	1.9	0.4

notes: 1. Percentage values; 2. Growth rates are averages over the
period 1973-1993

Table 10.2: Determinants of long-term employment

Table 10.2 attributes changes in employment as determined by the long-
term relations (10.1) to changes in output, prices and technology. In accordance with the reported elasticities, technological change is the major factor
behind the decrease in long-term employment of low- skilled workers. In
the exposed sector, a growth rate of output of at least 5.5 per cent is necessary to stabilise low-skilled employment at unchanged factor costs. In the
sheltered, sector the required growth rate would be 4.1 per cent. Changes in
factor prices play a less important role in explaining the shift in demand.

10.3.4 Comparison with other results

Hamermesh (1993) surveys (mainly American) studies on labour demand.
Although empirical work on price elasticities shows a wide range of estimates, some general conclusions can be drawn. Studies that do not distinguish between skill categories yield estimates somewhere between −0.15

and −0.75 for the own-demand elasticity of labour. Studies that allow for heterogeneous labour suggest that capital and skill are relative complements and that own-wage demand elasticities decrease with skill. Our results confirm this pattern. However, the own-demand elasticity is considerably lower than suggested by Hamermesh. The important role for technological progress in reducing demand for low-skilled labour is in line with Berman, Bound and Griliches (1992) and Shadman-Mehta and Sneessens (1995).

As mentioned above, empirical studies on the demand for heterogeneous labour in the Netherlands are scarce. Differences in methodology and data prevent a detailed comparison of the results across studies. However none of the studies on the Netherlands are fully consistent with the 'stylized facts' as reported in Hamermesh. In line with our results, both Broer and Jansen (1990) and Hebbink (1992) find that labour-saving technological change plays a major role in reducing the demand for unskilled labour. Regarding the price elasticity of labour demand, the evidence is mixed. To illustrate, Gelauff, Haan and Okker (1986) report an elasticity close to zero. Draper (1989), however, finds a value of −0.2. Hamermesh suggests that measurement error in the price of capital biases the estimate of the elasticity of labour demand towards zero. This bias might explain why studies that use the real price of labour as a determinant (Draper, 1989) find a higher demand elasticity than studies that adopt the price ratio of labour and capital as an explanatory variable (Gelauff, Haan and Okker, 1986).

10.4 Wage determination and matches

The wage block determines wage levels for high and low-skilled labour, macro contract wages and the macro wage drift. The basic specification for wages is formed by the wage curve, linking the level of wages to prices, productivity, a wedge term, the replacement rate and the unemployment rate. Most of these explanatory elements are standard practice in the wage curve literature, although recently there has been a debate on the effect of the wedge on wage formation.

Policy makers often point to the wedge as an important cause of unemployment because the wedge is an important component of total labour costs. It is, however, not clear that this argument is valid, because the incidence of the wedge may be fully born by the workers in the form of lower net real wages. Many bargaining models of wage formation suggest that this is indeed the case. Indeed, in their influential books, Layard, Nickell and Jackman (1991, 1994) argue more broadly that the tax wedge should have no long-term influence on wage costs and is, therefore, fully born by workers.

If so, the wedge should not have any effect on unemployment, at least in the long run. Most of the empirical literature, however, seems to indicate that the wedge does have a long-run effect on wage formation, and moreover, that different subcomponents of the wedge affect labour costs differently.[4]

In addition to these macro issues, our focus naturally lies in the disaggregation of wages by skill level. Macro wage equations are generally generated by a bargaining model, at least in a European context. These models assume homogeneous labour, and so if we maintain the bargaining model, we have to disaggregate the union gain function. It is not clear how this is to be done. Contrary to a firm, who may be thought of as a single agent, the union is in essence a collection of individual members, each of whom may decide to leave the union/firm if he thinks he can get a better deal elsewhere. As a result, a union utility gain function consisting simply of total gain for all its members seems rather implausible, because that would make the union be indifferent between equal pay for all members and the total wage bill going to a single member. The latter outcome clearly is not consistent with optimization at the level of the individual union members.

One way of thinking about this problem intuitively may be that it takes place in two stages. First, the union and the firm bargain about the total wage bill, and then the individual members bargain about the division of this wage bill. The solution to the second stage would set wages by maximizing the product of the individual gain functions subject to the constraint that the total wage bill be distributed.

Based on this idea, we define the union gain function as the product of the individual gain functions, and we state the overall bargaining problem as follows[5]:

$$
\underset{w_i}{\text{argmax}} \left(\frac{p_y y - \sum_{i=L,H} n_i w_i \left(1 + se_i\right)}{p_c} \right) \cdot \prod_{i=L,H} \left(\frac{w \left(1 - sl_i\right) \left(1 - td_i\right)}{p_c} - F_i \right)^{\frac{\alpha_i n_i}{n}}
$$

$$(10.4)$$

where p_y denotes the value added price deflator, p_c the consumer price deflator, y total output, w the gross wage rate, se the rate of employers' social security contributions and other labour costs (such as pension costs), sl the premium rate for the social security contributions paid by labour, td the rate of direct taxes paid by labour, F the fallback position of a worker, and α a measure of relative bargaining power. The subscript i denotes the skill level (either L for low-skilled or H for high-skilled), n_i the number of workers of skill level i and n the total number of workers. If n equals 1 or if all workers are alike, the above problem reduces to the standard bargaining problem with homogeneous labour.

In order to derive some simple expressions for the first order condition, we define some additional elements. Let wy_i denote the real product wage, defined as $wy_i = w_i(1 + se_i)/p_y$. Let wc_i denote the real consumption wage, defined as $wc_i = wi(1 - sl_i)(1 - td_i)/p_c$. Let Λ denote the wedge, defined as $\Lambda = wy_i/wc_i$. Define rp_i^* as $rp_i^* = F_i/wc_i$. rp_i^* is the ratio of the utilities of unemployed and employed workers of skill level i and can be thought of as a replacement rate. Also define z_i as $z_i = \alpha_i/(1 - rp_i^*)$. z_i is a measure of the total bargaining strength of workers of skill level i, combining the measures αi and rp_i.

The first-order conditions to the bargaining problem may now be stated as:

$$wy_i = \frac{z_i}{1 + \frac{1}{n}\sum_{i=L,H} n_i z_i} \frac{y}{n} = \frac{z_i}{1 + z}h, \quad i = L, H \qquad (10.5)$$

where h denotes average labour productivity and z average z_i. The relative wages for workers i and j are equal to the ratio of their relative bargaining strengths:

$$\frac{w_i}{w_j} = \frac{wy_i}{wy_j} = \frac{z_i}{z_j} \qquad (10.6)$$

and the average or macro real product wage, wy, is given by:

$$wy = \frac{z}{1 + z}h$$

These equations indicate that if productivity h and the coefficients α_i are exogenous, the various product wages only depends on the rp_i^*. So, if the wedge affects labour costs, it does so only indirectly, by altering rp^*, the ratio of the utilities of the unemployed and employed workers. Even if this is the case, the wedge does not enter the equation for wy once we include rp^*. Therefore, if we find an independent effect of the wedge next to a measure of the replacement rate, this must be because that measure of the replacement rate is not equal to rp^*.

10.4.1 The fallback position and the informal market

This may be the case if we include the informal sector or household production into the analysis. For instance, suppose that if negotiations break down, the worker becomes unemployed and tries to find work again.[6] If he succeeds he gets a wage which we assume to be equal to the wage he would have gotten if the first negotiations had resulted in agreement, namely w_i.

If he cannot find another job, he gets unemployment benefits equal to the replacement wage times that wage. While unemployed, he may also engage in activities in the informal sector. We assume that these activities are not taxed. This may be because these activities simply are not subject to taxation, such as household production, or because taxes are evaded (black market activities). In either case we think of the worker selling his labour directly to himself or to another consumer. The (implicit) wage is a fraction γ of the going total cost of his labour, including all direct and indirect taxes and premiums costs.

Letting p denote the probability of finding another job, rp the official replacement rate, and wn the net wage, defined as $wn_i = w_i(1 - sl_i)(1 - td_i)$, the formula for the fallback position of the worker is:

$$F_i = \frac{p_i wn_i + (1 - p_i)\left(rp_i wn_i + \gamma_i w_i\left(1 + se_i\right)\left(p_c/p_y\right)\right)}{p_c} \quad (10.7)$$

Note that the ratio of p_c over p_y equals 1 plus the indirect tax rate.[7] It makes sense that the probability to find another job, p, depends negatively on the unemployment rate, ur. If we assume that $p_i = 1 - ur_i^{\varepsilon i}$, we get after some arithmetic[8]:

$$z_i = \frac{\alpha_i}{uw_i^\epsilon\left(1 - rp_i - \gamma_i\Lambda_i\right)} \quad (10.8)$$

Substituting this expression into (10.4), (10.5) and (10.6) we get equations for the wage levels by skill level, for the relative wage levels and for the average wage level. Note that if we disregard the informal sector, that is set $\gamma_i = 0$, $i = L,H$, the wedge again drops out of the analysis.

In all of the wage equations the wedge enters as a single variable, that is, all subcomponents have the same coefficient. An important assumption for this result is that prices are not sticky, which is a reasonable assumption for the long run. However, since most wage contracts specify gross wages, we may expect gross wages to be relatively fixed in the short run. An unexpected shock to employer social security payments then has to be absorbed by the employers in the short run, while unexpected shocks to employee taxes and social security payments must be born by workers. Many empirical studies have found that employer social security payments indeed have a stronger effect on wage costs than employee taxes and social security payments. So, this may well be an additional effect for the short run.

10.4.2 Estimation

In order to estimate wage equations for high and low-skilled workers, we

	low-skilled		high-skilled	
γ_1	.91	(8.29)	.91	(—)
γ_2	.25	(3.22)	.25	(—)
γ_3	.30	(3.19)	.07	(0.56)
γ_4	.86	(3.39)	.47	(1.31)
γ_5	.13	(2.11)	.36	(3.80)
α	-.66	(7.10)	-.57	(5.01)
β_1	.25	(5.09)	.14	(1.34)
β_2	.41	(7.47)	.36	(3.68)
β_3	-1.23	(5.37)	-2.40	(2.78)
β_4	.025	(2.59)	.039	(1.84)
const	-1.23	(5.37)	-.90	(2.51)
s.e.	.0051		.0072	
DW	1.88		1.73	

Estimation by 3SLS, with the change in lagged value added prices and current import prices as instruments. Sample: 1966-93. t-statistic in parentheses.

Table 10.3: Estimation results of the wage equations

substituted (10.8) into (10.4) and estimated the same general form for both skill levels, namely:

$$\log we_i^* = \log p_y + \log h_i + \beta_1 \log \Lambda + \beta_2 \log rp_i$$
$$+\beta_3 ur_i + \beta_4 ud_i \qquad (10.9)$$
$$\Delta \log we_i = \gamma_1 \Delta \log p_y + \gamma_2 \Delta \log h_i + \gamma_3 \Delta \log (1 + t_l)$$
$$+\gamma_4 \Delta \log (1 + se_i) + \gamma_5 \Delta \log rp_i$$
$$-\alpha \left(\log we_i - \log we_i^*\right) + \text{const}_i \qquad (10.10)$$

where we denotes total annual wage cost per fte (denoted $w(1+se_i)$ above), ur the unemployment rate, ud the average unemployment duration. The other variables are defined as above. The superscript * denotes long run, and the subscript i the type of labour, that is, high or low-skilled. The short-term influence of the wedge has been split into two components: social security contributions paid by employers and the rest. This is based on the idea that gross wages may be relatively fixed in the short run.

The estimation was performed with 3SLS, using the changes in current import prices and lagged value added prices as instruments for the current change in value added prices. We also restricted the coefficients of $\log p_y$ and $\log h$ to be the same in both equations, which was easily accepted. The estimation results are given in Table 10.3.

The estimation results for the two wage equations are very similar. The influence of the wedge is somewhat stronger on wages for low-skilled labour and the opposite holds for the influence of unemployment. The latter result may be explained by arguing that the opportunity cost for high-skilled labour is higher. It is interesting that the effect of unemployment duration on wages is positive. The intuition for this result may be that, given the level of unemployment, an increase in the average unemployment duration means that there are fewer short-term unemployed and more long-term unemployed. Since many studies indicate that the effect of long-term unemployment on wages is less than that of short-term unemployment, it is not surprising that the effect of unemployment duration on wages, given the level of unemployment, is positive.

The above equations describe the developments of total wage costs. We may decompose the rise in total wage costs for each period in a rise in contractual wages and a wage drift. This distinction is important because wage related benefits rise only with the macro contractual wage increase. To be able to make this subdivision, we also estimate an equation for the macro wage drift. We related it simply to productivity growth, arguing that higher productivity growth creates more room for bonusses, extra promotions, and so on. The results are:

$$\Delta \log (w_{drift}) \quad = \quad .23 \cdot \Delta \log h + \quad .005$$
$$(3.50) \qquad\qquad (2.18)$$
$$\text{s.e.} = .0077 \qquad \text{D.W.} = 1.28$$

10.5　The cumulative production structure

The final model block we want to discuss is the Cumulative Production Structure matrix (CPS). As indicated in Section 10.2, the CPS forms the link between sectoral value added and imports on the one hand and final demand on the other.[9] An aggregated version of the CPS is given in Table 10.4.

The CPS is the reduced form of an input output table. The columns are final demand categories and the rows input categories. The number 97 in the top left corner, for instance, indicates that 97 billion guilders worth of value added produced in the exposed sector is ultimately used for the production of export goods.

In modelling the CPS we distinguish between a CPS matrix for quantities and one for prices, linking respectively the quantities and prices of value added and imports to those of final demand. In estimation of these CPS matrices we have the difficulty that there is no real data for it. Therefore we proceeded in an indirect way. In both cases we specified equations for each of the inner cells of the CPS matrices, constructed the implied border totals

	exports	private cons.	gov't cons.	invest ment	total
value added exposed	97.0	45.5	5.0	16.3	163.9
value added sheltered	49.1	73.1	10.1	35.9	168.2
value added remainder	16.9	103.0	56.1	3.9	179.9
intermediate inputs	132.6	36.5	7.4	15.9	192.4
domestic production at factor costs	295.7	258.2	78.7	72.0	704.5
final imports	-	50.4	-	26.3	76.7
taxes minus subsidies	2.4	32.3	4.3	15.0	54.0
final demand at market prices	298.0	340.9	83.0	113.3	835.3

Table 10.4: Cumulative production structure for 1992, billions of guilders

by addition and estimated the coefficients of the matrix by estimating these implied border totals.

10.5.1 Prices

We will only deal with the CPS matrix for prices, the one for quantities follows a similar structure. Prices of imports in the CPS are exogenous. Prices of value added are assumed to be determined by the costs and capacity utilization rates of the sector of origin and by competitor prices for the final demand category for which the value added is used. Denoting sectoral indices by superscripts and final demand category indices by subscripts, the equations for value added prices are:

$$\Delta p_a^y = \Delta cy^y + \frac{\beta}{ycw_a}\Delta\left(pcom_a - cost_a\right) + \gamma\left(q^y - \bar{q}^y\right) \quad (10.11)$$

where:

p_a^y = log of added price of sector y for the production of final demand category a
cy^y = log of value added cost of sector y
q^y = capacity utilization rate of sector y
$\bar{q}^y$ = normal capacity utilization rate of sector y
$pcom_a$ = log of foreign competitor prices for final demand category a

$cost_a$ = log of total costs of final demand category a

ycw_a = cost share of total value added in final demand category a (a scaling factor used to facilitate the adding up of the specifications across columns)

The specifications of these individual value added prices are now used to construct prices for final demand categories and total value added by sector. These latter price equations are estimated simultaneously by SUR (leaving one of the equations out because all the equations together are dependent). The estimation results are presented in Table 10.5 below.

	exports		consumption			investment
	goods	services	market	non-market	government	equipment
β	.50	.25	.53	.25	.46	.83
	(11.7)	(6.1)	(14.6)	(4.5)	(11.4)	(22.0)
	exposed	sheltered				
γ	.28	.22				
	(4.4)	(4.4)				

Table 10.5: Estimation results for price equations in CPS, t-statistics in parentheses

The estimated coefficients on competing prices are fairly high. They are largely in line with previous research. However, other research by for instance Nieuwenhuis (this volume), suggests that should be lower and also argues that the feedback from the level of the capacity utilization rate to the rate of change in prices should be replaced by an equation completely in levels. This discussion appears to be very similar to the one about replacing the traditional Phillips curve equation with a wage level equation. This is a topic of current research.

10.6 Simulation analysis

To illustrate the properties of the model, we present the results of three simulations. These are:

1. an increase in world trade by 1 per cent
2. a reduction in the minimum wage rate and related benefits by 1 per cent;
3. an increase in the labour supply of high and low-skilled workers by 10 000 persons

The base run for these simulations is a dynamic historical simulation over the period 1981-1993. In each simulation, a single, one-time impulse is given to the rate of or absolute change of a certain variable. The response of the model is tracked by calculating the differences between the values (levels) of the model variables in the simulation run and in the base run. In all three simulations the impulse is given in 1981 and the simulation runs till 1993. Since the effects generally appear to stabilize in five to ten years, this time horizon is sufficient in most cases.

In general, a temporary shock to the growth rate of an exogenous variable will have a permanent effect, because the exogenous variable will be permanently different from its base run levels. A temporary shock to the growth rate of an endogenous variable will have no permanent effects, because none of the exogenous variables are affected and so, the long-run solution of the model is not affected either. Of the three simulations presented in this section, none involves a shock to an endogenous variable.

The results for each simulation are reported in a table in which the absolute or relative differences between the simulation run and the base run are presented for several key variables at the macro level, for the exposed and sheltered sectors, and for low and high-skilled labour.

10.6.1 An increase in world trade of one per cent

The impact effect of an increase in world trade is a proportional increase in the exports of goods (excluding energy). This generates a cyclical upturn, with higher production, investments, employment, labour productivity, utilization rates, (real) wages and prices. Over time, the capacity utilization rates return to their normal levels as increased investments and employment generate a faster growth rate of production capacity. The increase in prices causes a real appreciation, which partially offsets the growth in exports. The employment effects for high and low-skilled labour are nearly the same.

As is to be expected, initially the exposed sector benefits most. However, after a few years the growth of the sheltered sector dominates, both in terms of production and of employment. There are several reasons for this result. As the CPS matrix in Table 10.6 indicates, intermediate deliveries of the sheltered sector to exporting firms are substantial. In addition, the sheltered sector benefits from the increase in investment and, especially, consumption. Third, the shift in demand from exports to consumption which occurs over time due to the rise in real wages and the real appreciation favours the sheltered sector.

		1981	1982	1984	1986	1993
prices						
wage rate enterprises	%	0.06	0.55	0.93	0.52	0.45
gross value added private sector	%	0.07	0.41	0.70	0.42	0.36
quantities						
private consumption	%	0.04	0.23	0.42	0.25	0.19
private inv. excl. dwellings	%	0.02	1.40	0.98	−0.06	0.15
exports of goods excl. energy	%	0.93	0.66	0.11	0.17	0.20
imports of goods	%	0.59	0.65	0.41	0.26	0.28
net national income	%	0.21	0.42	0.43	0.25	0.26
gross value added exposed	%	0.42	0.30	0.06	0.07	0.09
gross value added sheltered	%	0.15	0.37	0.33	0.18	0.17
employment enterprises (fte)	%	0.02	0.09	0.19	0.17	0.14
empl. enterprises(fte) low skil	%	0.02	0.09	0.19	0.19	0.16
empl. enterprises(fte) hi skil	%	0.02	0.09	0.18	0.14	0.13
unemployed persons	D	−0.7	−3.2	−6.7	−6.2	−5.8
quotes						
capacity util. rate exposed	D	0.37	0.16	−0.11	−0.04	0.01
capacity util. rate sheltered	D	0.13	0.30	0.13	−0.06	−0.01
current account BOP (% NNI)	D	0.11	0.07	0.00	0.04	0.06
labour income share enterprises	D	−0.27	−0.17	0.13	0.10	0.05

Table 10.6: Cumulated effects of an increase in world trade by 1 per cent

10.6.2   A reduction of the minimum wage and related benefits by one per cent

The reduction in the minimum wage rate has two major effects. First, the income of the recipients of social benefits that are linked to the minimum wage is reduced. Second, the reduction in social benefits reduces the social security premiums and the replacement ratio, which reduce wage costs in the private sector. This in turn raises employment. In contrast to the previous simulation, a reduction in minimum wages has permanent effects because an institutional, and thus exogenous, variable has been changed.

10.6.3   An increase in the labour supply of low and high-skilled workers by 10 000 persons each (about 0.4 per cent of total labour force)

The increase in labour supply initially just raises unemployment. Reduction of the wage rate is the result, leading to lower prices. The competitive position improves, and exports increase. Consumption initially rises, because the

		1981	1982	1984	1986	1993
prices						
wage rate enterprises	%	−0.19	−0.36	−0.39	−0.31	−0.19
gross value added private sector	%	−0.10	−0.19	−0.19	−0.13	−0.04
quantities						
private consumption	%	−0.11	−0.20	−0.24	−0.20	−0.12
private inv. excl. dwellings	%	0.00	−0.10	0.21	0.41	0.28
exports of goods excl. energy	%	0.05	0.17	0.33	0.31	0.32
imports of goods	%	−0.04	−0.04	0.06	0.10	0.15
net national income	%	−0.06	−0.09	−0.04	0.02	0.12
gross value added exposed	%	0.00	0.04	0.10	0.10	0.12
gross value added sheltered	%	−0.04	−0.07	−0.02	0.02	0.07
employment enterprises (fte)	%	0.00	0.00	0.01	0.03	0.08
empl. enterprises(fte) low skil	%	0.00	0.01	0.06	0.11	0.18
empl. enterprises(fte) hi skil	%	0.00	−0.02	−0.05	−0.04	0.01
unemployed persons	D	0.0	0.1	−0.3	−1.3	−3.3
quotes						
capacity util. rate exposed	D	0.01	0.04	0.07	0.02	−0.01
capacity util. rate sheltered	D	−0.03	−0.05	0.01	0.03	0.02
current account BOP (% NNI)	D	0.02	0.06	0.09	0.07	0.09
labour income share enterprises	D	−0.06	−0.12	−0.19	−0.17	−0.12

Table 10.7: Cumulated effects of a minimum wage reduction and related benefits by 1 per cent

additional workers, though unemployed, still provide additional purchasing power. The drop in real wages that follow, however, reduces consumption later on. On balance, production rises throughout the simulation period. Net national income rises less because of terms-of-trade effects. The shift towards exports benefits the exposed sector most. The employment effects for high-skilled workers are faster because the feedback from unemployment to wages is stronger for high-skilled workers, and the original increase in their unemployment rate is larger as well.

At the end of the simulation period only one third of the increase in the labour supply has found work, although the increase in employment has by no means stopped at that point. In addition, the labour income share remains below base as well. The two are probably related and indicate a poor feedback mechanism in this area.

10.7 Summary and conclusion

This paper presented a preliminary version of a new macroeconometric model

		1981	1982	1984	1986	1993
prices						
wage rate enterprises	%	−0.02	−0.39	−0.48	−0.45	−0.27
gross value added private sector	%	−0.01	−0.20	−0.23	−0.17	−0.02
quantities						
private consumption	%	0.09	−0.01	−0.08	−0.08	−0.01
private inv. excl. dwellings	%	0.00	0.12	0.45	0.76	0.54
exports of goods excl. energy	%	0.00	0.10	0.38	0.43	0.52
imports of goods	%	0.05	0.05	0.20	0.27	0.34
net national income	%	0.03	−0.03	0.02	0.07	0.18
gross value added exposed	%	0.03	0.07	0.18	0.20	0.24
gross value added sheltered	%	0.04	0.02	0.08	0.13	0.17
employment enterprises (fte)	%	0.00	0.01	0.04	0.09	0.18
empl. enterprises(fte) low skil	%	0.00	0.01	0.00	0.02	0.18
empl. enterprises(fte) hi skil	%	0.00	0.02	0.09	0.15	0.18
unemployed persons	D	19.9	19.5	18.6	16.8	12.6
quotes						
capacity util. rate exposed	D	0.03	0.06	0.12	0.06	−0.01
capacity util. rate sheltered	D	0.04	0.01	0.05	0.06	0.00
current account BOP (% NNI)	D	−0.03	−0.03	0.02	0.02	0.08
labour income share enterprises	D	−0.05	−0.19	−0.29	−0.30	−0.21

Table 10.8: Cumulated effects of an increase in the labour supply of high and low-skilled workers by 10 000 persons each

of the Netherlands. The model is meant to be suitable for both short, medium and long-run analysis and to give a more elaborate description of the labour market. It builds on two existing models at the CPB Netherlands Bureau for Economic Analysis, FKSEC and MIMIC, and aims to integrate their relative strengths, namely an empirical base and a theoretical consistency. To be able to analyse issues related to low-skilled labour in more detail, we disaggregated the labour market into low and high-skilled labour.

To achieve consistent outcomes for the long run, we formulated and estimated models of production, wage formation and matches in levels and estimated these models using an error correction framework. The disaggregation into high and low-skilled labour entered naturally into the production framework. For wage formation and matches we simply applied the macro theory to both low and high-skilled labour separately.

The simulation results indicate that the model has a tendency to return to base when an endogenous variable such as the wage rate is perturbed, while there are long-run deviations from base when an exogenous or a policy variable is changed. The length of the adjustment path depends on the particular simulation, but in most cases, the effects appear to stabilize after around five to ten years. A notable exception is a shock to labour supply, which generates a very slow adjustment. We did not yet perform a thorough comparison with the FKSEC model, but a cursory glance indicates that the initial effects are fairly similar, but the final effects not. This is indeed what we would expect.

Overall, it appears that the model provides a consistent infrastructure to analyse issues with both short and long-run dimensions. Also, for the labour market, the model provides a framework to analyse issues related to labour skills and labour flows. The initial results are encouraging, and it appears promising to continue this type of research.

Acknowledgments

We want to thank A.J.G. Manders for agreeing to have part of his work with D.A.G. Draper on production functions included in this paper, R. van Stratum for his contributions in an earlier phase of the project, and participants of the Steering Committee for many helpful comments and suggestions. Netherlands Bureau for Economic Policy Analysis, Van Stolkweg 14, P.O.Box 80510, 2508 GM The Hague,The Netherlands. E-mail: FHH@CPB.NL

Notes

1. An alternative to this modelling strategy is provided by Real Business Cycle models, that are typically designed so that the dynamics follow from the optimal adjustments of agents to a changing economic environment. The empirical basis of these models is generally weak however (see Stadler, 1994).

2. For models of the Dutch economy that do contain elaborate financial sector, see *e.g.* the Freia model (CPB, 1983) and the MORKMON model (DNB, 1984).

3. This section appeared in similar form in Draper and Manders, (1996a). See also Draper and Manders, (1996b).

4. For instance, Knoester en Van der Windt (1987), testing a macro wage equation for ten OECD countries, find that the wedge has a significant effect on wage cost. They also find that the effect of employer and worker taxes and social security contributions is stronger that the effect of indirect taxes for Australia, Canada, Germany, Italy, Japan, The Netherlands, Sweden and the United States. For the United Kingdom, Layard and Nickell (1986) find that only the taxes and social security contributions paid by employers affect wages. For the Netherlands, most studies find that the effect of the social security contributions paid by employers is larger than that of the other components of the wedge (see, for instance, Fase *et al.*, 1990; CPB, 1992; Graafland, 1991, 1992; Graafland en Verbruggen, 1993).

5. The firm's gain function consists of operating profits. Total profits are given by operating

profits and some unspecified sunk or fixed costs. These could, for example, be hiring costs or capital costs. In equilibrium, these costs correspond to quasi rents that make the bargaining neccesary. The also generate unemployment as a neccesary force to reduce the union wage claims to a level compatible with general equilibrium, that is a level such that the firms' operating profits are high enough to cover the fixed costs.

6. One may also interpret the fallbeck positions F_i as the expected income to the worker during a strike. However, this would not affect the analysis.

7. Actually, it also includes a terms of trade effect. If the value of output in the informal market is not shielded from changes in the terms of trade, p_c should be replaced by the consumer price with the terms of trade effects taken out.

8. Note that if we had used the consumer price with the terms of trade effect taken out, as suggested in the previous note, the terms of trade would also disappear from this equation. Λ should then be interpreted as the wedge with the terms of trade effect taken out.

9. Our approach to linking macro final demand and sectoral value added was initiated by the Brookings model project (Fisher, Klein and Shinkai, 1965). It also existed in the Wharton Annual and Industrial Forecasting Model (Preston, 1975). For other models with similar approaches, see Bodkin, Klein and Marwah (1991). The modelling of their 'conversion' matrices was, however, rather rigid.

References

Allen, C.B. (1994), "A Supply Side Model of the UK Economy: An Application of Non-linear Cointegration," Centre for Economic Forecasting Discussion Paper.
Bean, C.R. (1994), "European Unemployment: A Survey," *Journal of Economic Literature*, 32:573-619.
Berman, E., J. Bound and Z. Griliches (1994), "Changes in the Demand for Skilled Labor within U.S. Manufacturing: Evidence from the Annual Survey of Manufacturers," *Quarterly Journal of Economics*:367-97.
Bodkin, R.G., L.R. Klein and K. Marwah (1991), *A History of Macroeconometric Model Building*, Edward Elgar, Brookfield.
Brayton,F. and E. Mauskopf (1985), "The Federal Reserve Board MPS Quarterly Econometric Model of the U.S. Economy," *Economic Modelling*, 2:170-292.
Broer, D.P. and W.J. Jansen, (1989), "Employment, Schooling and Productivity Growth," *De Economist* , 137:425-53.
CBS (1993), *Standaard onderwijsindeling SOI-1978, ed. 1993* ; codelijst van opleidingen.
CBS (1996), *Tijdreeksen arbeidsrekeningen 1969-1993*. Ramingen van het opleidingsniveau, een tussenstand, Statistics Netherlands, Voorburg.
CPB (1983), *FREIA, Een macroeconomisch model voor de middellange termijn* (FREIA, A Macroeconomic Model for the Medium Term), Distributiecentrum Overheidspublicaties.
CPB (1990), *ATHENA, Een bedrijfstakkenmodel voor de Nederlandse economie*, CPB monograph no. 30.
CPB (1992), *FKSEC, A Macroeconometric Model for the Netherlands*, Stenfert Kroese.
De Nederlandsche Bank (1984), *MORKMON, Een Kwartaalmodel voor Economische Analyse* (MORKMON, a Quarterly Model for Economic Analysis), Kluwer.
Diewert, W. and T. Wales (1987), "Flexible Forms and Global Curvature Conditions," *Econometrica*, 55:43-68.
Diewert, W. and T. Wales (1995), "Flexible Forms and Tests of Homogeneous Separability," *Journal of Econometrics*, 67:259-302.
Draper D. (1989), *Produktieblok voor de niet beschermde sector*, intern CPB memo I/89/14, CPB Netherlands Bureau for Economic Policy Analysis.
Draper D. and A. Manders (1996a), "Why Did the Demand for Dutch Low-skilled Workers Decline?," *CPB Report*, Quarterly Review of CPB Netherlands Bureau for Economic Pol-

icy Analysis, 1996, no 1.

Draper D. and A. Manders (1996b), *Structural Changes in the Demand for Labor*, CPB Research Memorandum No 128, CPB Netherlands Bureau for Economic Policy Analysis.

Fase, M.M.G., P. Kramer and W.C. Boeschoten (1990), *MORKMON II, Het DNB kwartaalmodel van de Nederlandse economie*, DNB, Monetaire monografieën

Fisher, F.M., L.R. Klein and Y. Shinkai (1965), "Price and Output aggregation in the Brookings Econometric Model," in: J.S. Duesenberry, G. Fromm, L.R. Klein and E. Kuh (eds.), *The Brookings Quarterly Econometric Model of the United States*, Rand McNally Company, Chicago.

Gelauff, G., A. de Haan and V. Okker, (1986), *Een macro produktiefunctie zonder jaargangen, voorlopige resultaten*, intern CPB memo I/86/14, CPB Netherlands Bureau for Economic Policy Analysis.

Gelauff, G.M.M, and J.J. Graafland (1994), *Modelling Welfare State Reform*, North Holland, Amsterdam.

Graafland, J.J. (1991), "Effecten van marginale belasting- en premiedruk op loonvorming," *Maandschrift Economie*, 55:442-55

Graafland, J.J. (1992), "From Phillips Curve to Wage Curve," *De Economist*, 140:501-14

Graafland, J.J. and F.H. Huizinga (1996), *Taxes and Benefits in a Non-linear Wage Equation*, CPB Research Memorandum No 125, CPB Netherlands Bureau for Economic Policy Analysis.

Graafland, J.J. and J.P. Verbruggen (1993), "Macro Against Sectoral Wage Equations for the Netherlands," *Applied Economics*, 25:1373-83.

Hall, S.G. (1995), "Macroeconomics and a Bit More Reality," *Economic Journal*, 105:974-88.

Hamermesh, D.S. (1993), *Labor Demand*, Princeton University Press, Princeton.

Hebbink, G.E. (1991), "Employment by Level of Education and Production Factor Substitutability," *De Economist*, 139:379-99.

Huigen, R, A. Kleiweg, G. van Leeuwen and K. Zeelenberg (1993), *A Microeconometric Analysis of Interrelated Factor Demands*, Research paper MOPS-30, Statistics Netherlands, Voorburg.

Jorgenson D. (1986), "Econometric Methods for Modelling Producer Behavior," in: Z. Griliches and M. Intriligator (eds.), *Handbook of Econometrics*, Vol. III, North Holland, Amsterdam.

Knoester, A. and N. van der Windt (1987), "Real Wages and Taxation in Ten OECD Countries," *Oxford Bulletin of Economics and Statistics*, 49:151-69.

Layard, P.R.G. and S.J. Nickell (1986), "Unemploment in Britain," *Economica*, 53:s121-69.

Layard, R., S. Nickell and R. Jackman (1991), *Unemployment, Macroeconomic Performance and the Labour Market*, Oxford University Press, Oxford.

Layard, R., S. Nickell and R. Jackman (1994), *The Unemployment Crisis*, Oxford University Press, Oxford.

Lucas, R.E. (1976), "Econometric Policy Evaluation, A Critique," in: K. Brunner and A. Meltzer (eds.), *The Phillips Curve and Labor Markets*, Carnegy-Rochester Conference Series on Public Policy.

National Institute (1994), *National Institute Model 12*, mimeo.

Nieuwenhuis, A. (1995), *Imperfect Competition and Aggregate Price Equations*, mimeo.

Ours, J. van and G. Ridder (1991), "Cyclical Variation in Vacancy Durations and Vacancy Flows," *European Economic Review*, 35:1143-55.

Ours, J. van, G. Ridder (1995), "Job Matching and Job Competition: Are Lower Educated Workers at the Back of Job Queues?," *European Economic Review*, 39:1717-31.

Preston, R.S. (1975), "The Wharton Long Term Model: Input-Output Within the Context of a Macro Forecasting Model," *International Economic Review*, 16:3-19.

Shadman-Mehta, F. and H. Sneessens (1995), *Skill Demand and Factor Substitution*, CEPR Discussion paper No. 1279, London.

Sims, C. (1980), "Macroeconomics and Reality," *Econometrica*, 48:1-48.

Stadler, G.W. (1994), "Real Business Cycles," *Journal of Economic Literature*, 23:1750-83.

PART VI
Intertemporal Modelling

11 Uses and Limitations of Public Debt

Willem H. Buiter and Kenneth M. Kletzer

11.1 Introduction

This paper addresses three related issues. First, in what sense is the government's ability to borrow limited by its capacity to tax? Second, when is Ponzi finance[1] feasible for an infinite-lived government? Third, when does the opportunity to engage in Ponzi finance enhance the government's ability to influence private resource allocation?

The standard answer to the first question, when the long-run (after-tax) rate of interest exceeds the long-run natural rate of growth, is based on the conventional government solvency constraint. This states that, in an infinite-lived economic system, the present discounted value of the 'terminal' government debt should be non-positive in the limit. This implies that the sum of the present discounted values of all future primary (non-interest) surpluses should be at least as large as the outstanding stock of public debt. As exhaustive government spending cannot be negative, the maximal amount of debt a solvent government can issue is constrained by the present discounted value of future taxes (net of transfers).

The traditional answer to the second question is that Ponzi finance is feasible only if the long-run after-tax rate of interest on the public debt is below the long-run natural rate of growth. If interest income is not taxed, Ponzi finance is only possible, in deterministic, competitive perfect foresight OLG models, if the economy is dynamically inefficient (see for example Gale, 1983, Tirole, 1985, O'Connell and Zeldes, 1988, Blanchard and Weil, 1992 and Azariadis, 1993). Under uncertainty, Blanchard and Weil (1992) argue that Ponzi schemes are feasible only if the competitive equilibrium allocation is not Pareto-efficient (because it fails to provide full intergenerational insurance), even if it is dynamically efficient. In answer to the third question, the existing literature points out that when the competitive equilibrium is Pareto-inefficient, Ponzi finance can be Pareto improving.[2]

We argue that convincing answers to all three questions require a careful specification of the government's 'capacity to tax,' that is, the richness of the set of lump-sum and/or distortionary tax-transfer instruments available to it. It is also necessary to make a novel distinction between *weak* and *strong* Ponzi finance. Under a weak Ponzi scheme, *aggregate* taxes are cut one period while spending is kept constant, additional debt is issued and in no subsequent period are aggregate taxes raised or is public spending reduced. Under a strong Ponzi scheme, taxes are cut for one generation during one period (again with spending constant), additional debt is issued and not only are *aggregate* taxes and public spending in all subsequent periods left unchanged, but the taxes and transfers of each generation are left unchanged.

We show that weak Ponzi finance may be feasible whether or not the competitive equilibrium is dynamically efficient or Pareto-efficient, and regardless of the long-run relationship between the interest rate and the growth rate. Strong Ponzi finance is possible only if the long-run (after-tax) rate of interest is below the long-run natural rate of growth. Key to our argument is our specification of the feasibility constraints on the government's fiscal-financial plan and of the set of available tax-transfer instruments. We prove that weak Ponzi finance is *always* feasible if the set of lump-sum taxes and transfers the government can choose from is unrestricted (across generations, and time[3]).

Our proof of the feasibility of weak Ponzi finance depends on our characterisation of feasible fiscal-financial plans for an infinite-lived government in an economy with overlapping generations of finite-lived households without private intergenerational gift motives. We express feasibility of the government's fiscal-financial plan as a set of three inequality constraints on admissible sequences of taxes, transfers, public debt and exhaustive public (consumption) spending. These are derived from the requirement that the private capital stock, private consumption by each generation and government consumption be non-negative in each period. In other words, feasibility for the plans of the infinite-lived government is derived from the (well-understood) requirement of solvency (or non-bankruptcy) for each of an infinite sequence of finite-lived households.

In the infinite-horizon OLG economy, our feasibility constraints can be met by government fiscal-financial plans that do not satisfy the conventional solvency constraint of an infinite-lived government. A central motivation for our paper was to *derive* this conventional solvency constraint from acceptable primitives. We find reasonable sufficient restrictions on taxes and transfers for our feasibility constraints to imply the conventional solvency constraint.

Consider equilibria in which the long-run real interest rate is above the long-run growth rate of efficiency labour. One sufficient condition for the conventional solvency constraint to be valid is that the maximal long-run growth rate of transfers to the young and of taxes on the old be less than the long-run interest rate. This condition is met if the long-run growth rate of taxes and transfers cannot exceed that of efficiency labour, or equivalently, if the ratio of taxes paid or transfers received in any one period by a generation to the value of the physical resources it owns is bounded. One reason may be that only (sufficiently) distortionary taxes and transfers are available. We further show that in the presence of such distortionary taxes and transfers the feasibility constraints are generally more restrictive than the conventional solvency constraint. They imply a bounded long-run ratio of debt to physical resources. The solvency criterion is a necessary but not sufficient condition for feasibility in this case. Another example of a sufficient condition is that the net transfer payments by the government to a generation cannot change sign over the life cycle of that generation.

With unrestricted lump-sum taxes and transfers, weak Ponzi finance is *inessential* in that it does not enhance the set of allocations that can be supported as competitive equilibria. Ponzi finance may also be feasible when lump-sum taxes and transfers are restricted. In such cases, Ponzi finance is *essential*; the ability to engage in Ponzi finance allows the government to support competitive equilibrium allocations that cannot be supported without it. This essential-inessential Ponzi finance distinction is extended to public debt in general.

We discuss three benchmark cases of restrictions on the set of fiscal instruments available to the government and state three equivalence results. The first is the well-known proposition (see for example Wallace, 1981, Chamley and Polemarchakis, 1984 and Sargent, 1987, Chapter 8) that with unrestricted time-, age- and state-specific lump-sum taxes and transfers, the ability to depart from budget balance does not permit additional equilibria to be supported. Specifically, any intergenerational redistribution and insurance that can be provided with government borrowing or lending can also be provided with a balanced government budget.

We prove a second equivalence result stating that, if the government is constrained to treat all overlapping generations the same during any given period, the ability to unbalance the budget permits it to support all equilibria that can be supported with completely unrestricted lumps-sum taxes and transfers.

We also establish a third equivalence proposition stating that even rather severe restrictions on the ability to vary taxes and transfers over the lifetime of a generation do not restrict the set of equilibria that can be supported, provided unbalanced budgets are permitted.

Alternative government financing policies not only effect redistribution among generations (giving rise to familiar 'financial crowding out' issues), in a stochastic environment they will also permit privately infeasible trades across states of nature or intergenerational insurance. Such intergenerational redistribution schemes as social security taxes and retirement benefits can provide insurance that either cannot be provided by the market or is provided inefficiently.[4] OLG models have incomplete market participation. Because individual households cannot enter into insurance contracts before they are born, there may be incomplete risk-sharing (Blanchard and Weil, 1992). Even in a dynamically efficient economy, the public provision of this insurance can have implications for Pareto-efficiency (see Enders and Lapan, 1982, Stiglitz, 1983, Fischer, 1983, Merton, 1984, Gordon and Varian, 1988, Pagano, 1988, Gale, 1990, Zilcha, 1990and Blanchard and Weil, 1992).[5] An investigation of the many interesting positive and welfare aspects of intergenerational redistribution and of the provision of intergenerational insurance through the government budget, is beyond the scope of this paper.

The outline of the rest of this paper is as follows. Section 11.2 develops the model. Section 11.3 concerns the feasibility of Ponzi finance. Conditions under which the feasibility constraints on government fiscal-financial plans imply the conventional solvency constraint are stated in Section 11.4. Section 11.5 concerns the usefulness of (weak) Ponzi finance and presents the equivalence results. Section 11.6 concludes.

11.2 The model

Consider the closed economy, one-good, two-period OLG growth model of Diamond (1965), with government borrowing or lending and lump-sum taxes or transfers.

11.2.1 The private sector

Individuals of the same generation are identical. Successive generations have the same utility functions and maximise expected utility. People live for two periods, work in the first period of life and retire in the second. There is no intergenerational gift or bequest motive. Labour supply is inelastic and scaled to unity for each young worker. There is no uncertainty.[6] The young have access to two stores of value, claims on real capital and one-period public debt.[7]

Population and labour force size in natural units are both denoted by L. It grows at the constant proportional rate $n > -1$, and L_0 is set equal to 1. There is labour-augmenting productivity growth, at a rate ω and the level of productivity is denoted by θ, so $1 + \omega_t \equiv \frac{\theta_t}{\theta_{t-1}}$. labour in efficiency units is therefore equal to θL. The growth rate of efficiency labour will be referred to as the natural rate of growth.

The optimisation problem of a competitive representative consumer born in period t is given in Equations (11.1), (11.2), (11.3) and (11.4). c_t^i and τ_t^i, $i = 1$, 2, are consumption, respectively taxes paid, by a member of generation t in the ith period of her life. w_t is the wage rate in period t. k_{t+1}^d and b_{t+1}^d are the amounts of capital, respectively bonds (measured for notational convenience per unit of efficiency labour of generation $t + 1$), demanded by a member of generation t, to be carried forward into period $t + 1$. The one-period real interest rate is denoted r and ρ is the rental rate of a unit of capital. In the absence of uncertainty, $r = \rho$. A member of generation t chooses c_t^1, k_{t+1}^d, b_{t+1}^d and c_t^2 to maximise (11.1)[8] subject to the sequence of constraints given in (11.2),(11.3) and (11.4).

$$v(c_t^1) + \beta v(c_t^2) \tag{11.1}$$

$$w_t - \tau_t^1 - c_t^1 \geq (k_{t+1}^d + p_t b_{t+1}^d)(1+n)\theta_{t+1} \tag{11.2}$$

$$c_t^2 + \tau_t^2 \leq (1 + r_{t+1})(k_{t+1}^d + b_{t+1}^d)(1+n)\theta_{t+1} \tag{11.3}$$

$$c_t^1, c_t^2, k_{t+1}^d \geq 0 \tag{11.4}$$

Since utility is strictly increasing in c^1 and c^2, (11.2) and (11.3) will hold with equality.

The interior first-order conditions[9] for a member of generation t are

$$v'(c_t^1) = \beta(1 + r_{t+1})v'(c_t^2) \tag{11.5}$$

Output Y is produced by a twice continuously differentiable production function with constant returns to capital K and labour in efficiency units θL and positive and diminishing marginal products: $\frac{Y_t}{\theta_t L_t} \equiv y_t = f(\frac{K_t}{\theta_t L_t}) = f(k_t)$; $f(0) = 0$; $f' > 0$; $f'' < 0$. It also satisfies the Inada conditions.

11.2.2 The public sector

The government imposes lump-sum taxes (transfers when negative) on the young and/or the old, spends on public consumption[10] and satisfies its single-period budget identity by borrowing or lending. B_t is the stock of government bonds at the beginning of period t and $G_t \geq 0$ the amount of public consumption spending. The single-period government budget identity is

$$B_{t+1} \equiv (1 + r_t)B_t + G_t - \tau_t^1 L_t - \tau_{t-1}^2 L_{t-1}$$

With $b_t \equiv B_t/(\theta_t L_t)$, and $g_t \equiv G_t/(\theta_t L_t)$, the single-period government budget identity can be rewritten as

$$b_{t+1}(1+n)(1+\omega_{t+1}) \equiv (1+r_t)b_t + g_t - \theta_t^{-1}[\tau_t^1 + \tau_{t-1}^2(1+n)^{-1}] \quad (11.6)$$

11.2.3 Factor and asset market equilibrium

The labour market and capital rental market are competitive and clear, so

$$w_t = \theta_t[f(k_t) - k_t f'(k_t)] \tag{11.7}$$

$$r_{t+1} = f'(k_{t+1}) \tag{11.8}$$

The economy-wide asset market equilibrium conditions are given by

$$B_{t+1} + K_{t+1} = L_{t+1}(b_{t+1}^d + k_{t+1}^d) \tag{11.9}$$

Substituting the asset market equilibrium conditions into equation (11.2) (assumed to hold with equality) yields:

$$(b_{t+1} + k_{t+1})(1 + n)(1 + \omega_{t+1}) = (w_t - \tau_t^1 - c_t^1)\theta_t^{-1} \tag{11.10}$$

11.2.4 Feasible fiscal-financial plans and government solvency

Solving the government single-period budget identity forward in time to $T > t$, we get for all $t \geq 0$

$$\delta_{t-1}b_t \equiv \sum_{s=t}^{T-1}(\theta_{s+1})^{-1}\left[\frac{\tau_s^1}{1+n} + \left(\frac{1}{1+n}\right)^2\tau_{s-1}^2 - \frac{\theta_s g_s}{1+n}\right]\delta_s + \delta_{T-1}b_T$$

$$(11.11)$$

$$\delta_s \equiv \prod_{j=0}^{s} \left(\frac{(1+n)(1+\omega_{j+1})}{1+r_j} \right) \text{ for } s \geq 0 \quad (11.12)$$

$$\equiv 1 \qquad \text{for } s = -1$$

We also define the market discount factor Δ as follows:

$$\Delta_s \equiv \prod_{j=0}^{s} [\frac{1}{1+r_j}] = \delta_s/(\theta_{s+1}L_{s+1}).$$

Note that δ is the 'natural rate of growth-adjusted' discount factor. Both Δ_s and δ_s are assumed to be positive for finite values of s and non-negative in the limit as $s \to \infty$.

The conventional government solvency constraint, given in (11.13) requires the discounted public debt to vanish in the long run for any realisation of the discounted debt sequence.

$$\lim_{T \to \infty} \delta_{T-1}b_T \equiv \lim_{T \to \infty} \Delta_{T-1}B_T = 0. \quad (11.13)$$

Equation (11.13) [11] implies (11.14)[12]

$$\delta_{t-1}b_t = \lim_{T \to \infty} \sum_{s=t}^{T-1} (\theta_{s+1})^{-1} \left[\frac{\tau_s^1}{1+n} + \left(\frac{1}{1+n} \right)^2 \tau_{s-1}^2 - \frac{\theta_s g_s}{1+n} \right] \delta_s$$
$$(11.14)$$

The solvency condition (11.13) has the *prima facie* attractive property of implying the same kind of present value or intertemporal budget constraint (11.14) for the infinite horizon case as for the finite horizon case. If $T-1$ is the finite terminal period, then the standard (and uncontroversial) government solvency constraint is $b_T \leq 0$. A rational household sector ensures that $b_T \geq 0$. From (11.11) these two weak inequalities imply, that the value of the current stock of debt is equal to the present discounted value of future primary surpluses. The imposition of (11.13) has been virtually automatic in modern macroeconomic analysis. For a small sample see Barro (1979), Buiter (1985), Pagano (1988), Blanchard et. al. (1990), Auerbach and Kotlikoff (1987) and Auerbach, Gokhale and Kotlikoff (1991). It has been the subject of extensive empirical testing (see for instance Hamilton and Flavin, 1986, Wilcox, 1989, Corsetti, 1990, Grilli, 1990, Trehan and Walsh, 1991and Buiter and Patel, 1992 and 1994), with mixed results.

282   *Willem H. Buiter and Kenneth M. Kletzer*

We believe that the analogy with the finite-horizon case is potentially misleading. It is by no means obvious what reasonable restrictions can be imposed, in an economy without a terminal date, on the debt strategies of an infinite-lived government facing an infinite sequence of finite-lived overlapping generations (see for example Shell, 1971 and Wilson, 1981). As we will show, without *a-priori* restrictions on taxes and transfers, our model has a wide range of feasible debt strategies, many of which allow for Ponzi finance. This contrasts with the case of an economy consisting of a finite number of infinite-lived consumers. In such an economy, the government can only run a Ponzi debt scheme if at least one infinite-lived household runs a Ponzi credit scheme. This would violate the (necessary) transversality condition for a household optimum, so that the conventional government solvency constraint is implied by feasibility and household rationality without prior restrictions on taxes and transfers (see McCallum, 1984, Cass, 1972, O'Connell and Zeldes, 1988 and Bohn, 1991).

It turns out to be necessary to distinguish between two kinds of Ponzi finance, which we shall refer to as *weak* and *strong* Ponzi finance respectively. This distinction was motivated by a recent paper by Ball, Elmendorf and Mankiw (1995). Formally, weak Ponzi finance, which corresponds to the conventional notion of Ponzi finance, is defined as follows for our model.

Definition 11.1   *A weak Ponzi scheme*

The government runs a weak Ponzi scheme if

$$B_{t+1} \geq (1 + r_t)B_t \text{ for all } t \geq 0 \qquad (11.15)$$
$$B_0 > 0$$

The government engages in weak Ponzi finance if, in each period, t, the value of the debt carried into the next period, $t + 1$, is at least as large as the cost of servicing the debt carried into period t. From the government's single-period budget identity it follows, that a government engages in weak Ponzi finance if $G_t - (\tau_t^1 L_t + \tau_{t-1}^2 L_{t-1}) \geq 0$ for all $t \geq 0$, for $B_0 > 0$, that is, if it never runs a primary budget surplus despite there being a positive stock of public debt outstanding. An example would be, starting from an initial debt of zero at time $t = 0$ and a balanced primary budget, a one-period primary deficit during period 0, say through a lump-sum transfer payment to either or both generations alive during that period, followed by a return to a balanced primary budget forever after. The debt issued during period 0 is rolled over in perpetuity.

In Section 11.5 we are also interested in sequences of new debt *minus* old debt service, $\{B_{t+1} - (1 + r_t)B_t\}_{t=0}^{\infty}$ that, while not themselves weak Ponzi schemes, possess infinite subsequences $\{B_{t_j+1} - (1 + r_{t_j})B_{t_j}\}_{t_j=0}^{\infty}$ that are weak Ponzi schemes.

A *strong* Ponzi scheme is most easily explained by reconsidering our weak Ponzi scheme example. In a *weak* Ponzi scheme, the government issues a transfer to one or both generations alive in any one period (say $t = 0$) and does not (concurrently or subsequently) raise *aggregate* taxes or reduce *aggregate* transfer payments in any period. The government's exhaustive spending program is also left unchanged. That is,

$$T_t = \tau_t^1 L_t + \tau_{t-1}^2 L_{t-1} \text{ is unchanged for } t > 0 \qquad (11.16)$$

$$G_t \text{ is unchanged for } t > 0 \qquad (11.17)$$

In a *strong* Ponzi scheme, the government issues a transfer to one generation during one period (say the young during period 0) and does not (concurrently or subsequently) levy additional taxes on, or reduce transfer payments to, *any generation* (including the beneficiaries of the transfer payment) in any period. The government's exhaustive spending program is also left unchanged. A strong Ponzi scheme is therefore characterised as follows.

Definition 11.2 *A strong Ponzi scheme.*

A strong Ponzi Scheme satisfies the conditions for a weak Ponzi Scheme (equation 11.15). In addition, if τ_0^1 is cut and (11.17) holds, it satisfies (11.18)

$$\tau_t^1 \text{ is unchanged for } t \;\; > \;\; 0 \qquad (11.18)$$
$$\tau_t^2 \text{ is unchanged for } t \;\; \geq \;\; 0$$

Every strong Ponzi scheme is therefore also a weak Ponzi scheme, but not *vice-versa*.

We proceed by investigating what kind of constraints the model of Section 11.2 imposes on the government's ability to issue debt. Equation (11.10), stating that the savings of the young in period t equal the sum of the capital stock and the stock of government debt carried into period $t + 1$, can be rearranged as Equation (11.19)

$$B_{t+1} + \tau_t^1 L_t = -K_{t+1} + [w_t - c_t^1]L_t \qquad (11.19)$$

Equation (11.3) (holding with equality), stating that the old consume all their after-tax resources, can be arranged as Equation (11.20)

$$(1 + r_{t+1})B_{t+1} - \tau_t^2 L_t = -(1 + r_{t+1})K_{t+1} + c_t^2 L_t \qquad (11.20)$$

It is immediately obvious from (11.19) that, *for given* K_{t+1}, w_t, c_t^1 *and* L_t, the value of the public debt issued in period t, B_{t+1} can be made arbitrarily large (positive or negative) by making matching large (positive or negative) period t transfers to the young, $-\tau_t^1 L_t$. Such an arbitrarily large (positive or negative) value of B_{t+1} is consistent with Equation (11.20) *for given* r_{t+1}, K_{t+1}, c_t^2 *and* L_t, as long as period $t + 1$ taxes on the old, $\tau_t^2 L_t$ are assigned a matching large (positive or negative) value.

Since c_t^1 and K_{t+1} are non-negative, the constraint on public debt implied by (11.19) is $B_{t+1} + \tau_t^1 L_t \leq w_t L_t$. There also is a lower bound on the amount of public debt that can be issued (or an upper bound on the stock of public credit extended to the private sector). It follows from non-negativity of consumption by the old in period t. From the resource constraint $K_{t+1} - K_t = w_t L_t + r_t K_t - c_t^1 L_t - c_{t-1}^2 L_{t-1} - G_t$ and $c_{t-1}^2 \geq 0$ it follows that $(w_t - c_t^1)L_t - G_t + (1 + r_t)K_t - K_{t+1} \geq 0$. From (11.19) this implies $B_{t+1} + \tau_t^1 L_t \geq G_t - (1 + r_t)K_t$.

These upper and lower bounds on the public debt in each period, together with the requirement that exhaustive public spending cannot be negative and cannot exceed the total physical resources available in any period, constitute our characterisation of feasible fiscal-financial plans.[13]

Definition 11.3   *Feasible fiscal-financial plans.*

A government fiscal-financial plan is feasible if and only if its debt, taxes, transfer payments and exhaustive spending satisfy, for all $t \geq 0$, the single-period government budget identity given in (11.7) and

$$B_{t+1} + \tau_t^1 L_t \leq w_t L_t \qquad (11.21)$$

$$B_{t+1} + \tau_t^1 L_t \geq G_t - (1 + r_t)K_t \qquad (11.22)$$

Equations (11.21) and (11.22) plus non-negativity of G_t imply:

$$0 \leq G_t \leq w_t L_t + (1 + r_t)K_t \qquad (11.23)$$

Note that this definition of feasibility of fiscal-financial plans can be generalised easily to all OLG models with finite household horizons. It relies only on the reasonable postulate that in the last period of its life, each household disposes of all real and financial assets (including public debt), pays off any debts it has carried into that period and does not purchase any new assets or incur any new debt.

Equations (11.21) and (11.22) can be rewritten as

$$(1 + r_t)B_t - \tau_{t-1}^2 L_{t-1} \le w_t L_t - G_t \tag{11.24}$$

and

$$(1 + r_t)B_t - \tau_{t-1}^2 L_{t-1} \ge -(1 + r_t)K_t \tag{11.25}$$

Since the government feasibility constraints (11.21) and (11.22) are derived from the requirement that c_t^1, c_{t-1}^2 and $K_t \ge 0$ for all $t \ge 0$, another way of interpreting them is that the government refrains from policies that will bankrupt the private sector: it does not select sequences for taxes, transfer payments, debt and exhaustive spending that will cause the non-negativity constraints on consumption by both generations and on the private capital stock to become binding as long as there exist alternative policies for which these constraints do not become binding.[14]

The feasibility conditions (11.21) and (11.25) can be rewritten as

$$B_{t+1} + \tau_t^1 L_t \le [f(k_t) - k_t f'(k_t)]\theta_t (1 + n)^t \tag{11.26}$$

$$\tau_{t-1}^2 L_{t-1} - (1 + f'(k_t))B_t \le [1 + f'(k_t)]k_t \theta_t (1 + n)^t \tag{11.27}$$

Equation (11.26) implies that the long-run growth rate of the total resource transfer from the young generation to the government (whether through (voluntary) purchases by the young of government debt or through (involuntary) taxes on the young) cannot exceed the long-run growth rate of efficiency labour. Note that there is no constraint on B_{t+1} or $\tau_t^1 L_t$ separately, only on their sum.

Equation (11.27) implies that the long-run growth rate of the total resource transfer from the old generation to the government cannot exceed the long-run growth rate of efficiency labour. Note that there is no constraint on $\tau_{t-1}^2 L_{t-1}$ or on $-(1 + r_t)B_t$ separately, only on their sum.

If the long-run interest rate exceeds the long-run growth rate of efficiency labour ($\lim_{t\to\infty} \Delta_t \theta_t (1 + n)^t = 0$) then the feasibility constraints (11.21) and (11.22) imply

$$\lim_{t'-\infty} inf_{t' \leq t \leq \infty}\{\Delta_t[B_{t+1} + \tau_t^1 L_t]\} \leq 0$$

$$\lim_{t'-\infty} \sup_{t' \leq t \leq \infty}\{\Delta_t[B_{t+1} + \tau_t^1 L_t]\} \geq 0$$

If the limit inferior and the limit superior are the same, we get

$$\lim_{t-\infty} \Delta_t[B_{t+1} + \tau_t^1 L_t] = \lim_{t\to\infty} [\tau_{t-1}^2 L_{t-1} - (1 + r_t)B_t] = 0 \quad (11.28)$$

Note how this constraint differs from the conventional solvency constraint $\lim_{t\to\infty} \Delta_t B_{t+1} = 0$. Equation (11.28) states that the present discounted value of the total resource transfer from the young to the government and the present discounted value of the total resource transfer from the old to the government should converge to zero. Without further restrictions on τ_t^1 and τ_{t-1}^2, Equation (11.28) does not constrain the behaviour of the public debt or the public credit in the long run.

11.3 The feasibility of weak and strong Ponzi schemes with unrestricted taxes and transfers

In this Section we show that with unrestricted lump-sum taxes and transfers, weak Ponzi schemes are both always feasible and inessential, that is, they do not enlarge the set of allocations that can be supported as competitive equilibria.

Consider two equilibria, the single-star equilibrium and the double-star equilibrium. Assume that (11.29) holds for all $t \geq 0$, and that the initial capital stocks and the sequences of exhaustive public spending in the two equilibria are identical.

$$\tau_t^{1**} + \frac{\tau_t^{2**}}{1 + r_{t+1}} = \tau_t^{1*} + \frac{\tau_t^{2*}}{1 + r_{t+1}} \quad (11.29)$$

For concreteness, let the single-star equilibrium have a balanced budget in each period and zero public debt, that is $\tau_t^{1*}L_t + \tau_{t-1}^{2*}L_{t-1} = G_t$ and $B_t^* = 0$ for all $t \geq 0$. We define τ_t^{1**} and τ_{t-1}^{2**} as follows: $\tau_t^{1**} = \tau_t^{1*} - \epsilon_t$ and $\tau_{t-1}^{2**} = \tau_{t-1}^{2*} + (1 + r_t)\epsilon_{t-1}$

It follows that[15]

$$B_{t+1}^{**} - (1+r_t)B_t^{**} \equiv G_t - [\tau_t^{1**}L_t + \tau_{t-1}^{2**}L_{t-1}] \qquad (11.30)$$

$$= G_t - [(\tau_t^{1*} - \epsilon_t)L_t + (\tau_{t-1}^{2*} + (1+r_t)\epsilon_{t-1})L_{t-1}]$$
$$= [(1+n)\epsilon_t - (1+r_t)\epsilon_{t-1}]L_{t-1}$$

Thus, by choosing appropriately growing values for ϵ_t, $t \geq 0$, (that is values such that $\epsilon_t/\epsilon_{t-1} > (1+n)^{-1}(1+r_t)$), we can raise the growth rate of public debt in any period to any level. Since the present discounted value of lifetime taxes is the same in the single star and the double star equilibrium, the equilibrium private consumption sequences are the same, and so will be the wage rate, capital stock, interest rate and debt price sequences. By making a larger transfer to the young of generation t, the government provides the young with the means for increasing their saving. By levying a large tax on that same generation when old, the government provides the young with an incentive to save in order to pay these higher taxes.

Since $\tau_t^{1**}L_t = \tau_t^{1*}L_t - \epsilon_t L_t$ and also $\tau_{t-1}^{2**}L_{t-1} = -\tau_t^{*1}L_t + (1+r_t)\epsilon_{t-1}L_{t-1} + G_t$, we note that when a weak Ponzi game is played ($\epsilon_t/\epsilon_{t-1} > (1+n)^{-1}(1+r_t)$), the total tax on the young, $\tau_t^{1**}L_t$, will ultimately become negative and increasingly large in absolute value, while the tax on the old, $\tau_{t-1}^{2**}L_{t-1}$, will become an increasingly large positive number. We will therefore see the lifetime pattern of taxes becoming one of ever increasing receipts of transfer payments when young and ever increasing tax payments when old. The lifetime pattern of taxes therefore has to change sign or *zig-zag*. It is obvious that this property generalises to any finite household horizon OLG model.[16]

Another way of interpreting this is that the debt can grow without bound (and at a rate higher than the interest rate) without affecting the equilibrium allocations for consumption and the capital stock, because the government can, effectively, tax the debt held by the old to pay for the servicing of the debt held by the old. Government debt held by the old increases the 'base' on which lump-sum taxes on the old can be levied.[17]

Note that the ability to engage in weak Ponzi finance when the long-run interest rate exceeds the long-run natural rate of growth is restricted to the government. While private agents can make transfer payments (unrequited non-negative payments to others) only the government can impose taxes (unrequited positive transfers to itself).

McCallum (1984), made this point in the context of an infinite-lived representative agent model[18] (see also Bohn, 1991). Spaventa (1987, 1988) also emphasizes the distinction between models in which only endowments can be taxed (such as Pagano, 1988) and models in which interest income too is taxable. He, however, does not make the distinction between taxes on the young, taxes on the old and aggregate taxes. As we shall see below, aggregate taxes can be zero, and therefore less than the endowment (wage income) and less that the sum of the endowment and interest income, while the debt grows at a rate at least equal to the rate of interest forever.

The foregoing discussion suggests the following Proposition.

Proposition 11.1 *With unrestricted age-and time-contingent taxes and transfers, any equilibrium for consumption by the young and the old and for the capital stock, can be supported with an infinity of weak Ponzi schemes.*

Proof: We assert that, if there exists an equilibrium (the single star equilibrium, say) c_t^{1*}, c_{t-1}^{2*}, w_t^*, k_t^*, r_t^*, τ_t^{1*}, τ_t^{2*}, b_t^* for $t \geq 0$, for a given feasible sequence of exhaustive public spending, g_t, $t \geq 0$, then there also exist, for the same sequence of exhaustive public spending, (infinitely many) equilibria (the double star equilibria) c_t^{1**}, c_{t-1}^{2**}, w_t^*, k_t^{**}, r_t^{**}, τ_t^{1**}, τ_t^{2**}, b_t^{**} for $t \geq 0$ such that $c_t^{1**} = c_t^{1*}$, $c_{t-1}^{2**} = c_{t-1}^{2*}$, $w_t^{**} = w_t^*$, $k_t^{**} = k_t^*$, $r_t^{**} = r_t^*$ and $(1+n)(1+\omega_{t+1})b_{t+1}^{**} - b_t^{**}(1+r_t^{**}) \geq 0$ for all $t \geq 0$

Given k_0 and b_0, a competitive equilibrium of our two-period OLG model satisfies Equations (11.31) to (11.34) for all $t \geq 0$.

$$v'(c_t^1) = \beta(1 + f'(k_{t+1}))v'(c_t^2) \qquad (11.31)$$

$$\theta_t[f(k_t) - k_t f'(k_t)] - \tau_t^1 - c_t^1 - \frac{c_t^2 + \tau_t^2}{1 + f'(k_{t+1})} = 0 \qquad (11.32)$$

$$(\theta_t[f(k_t) - k_t f'(k_t)] - \tau_t^1 - c_t^1)\theta_t^{-1} = (k_{t+1} + b_{t+1})(1+n)(1+\omega_{t+1}) \qquad (11.33)$$

$$b_{t+1}(1+n)(1+\omega_{t+1}) \equiv (1 + f'(k_t))b_t + g_t - \theta_t^{-1}[\tau_t^1 + (1+n)^{-1}\tau_{t-1}^2] \qquad (11.34)$$

For the double star taxes and debt to support the same consumption and capital stock equilibrium, it is necessary and sufficient that for all $t \geq 0$.

$$(1+n)(1+\omega_{t+1})^{-1}b_{t+1}^* + \theta_t^{-1}\tau_t^{1*} = (1+n)(1+\omega_{t+1})^{-1}b_{t+1}^{**} + \theta_t^{-1}\tau_t^{1**} \qquad (11.35)$$

$$(1+f'(k_t))b_t^* - (1+n)^{-1}\theta_t^{-1}\tau_{t-1}^{2*} = (1+f'(k_t))b_t^{**} - (1+n)^{-1}\theta_t^{-1}\tau_{t-1}^{2**}$$
$$\text{(11.36)}$$

For the debt to grow at least as fast as the rate of interest forever, it must be true that for all $t \geq 0$

$$(1+n)(1+\omega_{t+1})b_{t+1}^{**} - (1+f'(k_{t+1}))b_t^{**}$$
$$= g_t - \theta_t^{-1}[\tau_t^{1**} + \tau_{t-1}^{2**}(1+n)^{-1}] \geq 0 \qquad \text{(11.37)}$$

The two choice variables during period t in equation (11.37) are τ_t^{1**} and τ_{t-1}^{2**}. No matter what value is assigned to τ_{t-1}^{2**}, τ_t^{1**} can always be assigned a large enough negative value to ensure that (11.37) is satisfied: the debt grows at least as fast as the rate of interest.

From the single-period government budget identity, it follows that (11.35) and (11.36) are the same constraint. No matter what value is assigned to τ_t^{1**}, a value can be assigned to τ_{t-1}^{2**} that ensures that (11.35) and (11.36) are satisfied for any values of τ_t^{1*}, τ_{t-1}^{2*}, b_t^*, b_{t+1}^*, b_t^{**} and b_{t+1}^{**}. ∎

Another way of putting this is that, by increasing $-\tau_t^{1**}L_t$ for any given values of $\tau_{t-1}^{2**}L_{t-1}$ and G_t and for any inherited value of $(1+f'(k_t))B_t^{**}$, it is possible to raise the growth rate of the public debt to any positive level without affecting $B_{t+1}^{**} + \tau_t^{1**}L_t$, the term on the left-hand side of the fiscal-financial feasibility constraints (11.21) and (11.22). $\tau_{t-1}^{2**}L_t$ can then be chosen to ensure that $(1+f'(k_t))B_t^{**} - \tau_{t-1}^{2**}L_{t-1}$, the term on the left-hand side of (11.24) and (11.25) satisfies these inequalities. The government simply reshuffles a constant total resource transfer away from the young in period t, $B_{t+1}^{**} + \tau_t^{1**}L_t$, between borrowed resources, B_{t+1}^{**}, and explicit taxes, $\tau_t^{1**}L_t$. Appropriating for its own use an amount of resources equal to the value of exhaustive public spending, G_t, it pays out the remainder to the old, either as debt service, $(1+f'(k_t))B_t^{**}$ or as transfers $-\tau_{t-1}^{2**}L_t$. Again it is only the total, $(1+f'(k_t))B_t^{**} - \tau_{t-1}^{2**}L_t$ that matters for the consumption of the old.

Figure 11.1 illustrates how any equilibrium can be supported with an infinity of weak Ponzi schemes.[19] Consider an equilibrium life-time consumption bundle $(\hat{c}^1, \hat{c}^2)$ at E_2 in Figure 11.1. Without loss of generality, we assume that this consumption bundle is supported by the endowment bundle $(w, 0)$, that is, with zero taxes and transfers in both periods. It is clear that the same consumption equilibrium will also be supported by any endowment bundle $(w - \tau^1, -\tau^2)$ provided $\tau^1(1+r) + \tau^2 = 0$. In Figure 11.1, we illustrate this with the endowment bundle $(w - \tau_d^1, 0 + \tau_d^1[1+r])$ at E_1

290   *Willem H. Buiter and Kenneth M. Kletzer*

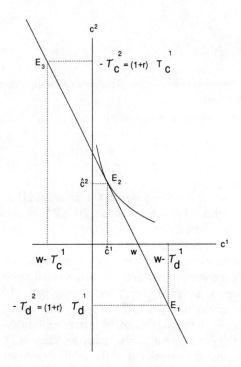

Figure 11.1: Equilibrium

where τ_d^1 is negative and with the endowment bundle $(w - \tau_c^1, 0 + \tau_c^1[1+r])$ at E_3 where τ_c^1 is positive (and sufficiently large in magnitude to make the after-tax period 1 endowment negative). The positive transfer to the young at E_1 could grow from any generation to the next at any rate. As long as the positive tax on the old grows at that same rate and thus continues to satisfy $\tau^1(1+r) + \tau^2 = 0$, the endowment point will move in a south-easterly direction along the intertemporal budget constraint through $E_3 E_2 E_1$, without the equilibrium consumption bundle (or any other equilibrium condition) being affected. Weak Ponzi games involving government debt corresponds to such movement along $E_3 E_2 E_1$ in a south-easterly direction. Without loss of generality, assume the equilibrium corresponding to the endowment $(w, 0)$ to have had zero public debt. In that case in all equilibria with non-zero lump-sum taxes and transfers, the stock of debt is equal to $B_{t+1} = -(1+n)^t \tau_t^1$. Weak Ponzi games involving government credit correspond to movements along $E_3 E_2 E_1$ in a northwesterly direction. In Figure 11.1, at E_3 there is a positive tax on the young $(\tau_c^1 > w > 0)$ and a positive transfer payment to the old of equal present value $(\tau_c^2 = -\tau_c^1(1 + r))$.

The Corollaries given below follow immediately from Proposition 11.1.

Corollary 11.1 *With unrestricted taxes and transfers, the competitive equilibrium model with the finite-lived OLG household sector does not require any bounds on the level or rate of growth of public debt. Weak Ponzi schemes are therefore always possible, regardless of the relationship between the interest rate and the growth rate and regardless of whether the economy is dynamically efficient.*

Corollary 11.2 *With unrestricted taxes and transfers, weak Ponzi finance is 'inessential,' that is, it does not permit additional equilibria to be supported.*

11.4 Restrictions on taxes and transfers and the conventional government solvency constraint.

Without restrictions on the government's ability to choose lump-sum taxes and transfers, the fiscal-financial feasibility constraints (11.21) and (11.22) do not imply the conventional government solvency constraint given by either (11.13) or (11.14) and allow weak Ponzi finance. In this section we consider restrictions on the government's set of available lump-sum taxes and transfers that imply that the conventional government solvency constraint applies. These restrictions rule out weak Ponzi finance whenever the interest rate exceeds the growth rate. The definition of strong Ponzi finance turns out to involve a restriction on the government's ability to vary lump-sum taxes and transfers over the life cycle that rules out strong Ponzi finance whenever the interest rate exceeds the growth rate. Other restrictions, including restrictions on the extent to which taxes and transfers can vary across generations alive at the same time, are shown in the next section to allow (essential) weak Ponzi finance, even if the interest rate exceeds the growth rate, but rule out strong Ponzi finance whenever the interest rate exceeds the growth rate.

The first restriction we consider is that not all taxes and transfers are lump-sum.

11.4.1 Distortionary taxes

If the taxes paid or transfers received by a generation are distortionary, it is unlikely that the long-run growth rate of per capita taxes or transfers can exceed the long-run growth rate of productivity. There will be some finite upper bound on the ratio of taxes and transfers per generation per period to the before-tax resources owned by that generation that period. Tax administration and collection costs that are strictly convex increasing functions of the amount of taxes raised will also put a finite upper bound on the ratio of taxes paid to real resources owned (see Barro, 1976, McCallum, 1984, Kremers, 1989and Bohn, 1991).

Assume that the young continue to be taxed in lump-sum fashion. The old in period $t + 1$ are taxed through a proportional tax at rate α_{t+1} on the gross return to their savings (their gross resources in old age), that is,

$$\tau_t^2 L_t = \alpha_{t+1}(1 + r_{t+1})(K_{t+1} + B_{t+1}) \tag{11.38}$$

This tax is equivalent to a proportional tax (at a rate $\alpha_{t+1}/(1 - \alpha_{t+1})$) on consumption by the old, since $\tau_t^2 = \frac{\alpha_{t+1}}{(1-\alpha_{t+1})}c_t^2$. By including (gross) debt income, $(1 + r_{t+1})B_{t+1}$, in the tax base, we do not prejudge the issue of whether taxes can grow faster than GDP forever, as we would if only the gross income from physical capital, $(1 + r_{t+1})K_{t+1}$, were taxed. We can safely restrict the analysis to the case where before-tax resources of the old in period $t + 1$ are non-negative and $\alpha_{t+1} \leq 1$. When $\alpha_{t+1} = 1$, $c_t^2 = 0$ and members of generation t don't save anything, so $K_{t+1} = 0$ and the economy collapses.

The first-order condition for the household can be written as

$$v'(c_t^1) = (1 - \alpha_{t+1})(1 + f'(k_{t+1}))\beta v'(c_t^2) \tag{11.39}$$

Substituting (11.38) and (11.39) into the life-time budget constraint of a member of generation t (assumed to hold with strict equality) yields

$$c_t^1 + \frac{c_t^2 \beta v'(c_t^2)}{v'(c_t^1)} = c_t^1 + c_t^2 D_{t+1} = w_t - \tau_t^1 \tag{11.40}$$

where $D_{t+1} \equiv [(1 - \alpha_{t+1})(1 + r_{t+1})]^{-1}$ is the (after-tax) market discount factor.

Note that (11.40) implies that, if τ_t^1 is positive (positive taxes are levied on the young), it cannot grow faster, in the long run, than w_t and therefore no faster than the natural rate of growth of the economy. With c_t^1, c_t^2 and v' non-negative, (11.40) implies $\tau_t^1 \leq w_t$. Weak Ponzi finance with growing public credit and long-run subsidies to saving ($\alpha_{t+1} < 0$) is infeasible if the *after-tax* long-run rate of interest (equal to the after-tax rate of return to capital in the deterministic model) exceeds the long-run growth rate of efficiency labour. Weak Ponzi finance with ever-growing transfers to the young (negative values of τ_t^1) is also infeasible, even with the tax rate α approaching 1, given our specification of the utility function, if the after-tax long-run interest rate exceeds the long-run growth rate of efficiency labour.

The proof is by contradiction. Suppose the government runs a financing scheme such that τ_t^1 is negative and growing in absolute value at a rate in excess of the natural rate. We rewrite (11.40) in growth rates as follows:

$$
a_t^1 \left(\frac{c_{t+1}^1 - c_t^1}{c_t^1} \right) + (1 - a_t^1) \left(\frac{c_{t+1}^2 - c_t^2}{c_t^2} \right) \tag{11.41}
$$

$$
+ (1 - a_t^1) \frac{c_{t+1}^2}{c_t^2} \left(\frac{D_{t+1} - D_t}{D_{t+1}} \right)
$$

$$
= \left(\frac{w_t}{w_t - \tau_t^1} \right) \left(\frac{w_{t+1} - w_t}{w_t} \right) + \left(\frac{-\tau_t^1}{w_t - \tau_t^1} \right) \left(\frac{-\tau_{t+1}^1 - (-\tau_t^1)}{-\tau_t^1} \right)
$$

where $a_t^1 \equiv \frac{c_t^1}{w_t - \tau_t^1}$ is the share of expenditure on c_t^1.

Since $-\tau^1$ grows at a rate in excess of the natural rate, the c^1 intercept of the household budget constraint $(w - \tau^1)$ asymptotically grows as fast as $-\tau^1$. If D grows fast enough to prevent c^2 from growing faster than the natural rate, the relative price of c^1 in terms of c^2 approaches zero. Our utility function is strictly concave with positive marginal utilities of both c^1 and c^2 for all non-negative consumption bundles. It follows that c^2 grows asymptotically at a rate exceeding the natural rate.[20] Note that even if c^2 were not to become unbounded when its price is zero, saving by the young would be zero and consequently the capital stock would be zero and the economy would collapse. Of course, if our economy were an endowment economy, zero saving by the young would not imply the collapse of the economy. Therefore, weak Ponzi finance is infeasible when the long-run after-tax interest rate exceeds the long-run natural rate of growth in this deterministic example.[21]

Figure 11.2 shows what would be required for weak Ponzi finance to be feasible under our period 2 consumption tax when the after-tax rate of interest exceeds the long-run natural rate of growth. For simplicity assume the natural rate of growth is zero. As the transfer to the young grows from $-\tau_t^1$ to $-\tau_t^{1'}$, the after-tax rate of return to saving has to fall (the tax-inclusive price of period 2 consumption has to rise). At the very least the (absolute value of the) slope of the intertemporal budget constraint $V'(w - \tau_t^{1'})$ has to be small enough relative to the slope of $V(w_t - \tau_t^1)$ to ensure that V' is

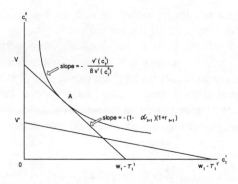

Figure 11.2: After-tax rate of interest exceeds long-run natural rate of growth

below V. If not, the new budget set would dominate the old one and the consumption of at least one good would be growing faster than the natural rate. However, as τ^1 continues to grow faster than the natural rate, the relative price of period 1 consumption falls to zero and c^2, the tax base of the old, declines to zero with it. Weak Ponzi finance is infeasible if the after-tax long-run interest rate exceeds the long-run growth rate of efficiency labour.

Note that by changing the specification of the utility function, it is possible to restore the possibility of weak Ponzi finance (when the after-tax interest rate exceeds the growth rate) even for the distortionary tax scheme considered here. Consider for instance the case where there is a positive minimum subsistence level of consumption in each period, $c_t^i > \underline{c}_t^i > 0$, $i = 1$, 2. The utility functions would have the property that $\lim_{c_t^i \downarrow \underline{c}_t^i} v'(c_t^i) = +\infty$, $i = 1, 2$. (The utility function continues to be increasing, strictly concave and twice continuously differentiable for $c^1 > \underline{c}^1$ and $c^2 > \underline{c}^2$). The limiting indifference curve going through the consumption bundle $(\underline{c}^1, \underline{c}^2)$ would be Leontief. In that case, even as the price of period 2 consumption rises without bound, period 2 consumption does not go to zero. Once period 2 consumption is at its lower bound, $\underline{c}^2$, (which will only happen if period 1 consumption is also at its lower bound as long as the relative price of period 1 consumption is non-zero) the period 2 consumption tax becomes a lump-sum tax, and weak Ponzi finance is possible. As there no longer is any substitution, further increases in the wealth tax impose no further distortions. As $-\tau_t^1$ grows from generation to generation, the entire increment in period 1 transfers is saved, despite the steadily worsening rate of return on saving, in order to keep period 2 consumption from falling below the subsistence level.[22]

The entire discussion thus far has been restricted to equilibria in which the after-tax rate of interest exceeds the natural rate of growth. In a recent paper Uhlig (1996) considers OLG models in which capital income taxation of the kind considered in this Section can drive the after-tax rate of interest (and the after-tax marginal product of capital) below the natural rate of growth[23] even when the before-tax rate of interest (marginal product of capital) is above the natural rate of growth and the economy is therefore not dynamically inefficient. Weak Ponzi schemes and strong Ponzi schemes are of course feasible whenever the after-tax rate of interest is below the natural rate of growth.

11.4.2 No 'sign reversal' in the net tax burden over the life cycle

The second restriction we consider is that net taxes in each of the two periods of a household's life must have the same sign:

$$\tau_t^2 = \lambda_t \tau_t^1 \qquad \lambda_t \geq 0 \text{ for all } t \geq 0. \qquad (11.42)$$

While this restriction may seem somewhat artificial, it covers many of the tax-transfer patterns actually found in the literature. These include (a) taxes on the young only (that is, $\lambda = 0$) and (b) common taxes on young and old (that is, $\lambda = 1$). It also covers the fiscal restrictions contained in the definition of a strong Ponzi scheme. We need just one of the weak inequalities of the fiscal-financial feasibility constraint in order to show that, under restriction (11.42), Ponzi finance is possible only if the long-run interest rate[24] is below the long-run growth rate of efficiency labour. Consider equations (11.21) and (11.24), rewritten for this case as

$$B_{t+1} \leq [f(k_t) - k_t f'(k_t)]\theta_t L_t - \tau_t^1 L_t \qquad (11.43)$$

$$(1 + f'(k_{t+1}))B_{t+1} \leq [f(k_{t+1}) - k_{t+1} f'(k_{t+1})]\theta_{t+1} L_{t+1} + \lambda_t \tau_t^1 L_t - G_{t+1} \qquad (11.44)$$

From (11.43), the only way for the debt to grow faster than the growth rate of efficiency labour forever, is for τ_t^1 to be negative and for $-\tau_t^1$ to grow at a rate higher than the growth rate of labour productivity. If the debt grows faster than the growth rate of efficiency labour forever, (11.44) can only be satisfied if $\lambda_t \tau_t^1 L_t$ is positive and has a growth rate higher than the growth rate of efficiency labour. That is impossible since $\lambda_t \geq 0$. We conclude that $-\tau_t^1$ can grow no faster than the growth rate of labour productivity and that the growth rate of the debt can therefore be no higher than the growth rate of efficiency labour. The debt can therefore grow faster than the interest rate forever only if the long-run interest rate is below the long-run growth rate of efficiency labour.

Blanchard and Weil (1992) considered the special case of the model with (11.42) where the labour force is constant $(n = 0)$, there is no productivity growth $(\theta_t = \theta$ for all $t \geq 0)$, there is no exhaustive public spending $(G_t = 0$ for all $t \geq 0)$ and there are no taxes or transfers $(\tau_t = 0$ for all $t \geq 0)$. They consider whether, starting from a zero initial public debt, a (small) increase in the stock of public debt can be rolled over forever. They therefore consider a *strong* Ponzi scheme. In their model, debt obviously cannot grow faster than wage income in the long run. In the deterministic version of their model, this means that only in a dynamically inefficient equilibrium (with the interest rate below the natural rate of growth) can there be viable weak Ponzi schemes (and strong Ponzi schemes), with the public debt growing forever at least as fast as the interest rate but no faster than the growth rate of labour income.[25]

Even if net taxes can change sign over the life cycle, if the long-run rate of interest exceeds the long-run growth rate of the *disposable* income of the young, that is, if $\lim_{t\to\infty}\{\Delta_t[\theta_t(f(k_t) - k_t f'(k_t)) - \tau_t^1](1 + n)^t\} = 0$, equations (11.21) or (11.24) imply

$$\lim_{t'\to\infty} inf_{t'\leq t\leq\infty}\{\Delta_t B_{t+1}\} \leq 0 \qquad (11.45)$$

From the public credit constraint (11.22) or (11.25) it follows that, if the long-run rate of interest exceeds the rate of growth of the *disposable* income of the old (capital income minus taxes on the old), that is, if $\lim_{t\to\infty}\{\Delta_t[(1 + f'(k_t))k_t\theta_t - (1 + n)^{-1}\tau_{t-1}^2](1 + n)^t\} = 0$, we have

$$\lim_{t'\to\infty} sup_{t'\leq t\leq\infty}\{\Delta_t B_{t+1}\} \geq 0. \qquad (11.46)$$

If $\lim_{t'\to\infty} inf_{t'\leq t\leq\infty}\{\Delta_t B_{t+1}\} = \lim_{t'\to\infty} sup_{t'\leq t\leq\infty}\{\Delta_t B_{t+1}\} = 0$, then we also have $\lim_{t\to\infty} \Delta_t B_{t+1} = 0$. This means that when the long-run rate of interest exceeds the long-run growth rate of the *disposable* income of the young and of the old, the conventional solvency constraint applies.

We summarise this discussion in the following Proposition.

Proposition 11.2 *Strong Ponzi schemes are feasible only if the long-run (after-tax) rate of interest is below the long-run natural rate of growth.*

Proposition 11.3 *The conventional government solvency constraint (which was* $\lim_{t\to\infty} \Delta_t B_{t+1} = 0$ *)* [26] *is implied by the fiscal-financial feasibility constraint (11.21), (11.22) and (11.23) if **(a)** The long-run interest rate exceeds the long-run growth rate of efficiency labour* ($\lim_{t\to\infty} \Delta_t \theta_t (1+n)^t = 0$) *and **(b)** Either, the net tax paid by a generation at any age cannot change sign over the lifetime of that generation, or the long-run growth rate of taxes paid or transfers received at any given age by a generation is less than the long-run natural rate of growth. Globally distortionary taxes* [27] *are sufficient to prevent the long-run growth rate of taxes and transfers from exceeding the long-run natural rate of growth.*

In order for the public debt to grow at least as fast as the rate of interest forever, when the long-run rate of interest is above the long-run growth rate of efficiency labour, it must be possible to make transfer payments to a generation when it is young and to tax it when it is old; in addition, the growth rate of these taxes and transfers must be at least as high as the interest rate.[28] Note that aggregate taxes net of transfers, $\tau_t^1 L_t + \tau_{t-1}^2 L_t$, need not grow at all and can indeed be equal to zero.

11.5 Inessential and essential Ponzi finance and public debt

We first restate, without a proof, a familiar equivalence or irrelevance result, due to Wallace (1981), Chamley and Polemarchakis (1984) and Sargent (1987). It generalises the familiar proposition that an equilibrium with positive public debt financed with taxes on the young is equivalent to a balanced budget, pay-as-you-go (or unfunded) social security retirement scheme in which lump-sum taxes on the young are paid out as lump-sum transfers to the old (see also Calvo and Obstfeld, 1988a,b). This Proposition implies that, in a deterministic model, any intergenerational redistribution that can be supported by debt, taxes and transfers can also be supported just with taxes and transfers and without debt. In the stochastic case it implies that any intergenerational insurance scheme supported with public debt, unbalanced budgets and unrestricted lump-sum taxes and transfers can also be provided with a balanced budget and without public debt.

Proposition 11.4 *Given initial values k_0 and b_0 and a feasible sequence g_t, $t \geq 0$, any equilibrium for k_t, c_t^1 and c_t^2 with arbitrary paths of debts b_t and of lump-sum taxes and transfers τ_t^1 and τ_t^2 for all $t \geq 1$ can be replicated without debt and deficits, that is, by using balanced-budget lump-sum taxes and transfers only.*

Proposition 11.4 implies that with unrestricted lump-sum taxes and transfers, public debt is redundant or 'inessential.'

We next consider three simple restrictions on how taxes and transfer can differ in any period across overlapping generations:

(a) Equal taxes per generation for all generations alive during any given period, that is, $\tau_t^1(1+n) = \tau_{t-1}^2$ for all $t \geq 0$.

(b) Equal per capita taxes for all generations alive during any given period, that is, $\tau_t^1 = \tau_{t-1}^2$ for all $t \geq 0$.

(c) Equal taxes per unit of efficiency labour for all generations alive during any given period, that is, $\tau_t^1 = (1 + \omega_t)\tau_{t-1}^2$ for all $t \geq 0$.

The next Proposition states that any intergenerational redistribution and intergenerational insurance supported with a balanced-budget and unrestricted lump-sum taxes and transfers, can also be supported with lump-sum taxes and transfers constrained to fall equally in per capita terms on all overlapping generations (case (b)), provided the public sector budget can be unbalanced. From Proposition 11.4, there is no loss of generality in taking the benchmark equilibrium of Proposition 11.5 to have a balanced budget and zero public debt. The proof of case (b) can be extended easily to the other two cases.

What drives these results is that even though the two generations (the young and the old) alive in any given period are treated in the same way during that period, we can still vary the present discounted value of lifetime taxes and transfers freely for each generation, making transfer payments to them while young and taxing them when old.

Proposition 11.5   *Given an initial value k_0 and a sequence of exhaustive public spending g_t, $t \geq 0$, any equilibrium for k_t, c_t^1 and c_{t-1}^2 for all $t \geq 0$ supported by unrestricted lump-sum taxes and transfers but without public debt and with balanced public sector budgets, can also be supported with equal per capita lump-sum taxes or transfers for both generations alive in any given period, provided unbalanced public sector budgets are allowed.*

Proof: Variables with single stars represent the benchmark balanced-budget policy with age-dependent taxes and transfers. Variables with double stars represent the age-independent tax/transfer case with an unbalanced budget. Note that $b_t^* = 0$, $\tau_t^{1*} = -(1+n)^{-1}\tau_{t-1}^{2*} + \theta_t g_t$ and $\tau_{t-1}^{2**} = \tau_t^{1**} = \tau_t^{**}$ for all $t \geq 0$.

From Equation (11.32) it follows that, if equivalence holds, it must be true that

$$b_{t+1}^{**} = (1 + f'(k_{t+1}))^{-1}\theta_{t+1}^{-1}[\tau_{t+1}^{1*} + \tau_{t+1}^{**}(1+n)^{-1} - \theta_{t+1}g_{t+1}] \quad (11.47)$$

From (11.47) and the government's single-period budget identity, (or equivalently from the economy-wide capital market equilibrium condition (11.33)), it follows that, if the double star regime supports the same equilibria as the single star regime, it must be true that (11.48) holds.

$$b_{t+1}^{**} = (1+n)^{-1}\theta_{t+1}^{-1}[\tau_t^{1*} - \tau_t^{**}] \tag{11.48}$$

For any τ_t^{1*} and τ_t^{**} it is clear that values of τ_{t+1}^{**} and $p_t b_{t+1}^{**}$ can be found to satisfy (11.47) and (11.48). The other equilibrium condition (11.31) is also satisfied under the double star regime. The solvency constraint under the single star regime is

$$\theta_t^{-1}\tau_t^{1*} \le f(k_t) - k_t f'(k_t) \tag{11.49}$$

and

$$\theta_t^{-1}\tau_t^{1*} \ge g_t - (1+\rho_t)k_t \tag{11.50}$$

Under the double star regime the solvency constraint is

$$b_{t+1}^{**} + (1+n)^{-1}\theta_{t+1}^{-1}\tau_t^{1**} \le f(k_t) - k_t f'(k_t) \tag{11.51}$$

and

$$b_{t+1}^{**} + (1+n)^{-1}\theta_{t+1}^{-1}\tau_t^{1**} \ge g_t - (1+\rho_t)k_t \tag{11.52}$$

It is clear from (11.48) that if the solvency constraint is satisfied under the single star regime ((11.49) and (11.50) hold), then it will also be satisfied under the double star regime ((11.51) and (11.52) hold). ∎

We have the following Corollary to Proposition 11.5:

Corollary 11.3 *Ponzi finance is essential when lump-sum per capita taxes and transfers are restricted be equal for overlapping generations.*

Proof: Equations (11.47) and (11.48) imply

$$\tau_{t+1}^{**} = -(1+r_{t+1})\tau_t^{**} + (1+r_{t+1})\tau_t^{1*} - (1+n)\tau_{t+1}^{1*} + (1+n)\theta_{t+1}g_{t+1} \tag{11.53}$$

Note that the homogeneous part of equation (11.53) changes sign each period (imparting a saw-tooth pattern) and grows at a proportional rate $1+r$

in absolute value. The saw-tooth pattern of tax receipts is passed on to the value of the per capita debt through the government budget identity under the double star policy, given in (11.54) below.

$$
\begin{aligned}
b_{t+1}^{**} = {} & (1+\omega_{t+1})^{-1}(1+n)^{-1}(1+f'(k_t))b_t^{**} \\
& +(1+\omega_{t+1})^{-1}(1+n)^{-1}g_t \\
& -(1+\omega_{t+1})^{-1}(1+n)^{-2}\theta_t^{-1}(2+n)\tau_t^{**}
\end{aligned} \tag{11.54}
$$

Equations (11.48) and (11.54) imply

$$
B_{t+1}^{**} = -(1+n)(1+f'(k_t))B_t^{**} - (1+n)G_t + (2+n)\tau_t^{1*}L_t \tag{11.55}
$$

The value of the public debt under the age-independent tax, unbalanced budget policy, B_{t+1}^{**}, is likely to zig-zag from a positive value in one period to a negative value in the next. If, for instance, τ_t^{1*} and L_t were constant over time, the saw-tooth pattern of the public debt, with debt in the homogeneous equation of (11.55) having a growth factor of $-(1+r_t)$ each period, is immediately apparent. Calvo and Obstfeld (1988b) noted such a pattern in an economy without population growth or productivity growth.

Over a two-period horizon, the public debt evolves according to

$$
\begin{aligned}
B_{t+2} = {} & (1+n)^2(1+f'(k_{t+1}))(1+f'(k_t))B_t \\
& -(1+n)[G_{t+1} - (1+n)(1+f'(k_{t+1}))G_t] \\
& +(2+n)[\tau_{t+1}^{1*} - (1+f'(k_{t+1}))\tau_t^{1*}]L_{t+1}
\end{aligned} \tag{11.56}
$$

Consider the simple example where $G_{t+1} = G_t = 0$ and $\tau_{t+1}^{1*} = \tau_t^{1*} = \tau^{1*}$ for all $t \geq 0$. Equation (11.56) simplifies to

$$
B_{t+2}^{**} = (1+n)^2(1+f'(k_{t+1}))(1+f'(k_t))B_t^{**} - (2+n)f'(k_{t+1})\tau^*L_{t+1}
$$

When τ^* is negative and r_{t+1} is non-negative, the public debt will, over a two-period horizon grow at a proportional rate at least equal to the sum of the real interest rate and the growth rate of population. If n is non-negative, the sequence of the public debt will therefore have infinite subsequences that are characterised by weak Ponzi finance. Public credit too will, over a two-period interval, grow at a rate asymptotically equal to the sum of the interest rate and the growth rate of population. This proves that 'subsequence weak Ponzi finance' is 'essential' in this case. ∎

Analogous results to Proposition 11.5 hold under the restriction that taxes per generation alive in any given period are equal and under the restriction that taxes per efficiency unit of labour in any given period are equal. Weak Ponzi finance is also essential in these cases.

The next proposition presents an example where public debt is essential, but weak Ponzi finance is infeasible, unless the long-run rate of interest is below the long-run growth rate of efficiency labour.

Proposition 11.6   *Given an initial value k_0 and a sequence of exhaustive public spending g_t, $t \geq 0$, any equilibrium for k_t, c_t^1 and c_{t-1}^2 for all $t \geq 0$ that can be supported with a balanced budget and unrestricted lump-sum taxes and transfers, can also be supported with taxes net of transfers that are required to have the same sign during the lifetime of each generation, provided unbalanced budgets are allowed.*

Proof: see Buiter and Kletzer (1996).

Proposition 11.4 states that public debt and deficits (and by implication weak Ponzi finance) are redundant policy instruments as long as the fiscal authority has unrestricted lump-sum taxes and transfers. Proposition 11.5 and Proposition 11.6 emphasize that a fiscal authority with a restricted tax-transfer instrumentarium may be able to use public debt and deficits as perfect substitutes for the missing age-specific taxes and transfers, provided the government's fiscal-financial feasibility constraint is specified as in (11.21), (11.22) and (11.23). Essential (subsequence) weak Ponzi finance may be a feature of these government borrowing and lending strategies.

11.6   Conclusion

Our results fall into two categories. The first contains a number of propositions that bring out how the government's ability to issue debt is constrained by its 'capacity to tax.' The second contains a number of propositions about how the government's ability to issue debt may expand the set of equilibria that can be supported when taxes are restricted.

Our approach to the feasibility of fiscal-financial plans has implications for the empirical approaches to testing for government solvency (see for example Hamilton and Flavin, 1986, Wilcox, 1989, Corsetti, 1990, Grilli, 1990, Trehan and Walsh, 1991and Buiter and Patel, 1992 and 1994). All these papers use variants of the conventional solvency criterion and study the long-run behaviour of the discounted public debt; typically, they test whether the (expectation of the) present discounted value of the future stock of public debt goes to zero in the limit. The conventional solvency criterion is neither necessary nor sufficient for our fiscal-financial feasibility conditions to be satisfied.

Our analysis suggests that the conventional test for government solvency are too weak if the long-run interest rate exceeds the long-run natural rate of growth and if the long-run growth rate of taxes and transfers cannot exceed the long-run growth rate of efficiency labour (which will be the case when there are distortionary taxes and/or transfers, or when there are strictly convex tax collection and transfer administration costs). While the conventional solvency constraint holds under these conditions, a more restrictive necessary condition for feasibility of the fiscal-financial plan also applies. This is that the growth rate of the public debt cannot indefinitely exceed the natural rate of growth. It would therefore be more informative to investigate the stationarity of, say, the public debt-GDP ratio than the stationarity of the discounted public debt.

The model considered in this paper has complete markets, even though, owing to the OLG demographic structure, there is incomplete market participation. Introducing incomplete markets, reflecting the existence of asymmetric information, would add an important dimension to the role of public debt. It would further qualify propositions about the redundancy of public debt. With asymmetric information about which private agents are constrained in their current consumption decision by current disposable resources rather than just by permanent income, government borrowing can be used as an 'information revelation device.' Because the government cannot differentiate between the two types of potential taxpayers, all private agents must be taxed equally. Only private agents that are not constrained in their current consumption by current disposable resources will *voluntarily* purchase additional government debt. Taxes (assuming they don't exceed currently disposable resources) will be paid (*involuntarily*) by both permanent income-constrained and current disposable resource-constrained agents. This distinction between taxes, that is, unrequited involuntary payments to the government, and government borrowing, that is, voluntary *quid-pro-quo* resource transfers to the government through the purchase of public debt by private agents, is central to an understanding of the cyclical stabilisation role of government borrowing.

Acknowledgments

The authors want to acknowledge financial support from Ford Foundation Grant Number 920-0109, 'Economics, Politics and Policies of Stabilisation, Structural Adjustment and Long-term Development,' administered by the

Economic Growth Center of Yale University. We are indebted to Pat Kehoe, other participants in the CEPR's Third International Macroeconomics Programme Meeting in Sesimbra Portugal, and to Costas Azariadis, Lans Bovenberg, Bill Brainard, Bill Nordhaus, Chris Sims, Nouriel Roubini and Harald Uhlig for helpful comments and suggestions. Willem H. Buiter is of the University of Cambridge, NBER and CEPR. Kenneth M. Kletzer is at University of California Santa Cruz.

Notes

1. According to the conventional definition, a government engages in Ponzi finance if, after some date, it never runs a primary (non-interest) budget surplus despite having a positive stock of debt outstanding. Equivalently, the value of the additional debt issued each period is at least as large as the interest payments made on the debt outstanding at the beginning of that period, or the growth rate of the debt each period is equal to or greater than the one-period interest rate on public debt.

2. With distortionary (non-lump-sum) taxes and transfers, real equilibrium allocations will almost always be affected by the ability, offered by unbalanced budgets, to vary the pattern over time of the excess burdens associated with the use of distortionary instruments. The same is true when there are tax collection or transfer administration costs that are increasing and strictly convex functions of marginal or average tax rates. See, for example, Barro (1979).

3. In the version of the model with uncertainty (see Buiter and Kletzer, 1996) taxes and transfers are also required to be unrestricted across states of nature.

4. Private intergenerational risk sharing motivated through altruism was analyzed by Hayashi, Altonji and Kotlikoff (1991).

5. Apart from the incomplete market participation that is intrinsic to OLG models without operative intergenerational gift motives or the institution of hereditary slavery, more standard types of insurance market failures can also create a potential welfare-improving role for taxes, transfer payments and public debt. (See Abel, 1988, Feldstein, 1989, Eaton and Rosen, 1980, Varian, 1980, Feldstein, 1988, Kimball and Mankiw, 1989 and Kaplow, 1991.)

6. The case with uncertain technology and taxes is considered in Buiter and Kletzer (1996).

7. Allowing for longer maturity debt would add notation but would not affect the equivalence results.

8. The single-period utility function v is twice continuously differentiable, strictly concave, increasing in c^1 and c^2 and satisfies the Inada conditions. Note that this utility function has the property that both c^1 and c^2 are normal goods.

9. The consumer's optimum will be turn out to be interior because (1) the utility function satisfies the Inada conditions, (2) the wage rate, the gross return on debt and the gross return on capital are positive and (3) government policy does not drive consumers to bankruptcy. Equation (11.5) anticipates point (3), which is introduced later in this Section.

10. Public consumption can be an argument in the private utility function. As long as it enters in an additively separable way, it will not affect the first-order conditions for private consumption. Public sector capital formation could be added in a straightforward manner.

304   *Willem H. Buiter and Kenneth M. Kletzer*

11. Often the weaker solvency criterion that (11.13) hold *in expectation*, is imposed for stochastic models, that is, $E_t \lim_{T \to \infty} \delta_{T-1} b_T \equiv E_t \lim_{T \to \infty} \Delta_{T-1} B_T = 0$. Bohn
(1990) argues quite convincingly, however, that the solvency criterion should apply to
each realisation of the discounted debt process, and not just to its mathematical expectation. See also Blanchard and Weil (1992).

12. When only the expectation of the discounted debt is required to go to zero in the limit,
Equation (11.14) is replaced by

$$\delta_{t-1} b_t = E_t \lim_{T \to \infty} [\sum_{s=t}^{T-1} (\theta_{s+1})^{-1} \left[\frac{\tau_s^1}{1+n} + \left(\frac{1}{1+n} \right)^2 \tau_{s-1}^2 - \frac{\theta_s g_s}{1+n} \right] \delta_s]$$

13. This is in the spirit of O'Connell and Zeldes (1988), who point out that in order for the
government to run a 'rational' Ponzi debt scheme, a rational private sector must be willing
to be at the receiving end of such a scheme, that is, willing to run a Ponzi credit scheme.

14. Kotlikoff (1989) points out that the ultimate constraint on the government is that it cannot
take, in present value, more from each household than the present value of its (pre-tax)
resources. Our criterion is more general than this, since, involving only period-by-period
physical (non-negativity) constraints, it does not rely on equilibrium prices and interest
rates. It therefore works even if markets are incomplete. Where present values can be
defined properly, our criterion implies that suggested by Kotlikoff. Note, however, that
Kotlikoff (1989) continues to impose the conventional government solvency constraint.

15. Note that the total transfer to the government by the young during period t evolves ac-
cording to $B_{t+1}^{**} + \tau_t^{1**} L_t = (1+r_t)(B_{t-1}^{**} + \tau_{t-1}^{1*} L_{t-1}) + [(1+n)\tau_t^{1*} - (1+r_t)\tau_{t-1}^{1*}] L_{t-1}$

16. When there are overlapping generations of infinite-lived households, (that is, asymptotically, there is an infinite population of infinite-lived households), the same zig-zag pattern
also emerges.

17. Note that such a tax is not perceived by those investing in government debt as a 'tax on
debt' affecting expected returns from holding debt, even if the tax is fully anticipated.
The amount paid is perceived by each individual to be independent of her actions. Aggregate tax collections, however, can move systematically with the total amount of debt
outstanding.

18. McCallum's specification of the private sector solvency constraint still implies that the
long-run growth rate of the public debt must the less than the private rate of time preference. If the time preference rate exceeds the growth rate of efficiency labour, the stock
of debt per unit of efficiency labour can grow without bound in McCallum's model.

19. We owe this picture to Bill Nordhaus.

20. If the value of the household's lifetime resources is positive in terms of both c^1 and c^2,
c^1 becomes unbounded as its relative price goes to zero for the more general case of
strictly quasi-convex preferences (see Arrow & Hahn, 1971, Chapter 4, Section 6). Our
further assumption of positive marginal utilities for non-negative consumption bundles
takes care of the case where the intercept of the budget constraint on the c^2 axis is zero.

21. With logarithmic single-period utility, $v(c) = \ln(c)$, the proof that transfer payments
to the young cannot grow faster than the natural rate of growth forever is immediate,
since in that case (11.40) becomes $\tau_t^1 = w_t - (1+\beta)c_t^1$. Since $w_t \geq 0$, this implies
$-\tau_t^1 \leq (1+\beta)c_t^1$.

22. We owe this point to Bill Brainard.

23. As shown in Diamond (1970), an increase in the capital income tax rate need not necessarily lower the after-tax rate of return on capital in the class of OLG models under

consideration in this paper. For a high enough intertemporal elasticity of substitution, an increase in the capital income tax rate (with tax revenues paid out to the young as a lump-sum transfer), could raise the before-tax rate of return to capital by enough to also raise the after-tax rate of return to capital.

24. We no longer consider distortionary taxes. There is therefore no need to distinguish between before- and after-tax rates of return.

25. In Tirole's extension of Diamond's OLG model, a speculative bubble has many of the features of public debt in our version of the Diamond model. However, his bubble is a private asset whose behaviour is restricted by private agents' budget constraints and voluntary exchange. Our potentially explosive debt bubbles rely on the government's unique capacity to tax. In Tirole's case, '*As bubbles crowd out productive savings and cannot grow faster than the economy, their existence is naturally shown to rely on the comparison between the asymptotic rates of growth and interest in the bubbleless economy.*' (Tirole, 1985, p. 1500).

26. Strictly speaking, this should be

$$\lim_{t'-\infty} inf_{t' \le t \le \infty} \{\Delta_t B_{t+1}\} \le 0$$

and

$$\lim_{t'-\infty} sup_{t' \le t \le \infty} \{\Delta_t B_{t+1}\} \ge 0$$

If the *lim inf* and the *lim sup* are both equal to zero then $\lim_{t \to \infty} \Delta_t B_{t+1} = 0$.

27. That is, taxes that remain distortionary for all feasible values of the tax base. This excludes the period 2 consumption tax when period 2 consumption is at the minimum subsistence level.

28. What we require, strictly speaking, is that the sequences of taxes and transfers per generation, $\{\tau_t^1 L_t\}_{t=0}^{\infty}$ and $\{\tau_{t-1}^2 L_{t-1}\}_{t=0}^{\infty}$ have infinite subsequences $\{\tau_{t_j}^1 L_{t_j}\}_{t_j=0}^{\infty}$ and $\{\tau_{t_j-1}^2\}_{t_j=0}^{\infty}$ whose elements have a growth rate at least as high as the interest rate.

References

Abel, A.B. (1988), "The Implications of Insurance for the Efficacy of Fiscal Policy," *NBER Working Paper* 2517, February.

Arrow, K.J. and F.H. Hahn (1971), *General Competitive Analysis*, North Holland, Amsterdam.

Auerbach, A.J. and L.J. Kotlikoff (1987), *Dynamic Fiscal Policy*, Cambridge University Press, Cambridge.

Auerbach, A.J., J. Gokhale and L.J. Kotlikoff (1991), "Generational Accounts- A Meaningful Alternative to Deficit Accounting," *NBER Working Paper* 3589, Januari.

Azariadis, C. (1993), *Intertemporal Macroeconomics*, Blackwell Publishers, Oxford.

Barro, R.J. (1976), "Reply to Feldstein and Buchanan," *Journal of Political Economy*, 84:343-49.

Barro, R.J. (1979), "On the Determination of the Public Debt," *Journal of Political Economy*, 87:940-971.

Ball, L., D.W. Elmendorf and N.G. Mankiw (1995), "The Deficit Gamble," *NBER Working Paper* 5015, February.

Blanchard, O.J., J.C. Chouraqui, R.P. Hagemann, and N. Sartor, "The Sustainability of Fiscal Policy: New Answers to an Old Question," *OECD Economic Studies*, No. 15:7-36 (Autumn)

Blanchard, O. J. and P. Weil (1992), "Dynamic Efficiency, The Riskless Rate, and Debt Ponzi

Games Under Uncertainty", *NBER Working Paper* 3992, February.

Bohn, H. (1990), "Sustainability of Budget Deficits in a Stochastic Economy", *mimeo*, Department of Finance, The Wharton School, University of Pennsylvania.

Bohn, H. (1991), "The Sustainability of Budget Deficits with Lump-Sum and with Income-Based Taxation," *Journal of Money, Credit and Banking*, 23(3), Pt. 2, August: 580-604.

Buiter, W.H. (1985), "A Guide to Public Sector Debt and Deficits," *Economic Policy*, 1:13-79.

Buiter, W.H. (1990), "Debt Neutrality, Redistribution and Consumer Heterogeneity: A Survey and Some Extensions," in: William C. Brainard, William D. Nordhaus and Harold W. Watts (eds.), *Money, Macroeconomics and Economic Policy*, MIT Press, Cambridge MA; also in W.H. Buiter, *Principles of Budgetary and Financial Policy*, MIT Press, Cambridge, MA: 183-222.

Buiter, W.H. and K.M. Kletzer (1990), "Fiscal Policy, Interdependence and Efficiency," *NBER Working Paper* 3328, April.

Buiter, W.H. and K.M. Kletzer (1991), "The Welfare Economics of Cooperative and Noncooperative Fiscal Policy," *Journal of Economic Dynamics and Control*, 15:215-244.

Buiter, W.H. and K.M. Kletzer (1996), "Uses and Limitations of Public Debt: Ponzi Finance, Dynamic Efficiency and Government Solvency," *mimeo*, University of Cambridge, October.

Buiter, W.H. and U.R. Patel (1992), "Debt, Deficits and Inflation: an Application to the Public Finances of India," *Journal of Public Economics*, 47:171-205.

Buiter, W.H. and U.R. Patel (1994), "Indian Public Finance in the 1990s: Challenges and Prospects," Yale University Department of Economics, *mimeo*, January.

Calvo, G.A. and M. Obstfeld (1988a), "Optimal Time-Consistent Fiscal Policy with Finite Lifetimes," *Econometrica* 56:411-32.

Calvo, G.A. and M. Obstfeld (1988b), "Optimal Time-Consistent Fiscal Policy with Finite Lifetimes: Analysis and Extensions," in: E. Helpman, A. Razin and E. Sadka (eds.), *Economic Effects of the Government Budget*, MIT Press, Cambridge, MA:163-198.

Cass D. (1972), "Distinguishing Inefficient Competitive Growth Paths: A Note on Capital Overaccumulation and Rapidly Diminishing Future Value of Consumption in a Fairly General Model of Capitalistic Production," *Journal of Economic Theory* 4:224-40.

Chamley, C. and H. Polemarchakis (1984), "Assets, General Equilibrium and the Neutrality of Money," *Review of Economic Studies*, 51:129-138.

Corsetti, G.o (1990), "Testing for Solvency of the Public Sector: An Application to Italy," Yale University, *Economic Growth Center Discussion Paper* 617, September.

Diamond, P.A. (1965), "National Debt in a Neo-Classical Growth Model," *American Economic Review* 55:1126-50.

Diamond, P.A. (1970), "Incidence of an Interest Income Tax," *Journal of Economic Theory*, 2:211-24.

Eaton, J. and H. Rosen (1980), "Taxation, Human Capital and Uncertainty," *American Economic Review*, 70:705-715.

Enders, W. and H.E. Lapan (1982), "Social Security Taxation and Intergenerational Risk Sharing," *International Economic Review*, 23:647-658.

Feldstein, M (1988), "The Effects of Fiscal Policies When Incomes are Uncertain: A Contradiction to Ricardian Equivalence," *American Economic Review*, 78:14-23.

Feldstein, M. (1989), "Imperfect Annuities Markets, Unintended Bequests, and the Optimal Age Structure of Social Security Benefits," *NBER Working Paper* 2820, January.

Fischer, S. (1983), "Welfare Aspects of Government Issue of Indexed Bonds," in: R. Dornbusch and M. H. Simonsen, *Inflation, Debt, and Indexation*, The MIT Press, Cambridge, MA: 223-46.

Gale, D. (1983), *Money: In Disequilibrium*, Cambridge University Press, Cambridge, MA.

Gale, D. (1990), "The Efficient Design of Public Debt," in: R. Dornbusch and M. Draghi (eds.), *Issues in Debt Management*, forthcoming.

Gordon, R.H. and H.R. Varian (1988), "Intergenerational Risk Sharing," *Journal of Public Economics*, 37:185-202.

Grilli, V. (1989), "Seigniorage in Europe," in: M. de Cecco and A. Giovannini (eds.), *A European Central Bank?*, Cambridge University Press, Cambridge U.K.: 53-79.

Hamilton, J.D. and M.A. Flavin (1986), "On the Limitations of Government Borrowing: A Framework for Empirical Testing," *American Economic Review*, 76:808-19.

Hayashi, F., J. Altonji and L. Kotlikoff (1991), "Risk-Sharing, Altruism, and the Factor Structure of Consumption," *NBER Working Paper* 3834, September.

Kaplow, L. (1991), "A Note on Taxation as Social Insurance for Uncertain Labour Income," *NBER Working Paper* 3708, May.

Kimball, M.S. and N.G. Mankiw (1989), "Precautionary Saving and the Timing of Taxes," *Journal of Political Economy*, 97:863-879.

Kotlikoff, L.J. (1989), "From Deficit Delusion to the Fiscal Balance Rule: Looking for an Economically Meaningful Way to Assess Fiscal Policy", *NBER Working Paper* 2841, February.

Kremers, J. (1989), "U.S. Federal Indebtedness and the Conduct of Fiscal Policy," *Journal of Monetary Economics*, 23:219-38.

McCallum, B.T. (1984), "Are Bond-Financed Deficits Inflationary? A Ricardian Analysis," *Journal of Political Economy*, 92:123-35

Merton, R.C. (1984), "On the Role of Social Security as a Means for Efficient Risk-Bearing in an Economy where Human Capital is not Tradable," in: Zvi Bodie and John Shoven (eds.), *Financial Aspects of the United States Pension System*, University of Chicago Press, Chicago, Ill.

O'Connell, S.A. and S.P. Zeldes (1988), "Rational Ponzi Games," *International Economic Review*, 29:431-50.

Pagano, M. (1988), "The Management of Public Debt and Financial Markets," in: F.Giavazzi and L. Spaventa (eds.), *High Public Debt: The Italian Experience*, Cambridge University Press: 135-66.

Samuelson, P.A. (1958), "An Exact Consumption-Loan Model of Interest With or Without the Social Contrivance of Money," *Journal of Political Economy*, 66:467-482.

Sargent, T.J. (1987), *Dynamic Macroeconomic Theory*, Harvard University Press, Cambridge.

Shell, K. (1971), "Notes on the Economics of Infinity," *Journal of Political Economy*, 79: 1002-11.

Spaventa, L. (1987), "The Growth of Public Debt: Sustainability, Fiscal Rules, and Monetary Rules," *IMF Staff Papers*, 34.

Spaventa, L. (1988), "Discussion," in: F.Giavazzi and L. Spaventa (eds.), *High Public Debt: the Italian Experience*, Cambridge University Press, Cambridge, MA:173-176.

Stiglitz, J. (1983), "On the Relevance or Irrelevance of Public Financial Policy: Indexation, Price Rigidities, and Optimal Monetary Policies," in: R. Dornbusch and Mario Henrique Simonsen (eds.), *Inflation, Debt and Indexation*, MIT Press, Cambridge, MA: 183-222.

Tirole, J. (1985), "Asset Bubbles and Overlapping Generations: A Synthesis," *Econometrica*, 53:1071-100.

Trehan, B. and C.E. Walsh (1991), "Testing Intertemporal Budget Constraints: Theory and Applications to the US Federal Budget and Current Account Deficits," *Journal of Money, Credit and Banking*, 23:206-23.

Uhlig, H. (1996), "Capital Income Taxation and the Sustainability of Permanent Deficits," *mimeo*, CentER for Economic Research, Tilburg University, May.

Varian, H.R. (1980), "Redistributive Taxation as Social Insurance," *Journal of Public Economics*, 17:49-68.

Wallace, N. (1981), "A Modigliani-Miller Theorem for Open-Market Operations," *American Economic Review*, 71: 267-74.

Wilcox, D.W. (1989), "The Sustainability of Government Deficits: Implications of the Present Value Borrowing Constraint," *Journal of Money, Credit and Banking*, 21:291-306.

Wilson, C. A. (1981), "Equilibrium in Dynamic Models with an Infinity of Agents," *Journal of Economic Theory*, 24: 95-111.

Zilcha, I. (1990), "Dynamic Efficiency in Overlapping Generations Models with Stochastic Production," *Journal of Economic Theory*, 52: 364-379.

12 Capital Income Taxation and the Sustainability of Permanent Primary Deficits

Harald Uhlig

I place economy among the first and most important virtues and public debt as the greatest of dangers to be feared. To perserve our independence, we must not let our rulers load us with perpetual debt. If we run into such debts, we must be taxed in our meat and drink, in our necessitites and in our comforts, in our labour and in our amusements. If we can prevent the government from wasting the labour of the people, under the pretense of caring for them, they will be happy. (Thomas Jefferson)

12.1 Introduction

Arguably the most visible part of Reagan's legacy is the budget deficit. The reasons for why it is of concern to many need not be repeated here. They have been discussed already by, say, Krugman, 1990, Buiter and Kletzer (1992b), Eisner (1992), Friedman (1992), Vickrey (1992). Auerbach, Gokhale and Kotlikoff (1994) use generational accounting to evaluate this issue, which in turn has been critized by Buiter (1993). Several of the arguments brought forward, however, state, that we need not worry. For example, some argue that a good part of the deficit corresponds to government investment and may be therefore a good thing after all: the deficit, corrected for this fact, is then actually quite a bit smaller (see Eisner, 1986). Some claim, that because parents care about their children, it does not matter whether government expenditures are financed by taxes or by debt (Barro,1974).

Finally, it is sometimes heard that we may just grow out of the deficit over time. Suppose, there is a government deficit and the interest rate is not 'too high.' Then, over time, even though the real value of the debt rises, the output of the economy may rise even faster, trivializing the debt through the enlarged tax base. Indeed, in this case, there is even room for a Pareto-improving redistribution which makes everybody better off, since providing each generation with a 'free,' deficit-financed lunch and then simply rolling over this debt forever is feasible. The crucial issue then is, whether indeed a permanent deficit is sustainable. For a recent book-length contribution regarding this argument and the related literature, see Carlberg (1995). An empirical investigation is in Bovenberg and Petersen (1992).

We reexamine this last point of view in the context of four versions of a basic overlapping generations model, adding one by one three relevant features: capital, investment and depreciation, and capital income taxation. This paper is thus an exercise in model engineering: by moving from a simple to a complicated framework step by step, it becomes transparent how the individual parts in the final machinery interact and contribute to the analysis.

In the very basic endowment economy, the government can run a deficit forever if the deficit as a fraction of GNP is not too high. Intuitively, debt here fulfills the role of money in other overlapping generations models. The result here corresponds to standard results about seignorage (see Wallace, 1980). For the second of the four models, we add a fixed capital stock which is traded from generation to generation. Since the value of the capital stock rises with the growth rate, so too must the value of government bonds. The total, outstanding deficit then explodes, thereby eliminating the possibility of a sustainable deficit.

We then add the possibility for capital income taxation (or, equivalently, savings taxation) to the instruments of the government, where the returns on private capital are taxed. The emphasis here is on this distortionary aspect of taxation, in contrast to the analysis in, say, Buiter and Patel (1992a) or Buiter and Kletzer (1994). A higher capital income tax drives down the realized rate of return on capital, possibly rendering the deficit sustainable. The necessary tax rate for accomplishing this is typically quite sizeable: even as the deficit-to-GNP ratio becomes negligible, the capital income tax does not. The intuition behind this result is that the capital income tax needs to drive the interest rate down to at most the growth rate of the economy in order to have sustainable deficits at all. It turns out that for most tax rate, there is a 'good' and a 'bad' steady state equilibrium. Though there is some choice for the tax rate, the tax rate can only be chosen so that the good equilibrium becomes better when the bad equilibrium becomes worse at the same time.

Finally, we make the capital stock variable by introducing investment and depreciation. Since the model is a neoclassical growth model in nature, there will not be any growth effects. However, the level effects resulting from the different capital income taxes which make a government deficit sustainable, can be quite dramatic. The welfare effects are much less clear cut. Furthermore, a positive capital income tax may not be necessary for sustainability, if the economy without the government is already dynamically inefficient: this is demonstrated in a numerical example.

This paper is a variation of Diamond's (1965) celebrated analysis. Since rolling over the debt amounts to the creation of a bubble, this paper can also be viewed as an application of the bubble literature as in Tirole (1985) or Blanchard and Fisher (1989). However, while the focus there and in Diamond (1973) as well as Atkinson and Sandmo (1980) is on the normative aspects of government policy, the focus here is on the positive aspect. The question is not, whether government should run a deficit forever, but whether it can. Note that sustainability of a permanent deficit means that the interest rate is below the growth rate of the economy and that therefore the economy is dynamically inefficient: from a welfare perspective, there is always a Pareto improving redistribution, which, depending on the structure of the model, may or may not require a deficit (see Cass, 1972, Balasko and Shell, 1980, Sargent, 1987, and Abel *et al.*, 1989). The sad fact is that permanent deficits seem to be politically attractive. The point of this paper is then to analyze what happens, given that a permanent deficit needs to be sustained.

This paper is related to Sargent and Wallace (1981), Darby (1984), Miller and Sargent (1984) and Aiyagari (1985). All of these papers, however, consider at most a savings technology with a fixed rate of return instead of a productive capital stock with a rate of return calculated from equilibrium conditions and taxes are lump sum, if introduced at all. In that respect, Chari (1988), Lucas (1990) and Bohn (1990) are more closely related, but they use an infinite-lived agent framework.

Finally, it should be emphasize, that the entire analysis proceeds in the context of a closed economy. Open economy issues make capital income taxation a much more tricky issue, and many additional problems may arise. For some of the related literature, see Bovenberg (1989, 1992) and Broer, Westerhout and Bovenberg (1994).

12.2 Model 1: No capital

In each period t, $t = 1, 2, ..$, a new generation of N two-period lived agents is born. There also is a generation of N initially old agents alive at date 1. The effects of population growth, general excess demand functions or distributional issues are not examined here. N is chosen to equal one, keeping in mind that each agent is meant to be representative of his generation and therefore does not act strategically.

There is one consumption good each period. An agent born at t cares about consumption c_{1t} when young and c_{2t+1} when old according to the utility function

$$u(c_1, c_2) = \log(c_1) + \log(c_2).$$

The specific form has been chosen to make the results easy and tractable. Observe that a discount factor is not included: again, this keeps the algebra simple. The special form for the utility function implies a vertical savings line in Diamond's (1965) diagram 1, thereby ruling out his 'perverse case.'

The agent is endowed with one unit of labour when young, which he can use to produce the consumption good according to the production function

$$y_t = \zeta_t n_t,$$

where ζ_t is the productivity parameter at time t. The productivity parameters are assumed to be

$$\zeta_t = \zeta^t,$$

where

$$\zeta > 1$$

is some given constant. Since labour is supplied inelastically, it follows that the growth rate γ of the economy is given by $\gamma - 1$, where

$$\gamma = \zeta$$

(the symbol γ is introduced to keep the notation consistent throughout the paper).

There is a government, who tries to finance a deficit in each period by rolling over its debt. We assume that the governmental deficit is a constant fraction α of total output,

$$g_t = \alpha\, y_t,$$

where $\alpha > 0$. If R_t is the return (that is, one plus the interest rate) from period $t - 1$ to t, the government budget constraint is given by

$$b_t = g_t + R_t b_{t-1}, \qquad (12.1)$$

where b_t are the one-period bonds issued by the government in time t. Note that the deficit is financed entirely by rolling over the debt. There are no income taxes and the like, since they are not the issue here (it is easy to append the model by having some kind of income tax, financing some government expenditures in excess of the deficit described above: in that case, the output y_t is to be read as the after — tax income). The results stay the same.

A *steady state equilibrium* is given by numbers $\beta > 0, \sigma > 0$ and $R > 0$, so that for

$$
\begin{aligned}
y_t &= \zeta^t \\
b_t &= \beta\, y_t \\
s_t &= \sigma\, y_t \\
R_t &= R,
\end{aligned}
$$

each agent maximizes its utility at savings s_t, given the gross return $R_{t+1} = R$, the government budget constraint is satisfied and markets clear:

1. the consumption goods market

$$c_{1t} + c_{2t} + g_t = y_t \qquad (12.2)$$

2. the bond market

$$s_t = b_t. \qquad (12.3)$$

It is easily shown that the savings of a young agent are given by

$$s_t = y_t/\, 2,$$

independently of the interest rate (which makes the logarithmic specification of the utility function so convenient for our purposes). Thus, the remaining constants β and R can be calculated from (12.1) or (12.2), given α : one equation suffices by Walras' law. The result is given by

$$
\begin{aligned}
R/\gamma &= 1 - 2\alpha \\
\beta &= 1/2.
\end{aligned}
$$

Since $R > 0$ is required for the steady state equilibrium, it follows, that $\alpha < 1/2$ is necessary and sufficient for a steady state equilibrium to exist. These results are summarized by

Proposition 12.1 *If there is no capital, any permanent deficit up to 50 per cent of total output each period is sustainable by rolling over the debt.*

Note, that the number of 50 per cent is simply the total savings of the agent in the model economy. This number is not meant to be interpreted as describing the actual situation in any particular country and depends critically on the specification of the utility function. A result of this type, however, probably holds for a wide variety of utility functions. The proposition seems like good news for politicians: optimality questions aside, it is at least possible to sustain a sizeable deficit forever. The question, of course, is, whether a crucial element is missing in deriving this answer to the sustainability question by making the model possibly too simple. That this is probably so should already be indicated by the following observation in the model.

Proposition 12.2 *If there is no capital, the size of the total outstanding debt is independent from the government deficit, as long as it is sustainable.*

This proposition simply follows, because $\beta = 1/2$ is independent of α (or R, for that matter). This proposition runs counter to the intuition one usually has about the size of a government deficit: one would think that a larger yearly deficit implies a larger outstanding stock of debt. The reason that the model here does not deliver such a result is simple: government bonds are the only means of cross-generational trade in this model. Government bonds act like money and the government deficit like seignorage or an inflation tax: while these are disturbing the amount an old agent will receive, it will not change the amount a young agent wants to save due to the logarithmic specification of the utility function. Thus, savings and not the size of the budget deficit is what determines the amount of outstanding debt (compare to Sargent, 1987).

It can be concluded that this model is indeed too simple to give a reliable insight into the question of the sustainability of permanent deficits. Therefore, another element needs to be added: a different vehicle for saving. More precisely, a privately owned capital stock is added as a feature of the model in the next section.

12.3 Model 2: fixed capital stock, no capital income tax

Let there be a fixed capital stock $k > 0$, which does not depreciate over time. Production is now given by the Cobb-Douglas production function

$$y_t = \zeta_t k^\rho n^{1-\rho},$$

where $\rho \in (0\,,1)$ is the share of capital, a constant. The capital is owned by the old, who sell it to the young for a total price of q_t. The young receive wage for their labour, spend part of it on consumption c_{1t}, part of it on saving in capital s_{kt} and part of it in saving in governmental bonds s_{bt}. When old, they receive the dividends from their capital holdings as well as the resale price and they are paid the interest on their bonds.

All markets are competitive. In particular, in order for any government bonds to be hold, it must be the case that the return on government bonds and on capital are the same in equilibrium. Furthermore, it is straightforward to calculate that the wage income is given by $(1-\rho)y_t$ and the dividend income by ρy_t, which we will substitute into the definition.

An *equilibrium* is given by sequences $(c_{1t}, c_{2t}, s_t, q_t, R_{t+1}, b_t)$, so that for

$$y_t = \zeta^t k^\rho,$$

it is the case that

1. for each t, $t = 1, 2, \ldots$, the agent born at t, maximizes his utility at c_{1t} and $c_{2t+1} = R_{t+1}s_t$, given the budget constraint:

$$c_{1t} + s_t = (1 - \rho)y_t,$$

2. the government budget constraint is satisfied:

$$b_t = g_t + R_t b_{t-1}$$

3. markets clear:

   (a) the consumption goods market

   $$c_{1t} + c_{2t} + g_t = y_t$$

   (b) the capital market

   $$s_t = q_t + b_t,$$

4. no arbitrage:

$$R_t = (\rho\, y_t + q_t)/q_{t-1}, \qquad (12.4)$$

where $q_t > 0$ and $R_t > 0$.

The condition $b_0 \geq 0$ ensures that the government cannot start up the economy by handing a liability to the old agents, which they may trade from generation to generation. The restrictions $q_t > 0$ and $R_t > 0$ are the usual positivity restrictions on prices. Finally, (12.4) is the restriction that the return on government bonds and capital must be equal (since the deficit is assumed to be strictly positive, this restriction must hold except for degenerate cases). This is called a no arbitrage condition, because that is its economic interpretation. It is of course possible to write the definition of an equilibrium without this condition and derive it from a more elaborate description of the maximization problem for the agents. Since this step is straightforward, the version of the definition above and in similar spirit everywhere below was chosen.

In contrast to the model without capital, the following result is obtained.[1]

Proposition 12.3   *If there is fixed capital stock and no capital income taxation, the government cannot sustain a permanent deficit of a constant fraction of total output.*

Proof.   Suppose, there was an equilibrium. Market clearing in the consumption goods sector implies

$$(1 - \rho)y_t/2 + R_t(1 - \rho)y_t/(2\gamma) + \alpha\,y_t = y_t, t \geq 2,$$

where $\gamma = \zeta$ is the growth rate of the economy, as before. Thus the return

$$R_t \equiv R, t \geq 2$$

has to be a constant. Define the fraction of saving which is capital by

$$\varphi_t = q_t/\,s_t,$$

and note that then $\varphi_t \in (0\,,1)$, since $q_t > 0$ and $b_t > 0, t \geq 1$. The condition (12.4), which guarantees an equal return on capital and government bonds can now be rewritten as

$$R/\gamma = \frac{\varphi_t + 2\frac{\rho}{1-\rho}}{\varphi_{t-1}}$$

or, equivalently,

$$\varphi_t = \frac{R}{\gamma}\varphi_{t-1} - 2\frac{\rho}{1 - \rho}$$

for $t \geq 2$. Note that $\frac{R}{\gamma} > 0$. Consider the following three cases.

1. Suppose, that $\frac{R}{\gamma} < 1$. Then

$$\varphi_t \to -2\frac{\rho}{1 - \rho}\frac{1}{1 - \frac{R}{\gamma}} < 0,$$

   in contradiction to the positivity of φ_t.
2. Suppose, that $\frac{R}{\gamma} = 1$. Then

$$\varphi_t \to -\infty,$$

   in contradiction to the positivity of φ_t.
3. Suppose then, that $\frac{R}{\gamma} > 1$. But then the outstanding government debt outgrows the economy and is therefore not sustainable: let

$$\beta_t = b_t/\,y_t$$

· be the debt-to-GNP ratio. The government budget constraint can be rewritten as

$$\beta_t = \alpha + \frac{R}{\gamma}\beta_{t-1}.$$

Since $\beta_{t-1} \geq 0$, $t \geq 1$, $\alpha > 0$ implies $\beta_t \to \infty$. This is impossible, since by capital market clearing and $q_t > 0$, we need to have $\beta_t \leq 1$.

Since these three cases exhaust all possibilities, an equilibrium cannot exist. ∎

After some thought, the result is actually not that surprising: the value of the capital as well as the value of labour keep growing at the rate of the overall growth rate of the economy. But that means that the rate of return on capital must be even higher, that is, it must be the case that

$$R > \gamma.$$

But then the outstanding debt grows faster than the economy and there is no way that output can catch up any more.

Given that intuition, the return on capital is somehow too high to make a deficit sustainable. So why not give the government some instrument to lower the return on capital. That will ease the debt problem as well! It is therefore natural to consider a capital income tax or savings tax.

12.4 Model 3: fixed capital stock and capital income tax

A capital income tax in this model is a tax on the net return on capital. In order to keep the notation simple, a tax rate τ on the entire return on capital is introduced. Both formulations are equivalent, if capital income tax rates are allowed to exceed 100 per cent (It turns out that they would need to for the numerical examples presented below. Whether this is reasonable will be discussed in the last section before the conclusion.)

Since the focus in this paper is on the sustainability of a permanent deficit and therefore the effects of the elements in our model with respect to that, the tax is not used towards reducing the deficit, but simply increases government consumption. Also, the tax is not imposed on the return on government bonds for convenience. Otherwise, let the government use the tax on the bond returns to repay its bonds: the result is equivalent to the economy below except that the return on the government bonds is simply higher by the tax rate. Taxing government bonds just amounts to rewriting the government budget constraint in another way by doing the accounting differently.

A *steady state equilibrium* is given by numbers $\beta > 0, \sigma > 0, R > 0,$

$\theta > 0$ and a tax rate $\tau \geq 0$, so that for

$$
\begin{aligned}
y_t &= \zeta^t k^\rho \\
b_t &= \beta\, y_t, \\
s_t &= \sigma\, y_t, \\
R_t &= R, \\
q_t &= \theta\, y_t,
\end{aligned}
$$

it is the case that

1. for each t, $t = 1, 2, ...$, the agent born at t, maximizes his utility at c_{1t} and $c_{2t+1} = R_{t+1}s_t$, given the budget constraint:

$$c_{1t} + s_t = (1 - \rho)y_t,$$

2. the government budget constraint is satisfied:

$$b_t = g_t + R_t b_{t-1} \tag{12.5}$$

3. markets clear:

   (a) the consumption goods market

   $$c_{1t} + c_{2t} + g_t + \tau(\rho y_t + q_t) = y_t \tag{12.6}$$

   (b) the capital market

   $$s_t = q_t + b_t, \tag{12.7}$$

4. no arbitrage:

$$R_t = (1 - \tau)(\rho\, y_t + q_t)/q_{t-1}. \tag{12.8}$$

Using the decision rules of the agent resulting from his maximization problem as well as substituting b_t by βy_t, and so on, equations (12.5) through (12.8) can be rewritten as

$$\beta = \alpha + \frac{R}{\gamma}\beta \tag{12.9}$$

$$\frac{1-\rho}{2} + \frac{R}{\gamma}\frac{1-\rho}{2} + \alpha + \tau(\rho + \theta) = 1 \tag{12.10}$$

$$\frac{1-\rho}{2} = \theta + \beta, \tag{12.11}$$

$$\frac{R}{\gamma} = (1 - \tau)(\frac{\rho}{\theta} + 1), \tag{12.12}$$

which, by Walras law, must be dependent. Therefore equations (12.9), (12.11) and (12.12) can be used to solve for the unknown parameters β, θ, R and τ under the positivity restrictions. It turns out that there is one degree of freedom: ideally, one would then fix the tax rate τ and solve for the other three variables. It is more convenient to fix the return R instead and solve for β, θ and τ. Because $\beta > 0$, it must be the case that

$$0 < \frac{R}{\gamma} < 1 - \frac{2\alpha}{1 - \rho}$$

and for these values of R it follows that

$$\beta = \frac{\alpha}{1 - \frac{R}{\gamma}},$$

$$\theta = \frac{1 - \rho}{2} - \frac{\alpha}{1 - \frac{R}{\gamma}},$$

$$\tau = 1 - \frac{\frac{R}{\gamma}}{1 + \frac{\rho}{\frac{1-\rho}{2} - \frac{\alpha}{1-\frac{R}{\gamma}}}}.$$

Substituting these three formulas into (12.10) and checking that it holds for any value of R in the range described above can be used to verify the calculations. The qualitative insight is summarized by

Proposition 12.4 *With a fixed capital stock and capital income taxation, there is a range of interest rates*

$$0 < \frac{R}{\gamma} < 1 - \frac{2\alpha}{1 - \rho}$$

with corresponding capital income tax rates, so that the government deficit is sustainable forever.

The formula above allows for examining the behaviour of the capital income tax rate for various levels of α and R. As for the dependence on R, graphs are presented in Figures 12.1 and 12.2 with $\alpha = .10$, $\rho = .3$ and $\zeta = (1.03)^{25}$ to get results which are somewhat suggestive: a generation is thought of living 25 years, while young and accumulating wealth through labour, and 25 years, while old and consuming the returns to their investment.

320   *Harald Uhlig*

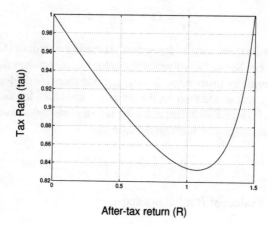

After-tax return (R)

Figure 12.1: The capital income tax rate τ in the case of a fixed capital stock in dependence of the equilibrium return R. Parameters are α = .1, ρ = .3 and ζ = (1.03)²⁵. Note, how there are two equilibrium returns R for any given τ in the appropriate range

Figure 12.1 (and all following Figures except for Figure 12.2) shows the relationship between the total tax rate on capital gains after 25 years and the return over 25 years. Figure 12.2 annualizes these numbers, determining the tax rate needed each year to compound to the total tax rate shown in Figure 12.1 in 25 years, likewise for the return. In Figure 12.2, we have shown only a part of the range of possible values for the annualized return $R^{1/25}$. Note in both figures, that the tax rate τ first falls and then rises again. This can be shown analytically to be correct. Furthermore, at a given tax rate, there will be typically two steady state equilibria (if at all), which are Pareto ordered.[2] These results are summarized in the next proposition.

Proposition 12.5   1. *The tax rate τ converges to 1 as the return R approaches its maximal or its minimal value supporting a steady state equilibrium with deficit α.*

2. *The tax rate τ first falls and then rises again as the gross interest rate is increased from its minimal to its maximal value. There is a unique minimum tax rate τ_{min}.*

3. *For each tax rate τ between τ_{min} and 1.0, there are two steady state equilibria, one with a lower return R than the other one.*

4. *The steady state allocations are Pareto ordered: Welfare increases as the return R increases.*

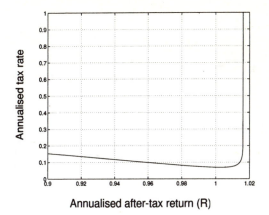

Figure 12.2: The annualized capital income tax rate $1-(1-\tau)^{1/25}$ in the case of a fixed capital stock in dependence of the annualized equilibrium return $R^{1/25}$. Parameters are $\alpha=.1$, $\rho=.3$ and $\zeta=(1.03)^{25}$. Note, how there are two equilibrium returns R for any given τ in the appropriate range

The proof for this proposition is in Appendix A. Note from the figures that the capital income tax necessary to sustain a government deficit forever is very large (especially taking into account that τ here is really the tax on total earnings on capital, not just on the gain). One might conjecture that the minimal capital income tax necessary to sustain a government deficit α converges to zero as α approaches zero. That this is not so is demonstrated in Figure 12.3 (for the same parameters as for Figure 12.1) and by the following proposition.

Proposition 12.6 *The minimal tax rate which sustains a permanent deficit is bound below by a number strictly bigger than zero, even as that deficit becomes arbitrarily small.*

Proof. Define

$$
\begin{aligned}
\tau^* &= \inf_{0<\alpha}\ \inf_{0<R/\gamma<1-\frac{2\alpha}{1-\rho}}\{\tau\} \\
&= \inf_{0<R/\gamma<1}\ \inf_{0<\alpha<(1-\rho)(1-R/\gamma)/2}\{\tau\}
\end{aligned}
$$

Since for fixed R, τ is increasing in α and defined for $\alpha=0$, it follows that

$$
\tau^* = \inf_{0<R/\gamma<1}\left\{1-\frac{1-\rho}{1+\rho}\frac{R}{\gamma}\right\}
$$

$$= \frac{2\rho}{1+\rho} > 0,$$

proving the claim. ∎

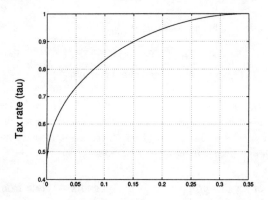

Primary deficit as a fraction of GNP (alpha)

Figure 12.3: The minimal capital income tax rate sustaining the permanent primary deficit αy_t

The intuition, that $\tau \to 0$ and $\alpha \to 0$ is wrong because, without τ, no government deficit is sustainable: R/γ ends up being strictly bigger than 1. In order to get sustainability, R/γ has to be suppressed strictly below one no matter how small the deficit is that is to be sustained.

The possibility for capital income taxation brings back the possibility for sustainable deficits by depressing the return on private capital sufficiently far. Capital income taxes are usually attacked by economists for their undesirable effects on the efficiency of an economic system (see, for example, Lucas, 1990), although they can potentially have beneficial effects in the context of overlapping generation models, see Uhlig and Yanagawa (1996). A closer look at this issue is taken in the next variation of our basic model in which now the temporary capital stock is the result of depreciation and investment.

12.5 Model 4: variable capital stock and capital income taxation

Let it be the case that

$$k_t = (1 - \delta)k_{t-1} + x_t,$$

where $\delta \in [0, 1)$ is the rate of depreciation, x_t is investment and k_t is the capital stock planted in period t and productive in period $t + 1$:

$$y_t = \zeta^t k_{t-1}^\rho n_t^{1-\rho}.$$

The steady state growth rate is now no longer $\zeta - 1$, since the capital stock will be growing as well. Also, it is necessary to calculate the value of the entire capital stock after dividends and depreciation, but before investment: the symbol q_t^{ante} is introduced for that. As before, the capital income tax revenues are used for additional government spending and not towards reducing the deficit.

A *steady state equilibrium* is given by numbers $\gamma > 0, \xi, \kappa > 0, \beta > 0, \sigma > 0, R > 0, \theta > 0, \theta^{ante} > 0$ and a tax rate $\tau \geq 0$, so that for

$$\begin{aligned}
y_t &= \zeta^t k_t^\rho, \\
x_t &= \xi\, y_t, \\
k_t &= \kappa\, y_t, \\
b_t &= \beta\, y_t, \\
s_t &= \sigma\, y_t, \\
R_t &= R, \\
q_t &= \theta\, y_t, \\
q_t^{ante} &= \theta^{ante} y_t,
\end{aligned}$$

it is the case that

1. for each t, $t = 1, 2, ...$, the agent born at t, maximizes his utility at c_{1t} and $c_{2t+1} = R_{t+1}s_t$, given the budget constraint:

$$c_{1t} + s_t = (1 - \rho)y_t,$$

2. the government budget constraint is satisfied:

$$b_t = g_t + R_t b_{t-1} \tag{12.13}$$

3. markets clear:

(a) the consumption goods market

$$c_{1t} + c_{2t} + g_t + \tau(\rho y_t + q_t^{ante}) + x_t = y_t \qquad (12.14)$$

(b) the capital market

$$s_t = q_t + b_t, \qquad (12.15)$$

4. no arbitrage:

$$R_t = (1 - \tau)(\rho\, y_t + q_t^{ante})/q_{t-1}. \qquad (12.16)$$

$$q_t^{ante} + x_t = q_t \qquad (12.17)$$

$$q_t = k_t \qquad (12.18)$$

5. the production function for capital holds:

$$k_t = (1 - \delta)k_{t-1} + x_t \qquad (12.19)$$

Equations (12.17) and (12.18) result from the definition of q_t^{ante} and the fact that the consumption good and the investment good are the same: both equations could be arrived at more fundamentally by focusing on the appropriate production technology, which transfers consumption goods into investment goods one for one and *vice versa*.

From the production function for output and the fact that capital is a constant fraction of output, the steady state growth rate $\gamma - 1$ is calculated as follows

$$\gamma = \frac{y_t}{y_{t-1}} = \frac{\zeta^t(\kappa y_{t-1})^\rho}{\zeta^{t-1}(\kappa y_{t-2})^\rho} = \zeta\,\gamma^\rho$$

or

$$\gamma = \zeta^{\frac{1}{1-\rho}}.$$

That is, the growth rate in this economy is a function of ζ, the growth rate of the underlying productivity parameter, and ρ, the capital share, and nothing else. Neither the budget deficit nor the capital income tax nor the interest rate have an impact on steady state growth. This is not the result of the particular utility function we used, but rather a standard result within models of the neoclassical growth variety, as can easily be seen from the derivation above: we should not expect the growth rate being changed by the capital income tax rate here. This would, of course, change in a model with endogenous growth.

However, the level effects of the capital income tax rate can be sizeable. They are derived now. Using the decision rules of the agent resulting from his maximization problem as well as substituting b_t by βy_t, *etc.*, we can rewrite Equations (12.13) through (12.18) as above as

$$\beta = \alpha + \frac{R}{\gamma}\beta$$

$$1 = \frac{1-\rho}{2} + \frac{R}{\gamma}\frac{1-\rho}{2} + \alpha + \tau(\rho + \theta^{ante}) + \xi \quad (12.20)$$

$$\frac{R}{\gamma} = (1-\tau)(\frac{\rho}{\theta} + \frac{\theta^{ante}}{\theta}), \quad (12.21)$$

$$\theta^{ante} + \xi = \theta$$

$$\theta = \kappa$$

$$\kappa = \frac{1-\delta}{\gamma}\kappa + \xi$$

which, again, by Walras law, must be dependent. Thus, leaving away Equation (12.20), the parameters $\xi, \kappa, \beta, R, \theta, \theta^{ante}$ and τ can be solved for under the positivity restrictions via the remaining equations. As before the solutions are parameterizable by the interest rate R, which can be chosen freely in a certain range

$$0 < \frac{R}{\gamma} < 1 - \frac{2\alpha}{1-\rho}.$$

The formulas for the other variables are

$$\beta = \frac{\alpha}{1 - \frac{R}{\gamma}}$$

$$\kappa = \theta = \frac{1-\rho}{2} - \frac{\alpha}{1 - \frac{R}{\gamma}},$$

$$\xi = (1 - \frac{1-\delta}{\gamma})\kappa,$$

$$\theta^{ante} = \frac{1-\delta}{\gamma}\kappa,$$

$$\tau = 1 - \frac{\frac{R}{\gamma}}{\frac{1-\delta}{\gamma} + \frac{\rho}{\frac{1-\rho}{2} - \frac{\alpha}{1 - \frac{R}{\gamma}}}},$$

which can be substituted into (12.20) to check the validity of the solution. The changes to the solution with the fixed capital stock are minor: for instance, the discount rate now enters the formula for the tax rate.

For $\alpha = \beta = \tau = 0$, we obtain a benchmark version of this model, in which there is no government. For the equilibrium return, we get from $\tau = 0$

$$\frac{R^*}{\gamma} = \frac{2\rho}{1-\rho} + \frac{1-\delta}{\gamma}.$$

and the steady state capital is given by $\kappa^* = \frac{1-\rho}{2}$. This benchmark economy is already dynamically inefficient, if the benchmark return R^* is smaller than the growth factor γ.

One immediate implication of this analysis is to figure out the effect of the capital income tax rate, which corresponds to a certain return, on the steady state path of output. For that, the solution above can simply be plugged into the formula for output in period 1

$$y_1 = \zeta(\kappa\, y_1/\gamma)^\rho$$

to find

$$y_1 = \zeta^{\frac{1}{1-\rho}} \Big(\frac{1-\rho}{2\gamma} - \frac{\alpha}{\gamma-R}\Big)^{\frac{\rho}{1-\rho}}.$$

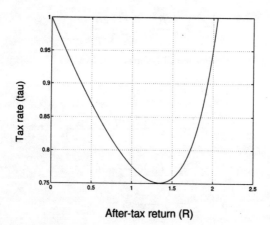

After-tax return (R)

Figure 12.4: The capital income tax rate τ in the case of a variable capital stock in dependence of the annualized equilibrium return R. The economy is dynamically efficient. Parameters are $\alpha = .1$, $\rho = .3$, $\delta = 1-(1-.1)^{25}$ and $\zeta = (1.03)^{25}$. Note, how there are again two equilibrium returns R for any given τ in the appropriate range

One can easily see the level effect: a higher α depresses the output in the first period and thus in all subsequent periods, because a higher α depresses the capital-to-output ratio κ.

Figure 12.4 repeats Figure 12.1 with the same parameters and additionally the parameter $\delta = 1 - (1 - .1)^{25}$ corresponding to a yearly depreciation of the capital stock of 10 per cent, where the capital stock is now variable (and the growth rate higher). It turns out, that the benchmark economy without a government is dynamically efficient for these parameters: $R^* > \gamma$. Note, that the solution for the supporting capital income tax did not change too much when compared to Figure 12.1. Figure 12.5 plots y_1 (in per cent of the level of first period output in the benchmark version) as a function of the return R in the relevant range. Figure 12.5 clearly shows, that output is a decreasing function of the return R, which can also be easily derived from looking at the equations above.

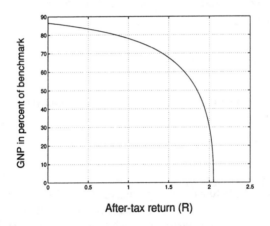

Figure 12.5: Equilibrium output in dependence of the annualized equilibrium return R in case of a variable capital stock. The economy is dynamically efficient. Parameters, as above, are $\alpha = .1$, $\rho = .3$, $\delta = 1 - (1 - .1)^{25}$ and $\zeta = (1.03)^{25}$

Thus, with variable capital, output is not maximized at the minimum capital income tax which makes the deficit sustainable, but rather at a capital income tax, which approaches 1 and a return which approaches zero. The intuition for this result is clear by looking at the algebra of how it is derived. A higher interest rate drives up the debt-to-GNP ratio β. This in turn drives down the amount of savings in the form of capital as a fraction of GNP, which is given by the parameter κ or θ : since total savings as a fraction of GNP remain constant, government debt crowds out capital. Finally, the level of output is increasing with the capital-output ratio κ : it is here, where the ratios expressing everything in terms of total GNP affect total GNP itself. To sum up, higher interest rates let government debt crowd out capital as a means of savings, thus lowering total output which needs capital as a productive factor. It is important to realize, that this is a steady state comparison. Nothing is said here about what will happen in these economies if the capital income tax is unexpectedly raised forever to a new level. The conclusion that a high capital income tax leads to maximal output is dangerous for another reason in this model too, of course: for a given capital income tax, there are typically two equilibria, one with a low and one with a high return, and the high return equilibrium for a capital income tax approaching one delivers the worst steady state output of all.

For welfare calculations, output is not the relevant measure, but rather utility, in which the return is of relevance. Up to a factor, which depends only on the time t, steady state welfare for each two-period lived generation is given by

$$W = 2\log(\frac{1-\rho}{2}y_1) + \log(R).$$

The welfare for the initally old generation is calculated as

$$W_0 = \log(R\frac{1-\rho}{2\gamma}y_1)$$

by stationarity. Both functions are plotted in Figure 12.6. It turns out, that the equilibria are no longer as nicely Pareto ordered as in the situation without investment. Anticipating Proposition 12.7, we can also plot the welfare-maximizing capital income tax rate as a function of the primary deficit parameter α, see Figure 12.7. As one can see, that tax rate can become quite substantial.

We now chose parameters so that the benchmark economy is dynamically inefficient: we chose $\rho = .15$. For small α, it then turns out, that a permanent deficit is sustainable even for a negative capital income tax (that is, for a savings subsidy) and furthermore, that welfare can improve due the deficit. This is shown in Figures 12.8, 12.9, 12.10 and 12.11, which correspond to the Figures 12.4, 12.5, 12.6 and 12.7 described above. Note, in particular, that welfare can even improve when compared to the benchmark economy with a government and without a government deficit. This is of course just a restatement of Diamond's (1965) insight.

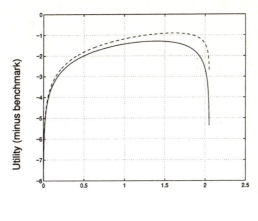

After-tax return (R)

Figure 12.6: Equilibrium welfare of the two-period lived as well as the initially old in dependence of the annualized equilibrium return R in case of a variable capital stock. The economy is dynamically efficient. Parameters, as above, are $\alpha = .1$, $\rho = .3$, $\delta = 1 - (1 - .1)^{25}$ and $\zeta = (1.03)^{25}$

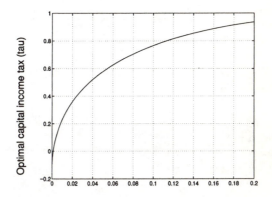

Primary deficit as fraction of GNP (alpha)

Figure 12.7: Welfare-maximizing capital income tax rate in dependence of the primary deficit αy_t in the case of a variable capital stock. The economy is dynamically efficient, except for very small values of α, where the optimal capital income tax rate is negative. Parameters, as above, are $\rho = .3$, $\delta = 1 - (1 - .1)^{25}$ and $\zeta = (1.03)^{25}$

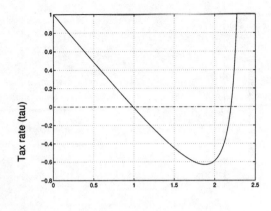

After-tax return (R)

Figure 12.8: The capital income tax rate τ in the case of a variable capital stock in dependence of the annualized equilibrium return R. The economy is dynamically inefficient. Parameters are $\alpha = .1$, $\rho = .15$, $\delta = 1 - (1 - .1)^{25}$ and $\zeta = (1.03)^{25}$. Note, how there are again two equilibrium returns R for any given τ in the appropriate range

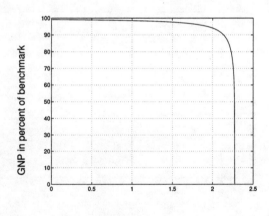

After-tax return (R)

Figure 12.9: Equilibrium output in dependence of the annualized equilibrium return R in case of a variable capital stock. The economy is dynamically inefficient. Parameters, as above, are $\alpha = .1$, $\rho = .15$, $\delta = 1 - (1 - .1)^{25}$ and $\zeta = (1.03)^{25}$

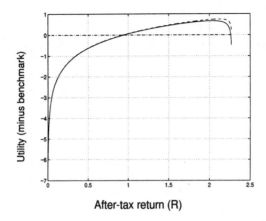

Figure 12.10: Equilibrium welfare of the two-period lived as well as the initially old in dependence of the annualized equilibrium return R in case of a variable capital stock. The economy is dynamically inefficient. Parameters, as above, are $\alpha = .1$, $\rho = .15$, $\delta = 1 - (1 - .1)^{25}$ and $\zeta = (1.03)^{25}$

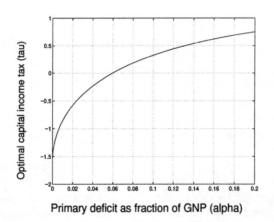

Figure 12.11: Welfare-maximizing capital income tax rate in dependence of the primary deficit αy_t in the case of a variable capital stock. The economy is dynamically inefficient, except for larger values of α, which imply an optimal positive capital income tax rate. Parameters, as above, are $\rho = .15$, $\delta = 1 - (1 - .1)^{25}$ and $\zeta = (1.03)^{25}$

Some of the theoretical facts are stated in the next proposition. It is important to keep in mind for these figures as well as for the following proposition, that this is a comparison across steady states: in particular, the endowment of the initial old is changing in this comparison. Furthermore, it is important to note, that different capital income taxes mean different total government expenditures. A complete welfare analysis will have to take that into account, if the goods purchased by the government enter the utility function of the agent.

Proposition 12.7 1. *The welfare of the initial old has a global maximum in the range of returns sustaining a deficit at a return $R_{\max,1}$. The welfare of the initial old is increasing in R for R smaller than this maximizing return and decreasing in R for R bigger than $R_{\max,1}$.*
2. The welfare of the two period lived agents has a global maximum in the range of returns sustaining a deficit at a return $R_{\max,2}$, which is smaller than the return $R_{\max,1}$. The welfare of the two period lived agents is increasing in R for R smaller than this maximizing return and decreasing in R for R bigger than $R_{\max,2}$.

The proof is in the appendix. The intuition behind this result should be clear, however. Consider again the model with a fixed capital stock and no investment. There, welfare was unambigously increasing in the return, because total output does not change, consumption and savings in the first period of the life stay the same, but second period consumption increases with the return on the constant savings. In the model here, where the capital stock is variable, this effect is counteracted by the decrease of output at higher returns for the reasons explained above. Overall then, the first effect is more important for very small returns: in essence, second period consumption rises faster than output declines. Eventually, however, the crowding-out effect starts to take over, resulting in declining welfare, as the return becomes too big. Associated with the return $R_{max,2}$, which makes all the two-period lived agents best off, is a capital income tax rate which can be calculated from the formulas above and the formulas given in the proof of proposition 12.7. Leaving out the initial old (which could be motivated with a social welfare function which aims at assigning the same weight to all generations), this capital income tax may be considered optimal, given that the deficit needs to be sustained forever. Note, that this tax will in general be quite substantial. This conclusion still holds true, if a social welfare function assigns weight to the initial old generation as well: in general, the maximizing tax rate will correspond to a return somewhere between $R_{max,2}$ and $R_{max,1}$. The welfare-maximizing tax rate has been plotted in Figures 12.7 and 12.11 as a function of the primary deficit parameter α for the two parameterizations used: note, how the tax rate can vary between a negative number in the case of a dynamically inefficient economy to a quite substantial positive number (with $\tau = 1$ corresponding to confiscation).

12.6 Interpretation

To what extent are the arguments brought forward in this paper relevant for the current situation of the Netherlands, say? It should be noted, that even though the numbers were picked to be suggestive, a more careful calibration would be necessary before drawing reliable conclusions. One insight from the numerical experiments is that the capital income tax necessary to sustain a deficit is very sensitive in particular to the capital share and to the deficit that needs to be sustained, when the benchmark economy is dynamically inefficient. The more reasonable choices for the parameters, however, seem to imply capital income taxes which are very high. This may give rise to worry about the applicability of the results presented here.

However, a few remarks may be in order in defense of these numbers. First of all, it actually may be the case that the realized capital income taxes in the Netherlands are very high indeed, simply because the taxation is not indexed by a price-deflator: if, say, inflation and not real appreciation has trippled the price of some asset, a capital gains tax of, say, 50 per cent amounts to a tax of 33 per cent on the original value of the asset. Furthermore, even if the government does not impose the extremely high capital income tax rates implied by the model, it certainly depresses the market interest rate, thereby easing the burden of debt, even though the government may not be able to sustain its deficit level forever. This point is possibly well understood by governments, which seem to have great difficulties in reducing the debt and at the same time are often reluctant to remove the capital income tax.

12.7 Conclusions

It was examined, whether a government can run a deficit forever as a fraction of total output in several overlapping generations models with growth by rolling over its debt. The answer is 'yes' for the model without capital, 'no' for the model with capital, but 'yes' for the model with capital and capital income taxation. The impact on steady state output and welfare by the capital income taxes that make a permanent government deficit sustainable was analyzed. It was shown that for a certain range of capital income taxes, there are two equilibria: one with a low return on savings and one with a high return on savings. For the low return equilibria, output is an increasing function of the capital income tax, whereas for the high interest rate equilibria, output is a decreasing function of the capital income tax. It was demonstrated, that output in the version of the model with investment was maximized at a capital income tax rate approaching 100 per cent, if the economy can be in the low return equilibrium. While welfare is not maximized at this point, it is still true that the welfare maximizing capital income tax may be very substantial, given that a government deficit needs to be sustained forever.

Appendix A

Proof of proposition 12.5

Part 1 is clear from looking at the formula for τ and letting R/γ approach 0 and $1 - \frac{2\alpha}{1-\rho}$, respectively.

Part 2. It is clear, that $\tau < 1$ for the range of admissable gross interest rates R. It thus remains to show that the derivative $\frac{d\tau}{dR}$ has a unique zero in this range. In order to prove this, it is convenient to change notation. Let

$$
\begin{aligned}
\nu &= \frac{1-\rho}{2\rho} \\
\mu &= \frac{\alpha}{\rho} \\
\lambda &= (\nu - \mu + 1) * \nu/\mu \\
\eta &= (1 - \mu/\nu) * (\nu - \mu) * \nu/\mu \\
z &= \frac{R/\gamma}{1 - \mu/\nu}
\end{aligned}
$$

noting that $\mu < \nu$ (and thus $\lambda > 1$) in order to have a nonempty range of interest rates to sustain a deficit in equilibrium to begin with. Then

$$
\begin{aligned}
\tau(z) &= 1 - \frac{z(1 - \mu/\nu)}{1 + \frac{1}{\nu - \frac{\mu}{1 - z(1 - \mu/\nu)}}} \\
&= 1 - \frac{\eta}{\frac{\lambda}{z} + \frac{1}{1-z}},
\end{aligned}
$$

where $z \in (0\,;1\,)$: it suffices to show, that $\tau'(z)$ has a unique zero in the unit interval. Note, that $\tau'(z) = 0$ on this interval is equivalent to

$$
0 = -\lambda(1-z)^2 + z^2, z \in (0\,;1)
$$

or

$$
z_{min} = \frac{\sqrt{\lambda}}{1 + \sqrt{\lambda}},
$$

proving the claim. Note that this z now corresponds to

$$R/\gamma = (1 - \mu/\nu)\frac{\sqrt{\lambda}}{1 + \sqrt{\lambda}},$$

and that one can calculate the location of the minimum to be

$$\tau(z_{min}) = 1 - \frac{\eta}{(1 + \sqrt{\lambda})^2}$$

Part 3 follows immediately from part 2. (d)
Part 4 follows from observing that

$$u(c_{1t}) + u(c_{2t+1}) = 2\log(y_t/2) + \log(R)$$

for $t \geq 1$ and

$$u(c_{20}) = \log(y_1/(2\gamma)) + \log(R)$$

for the initial old. ∎

Proof of proposition 12.7

Note, that the welfare function for both, the initially old and the two-period lived agents, is given by

$$W(R) = \log(R) + \varphi \log(y_1)$$

up to an additive term, where φ equals 1 for the inital old and φ equals two for the two-period lived agents (for a social welfare function, which weighs together all utility functions, φ will be somewhere between one and two). Taking the derivative with respect to R and using the formula for y_1 in the text, it follows that

$$
\begin{aligned}
W'(R) &= \frac{\gamma}{R/\gamma} - \frac{2\varphi\alpha/\gamma}{(1 - \frac{R}{\gamma})^2}\frac{1 - \frac{R}{\gamma}}{(1 - \rho)(1 - \frac{R}{\gamma}) - 2\alpha} \\
&= \frac{\frac{R^2}{\gamma^2} - (2 + (\varphi - 1)\frac{2\alpha}{1-\rho})\frac{R}{\gamma} + 1 - \frac{2\alpha}{1-\rho}}{R(1 - \frac{R}{\gamma})(1 - \frac{2\alpha}{1-\rho} - \frac{R}{\gamma})},
\end{aligned}
$$

where the quadratic in the numerator has the two roots

$$\frac{R}{\gamma} = 1 + (\varphi - 1)\frac{\alpha}{1 - \rho} \pm \sqrt{(1 + (\varphi - 1)\frac{\alpha}{1 - \rho})^2 - 1 + \frac{2\alpha}{1 - \rho}}.$$

Some more algebra then reveals that only the lower root lies within the admissable range, that the quadratic in the numerator is positive in this range for R/γ below this root and negative above it and that the lower root is a decreasing function of φ. This finishes the proof. ∎

Acknowledgments

An early draft of this paper was written while the author was an assistant professor at Princeton University. I am grateful in particular to Abhijit Baner-

336 *Harald Uhlig*

jee, Ben Bernanke, Timothy Besley, Henning Bohn, John Campbell, Glenn Donaldson, Jon Faust and Jim Hines for helpful comments. Harald Uhlig, CentER, Tilburg University, Postbus 90153, 5000 LE Tilburg, THE NETHERLANDS, UHLIG@KUB.NL, Fax: (+31) 13 - 4663066.

Notes

1. A result of this type can already be found in Scheinkman (1980), see also Tirole (1985).
2. Note, that this is a Pareto ordering of different steady states. In particular, the intially old agent is endowed differently in these different steady states, making this welfare comparison potentially misleading.

References

Abel, A.B., N.G. Mankiw, L.H. Summers and R.J. Zeckhauser (1989), "Assessing Dynamic Efficiency: Theory and Evidence," *Review of Economic Studies*, 56:1-20.
Aiyagari, S.R. (1985), "Deficits, Interest Rates, and the Tax Distribution," *Federal Reserve Bank of Minneapolis, Quarterly Review,* 9 (1):5-14.
Atkinson, A.B., and A. Sandmo (1980), "Welfare Implications of the Taxation of Savings," *Economic Journal,* 90:529-49.
Auerbach, A., J. Gokhale and L. Kotlikoff (1994), "Generational Accounting: A Meaningful Way to Evaluate Fiscal Policy," *Journal of Economic Perspectives,* 8:73-94.
Balasko, Y. and K. Shell (1980), "The Overlapping-Generations Model, I: The Case of Pure Exchange without Money," *Journal of Economic Theory,* 23:281-306.
Barro, R.J. (1974), "Are Government Bonds Net Wealth?," *Journal of Political Economy,* 82:1095-117.
Blanchard, Olivier J. and Stanley Fischer (1989), *Lectures on Macroeconomics*, MIT Press, Cambridge, Massachusetts
Bohn, H. (1990), "The Sustainability of Budget Deficits in a Stochastic Economy," *Rodney L. White Center for Financial Research Discussion Paper* 6-90, Wharton School of the University of Pennsylvania.
Bovenberg, L. and C. Petersen (1992), "Public Debt and Pension Policy," *Fiscal Studies*, 13:1-14.
Bovenberg, L. (1989), "The Effects of Capital Income Taxation on International Competitiveness and Trade Flows," *American Economic Review*, 79:1045-64.
Bovenberg, L. (1992), "Residence- and Source-Based Taxation of Capital Income in an Overlapping Generations Model," *Journal of Economics* (Zeitschrift für Nationalökonomie), 56:267-95.
Broer, D. Peter, Ed W. M. T. Westerhout and Lans Bovenberg (1994),"Taxation, Pensions and Saving in a Small Open Economy," *Scandinavian Journal of Economics,* 96 (3):403-24.
Buiter, W.H. and U.R. Patel (1992a), "Debt, Deficits, and Inflation: An Application to the Public Finances of India," *Journal of Public Economics,* 47 (2):171-205.
Buiter, W.H. and K.M Kletzer (1992b), "Who's Afraid of the Public Debt?," *American Economic Review,* 82:290-94.
Buiter, W.H. (1993), "Public Debt in the USA: How Much, How Bad and Who Pays?," *NBER Working Paper* 4362 .
Buiter, W.H. and K. M. Kletzer (1994), "Ponzi Finance, Government Solvency and the Redundancy or Usefulness of Public Debt," *Yale Cowles Foundation Discussion Paper* 1070.
Cass, D. (1972)., "On Capital Overaccumulation in the Aggregative Neoclassical Model of Economic Growth: A Complete Characterization," *Journal of Economic Theory*, 4:200-23.
Carlberg, Michael (1995), "Sustainability and Optimality of Public Debt," *Contributions to*

Economics, Springer, Physica, Heidelberg.

Chari, V.V .(1988), "Time Consistency and Optimal Policy Design," *Federal Reserve Bank of Minneapolis, Quarterly Review* 12 (4):17-31.

Darby, Michael R. (1985), "Some Pleasant Monetarist Arithmetic," *Federal Reserve Bank of Minneapolis, Quarterly Review* 9 (1):32-7, reprinted from Federal Reserve Bank of Minneapolis, Quarterly Review Spring (1984): 15-20.

Diamond, P (1965), "National Debt in a Neoclassical Growth Model," *American Economic Review*, 55:1126-50.

Diamond, P (1973), "Taxation and Public Production in a Growth Setting," in: J.A. Mirrlees and N.H. Stern (eds.), *Models of Economic Growth*, Macmillan, London.

Eisner, R. (1986), *How Real is the Federal Deficit*, Free Press, New York.

Eisner, R. (1992), "Deficits: Which, How Much, and So What?," *American Economic Review* 82:295-298.

Friedman, B. (1992), "Learning from the Reagan Deficits," *American Economic Review* 82:299-304.

Krugman, P. (1990), *The Age of Diminished Expectations — US Economic Policy in the 1990s*, MIT Press, Cambridge MA.

Lucas, R.E., Jr. (1990), "Supply-Side Economics: An Analytical Review," *Oxford Economic Papers* 42:293-316.

Miller, P.J. and T.J. Sargent (1985), "A Reply to Darby," *Federal Reserve Bank of Minneapolis, Quarterly Review* 9 (1):38-43, reprinted from Federal Reserve Bank of Minneapolis, Quarterly Review Spring(1984):21-26.

Sargent, T.J. (1987), *Dynamic Macroeconomic Theory*, Harvard University Press, Cambridge, Massachusetts.

Sargent, T. J. and N. Wallace (1985), "Some Unpleasant Monetarist Arithmetic," *Federal Reserve Bank of Minneapolis, Quarterly Review* 9 (1):15-31, reprinted from Federal Reserve Bank of Minneapolis, Quarterly Review Fall (1981): 1-17.

Scheinkman, J. (1980), "Note on Asset Trading in an Overlapping Generations Model," *mimeo*.

Tirole, J. (1985), "Asset Bubbles and Overlapping Generations," *Econometrica*, 53:1499-528.

Uhlig, H. and N. Yanagawa (1996), "Increasing the Capital Income Tax May Lead to Faster Growth," *European Economic Review,* 40:1521-40.

Vickrey, W. (1992), "Meaningfully Defining Deficits and Debt," *American Economic Review* 82:305-310.

Wallace, N. (1980), "The Overlapping Generations Models of Fiat Money," in: J. Kareken and N. Wallace (eds.), *Models of Monetary Economics*, Federal Reserve Bank of Minneapolis, Minneapolis.

PART VII
Policy
Implications

13 Policy Implications of Endogenous Growth Models

Jarig van Sinderen and Theo J.A. Roelandt

13.1 From growth accounting to endogenous technology growth

Up to the mid-eighties in traditional macroeconomic neoclassical growth theory exogenous technological development was presupposed. It was assumed that economic actors behaviour barely influenced technological renewal. Classical growth theory treated technology like manna from heaven, positively influencing per capita income growth in countries. The unexplained residual in traditional growth accounting estimates was attributed to *exogenous* technological progress. In other words, a statistical 'measure of our ignorance,' has been interpreted as an indicator of the impact of technology on growth. This kind of reasoning leaves little room for an analysis of the question how firms' strategic behaviour as well as government intervention may affect long-run growth through its impact on technological progress apart from exogenously stimulating general technology development. Also the influence of profit motivated R&D within firms could not be explained by this growth theory. From this traditional perspective technology policy primarily concentrates on enhancing competition (competition policy) and on stimulating fundamental scientific research (science policy). The actual measurement of the effectiveness of such policies is quite difficult. As a consequence, in economic theory the role of the government was underestimated. In this traditional view the influence of investment in fysical and technological infrastructure on growth is absent. However, in practice economic policy making has concentrated on infrastructure and also on investments in technology and human capital and is today motivated by market failure and microeconomic notions thus neglecting old fashion growth theory.

Dissatisfaction with the gap between actual policies pursued and the failure of macroeconomic growth theory to underpin such policies in an era in which neoclassical economics submerged and supply-side notions became important, stimulated economists in the eighties to develop models analysing technological progress as a result of economic decision making by firms, consumers and governments. In this new approach investments in technology and in research and development are expected to improve firms competitiveness and economic performance. At the core of these models is the proposition that investment in human as well as physical capital and the production of new processes and products, is central to growth which can be driven on without being halted by diminishing returns (Crafts, 1996). Productivity growth results from an active search process for innovations in which the ability to appropriate profits determines the resources devoted to innovative activity (OECD, 1992, Crafts, 1996). Some new growth economists emphasise the role of economic institutions and profit-motivated investments in searching for new products and processes. The key feature of endogenous growth models is that growth depends on the incentives to invest in improving technology. Endogenous growth models allow institutions and government policy to play a central role in long-run growth outcomes (Crafts, 1996). Government policy has an impact on several growth inducing factors such a infrastructure, human capital formation and enhancing the functioning of markets.

Especially the work of Romer (1986) has given rise to developing new growth models including knowledge as a substantive means of production in generating economic growth. Table 13.1 summarizes some of the variables at stake in modern growth models that are key variables to government policies. The studies included in this table illustrate that modern growth theory covers a broad spectrum of the possible relations between economic growth and government policy like investment in technological and physical infrastructure, profit induced R&D and the spill-over effects of R&D from one sector to another. In modern growth theories firms' research and development efforts have an important impact on economic performance by influencing labour and capital productivity, by increasing the knowledge intensity of products and by improving value adding economic activities. Firms' research and development efforts and immaterial investments also have spill-over effects in the sense that technology developed in one sector can improve the productivity and knowledge intensity of labour and capital held by others. Firms invest in technology development when private profit opportunities and a substantive private return on R&D-capital can be expected. Governments have an important role to play when clear external spill-over effects exists (social return on R&D- investments), achieving the right balance between patent regulation (knowledge protection and disclosure), competition policy (enhancing competition) and technology policy (public investments in education and R&D and improving the interaction between knowledge producing and using agents).

Study	Variable
Romer (1986)	Increasing marginal productivity of knowledge (learning by doing)
Griliches (1986)	R&D-capital, investment in basic research
Fagerberg (1987)	Growth in patent applications
Lucas (1988)	Human capital in production function
Englander & Mittelstädt (1988)	R&D capital growth.
Romer (1990)	Human capital stock
Rebelo (1991)	Spill overs (innovation by imitation and adoption). Private and social return on R&D-investments
Grosman & Helpman (1990)	Product innovation.
Romer (1990) Grosman & Helpman (1991)	Seperate technology sector
Aghion and Howitt (1990)	Learning by designing
Barro (1990)	Public R&D and investment in infrastructure
Van Sinderen (1990) (1993)	Infrastructure in production function; direct impact of taxation on R&D
Verspagen (1994)	R&D capital growth in private and government sector
Ben-David & Loewy (1996)	Knowledge accumulation and liberalised international trade
Sakurai *et al.* (1995)	(Inter)national embodied technology flows (absorptive capacity)
Sources: Fagerberg (1994); Schneider and Ziesemer (1994), Ministry of Economic Affairs, (1995); Roelandt *et al.* (1996)	

Table 13.1: Technology related sources of growth in new growth models

In 1986, Romer developed a model in which economic growth is explained by economic decisions at the micro level. Private agents respond to perceived profit opportunities when investing in human capital or R&D, while governments provide public goods or stimulate private investment through subsidies. This has been seen as the starting point of endogenous growth theory. Romer endogenizes the growth rate by including knowledge as a production factor.

13.2   Endogenous technological progress and public policy

New growth economists have explored various mechanisms through which economic policy may affect economic growth. Lucas (1988), for example, incorporates human capital in the production function. Private firms and individual households have incentives to accumulate human skills and knowledge because it improves their productivity. Moreover, the productivity of each worker increases with the total stock of human capital in the economy. This constitutes a positive externality related to the individual accumulation of human capital and, as such, provides a motivation for government intervention. Rebelo (1990) stresses the importance of positive spill-over effects. Some activities, undertaken by agents and induced by private motives, may have positive external effects to other agents in the economy. For example, firms that invest in R&D may discover a more efficient production process or invent a new product. Other firms may adopt the same technology or imitate the product when information becomes available, for example, when a patent expires and no longer offers protection against imitation. Hence, when a firm has successfully invested in R&D, not only the individual firm itself will benefit, but other firms (both domestically and outside the country) do as well. Consequently, the social benefits of R&D expenditures are larger than the private benefits so that there may be a role for government intervention in stimulating R&D. Grossman and Helpman (1990) also recognise that new product varieties or new technologies may be important engines of growth. Hence, public policies that encourage product and process innovations may improve the performance of the economy. Other examples of such spill-overs are learning-by-doing effects from investment in physical capital (Romer, 1986 and De Long and Summers, 1991), learning-by-designing effects induced by R&D activities (Aghion and Howitt, 1990), or the effects of complementary public services (such as education, public R&D activities or infrastructural projects) which raise the productivity of economic activities in the private sector (Barro, 1990; Toen-Gout and Van Sinderen, 1995). Indeed, in each of these cases, public policies can stimulate economic growth by internalising the external effects connected to certain activities. The government can accomplish this either by subsidising private activities with positive external effects or by providing public services with positive external effects itself. However, the actual effect on economic growth depends also on the way these expenditures are financed. Whether the government expenditure is financed by taxation, on the capital market or by means of a reshuffeling of government outlays will influence the effect on economic growth. Taxation on capital or on labour will work out differently on the growth rate and on the labour intensity of growth.

In a recent article Ben-David and Loewy (1996) considered an open economy endogenous growth model with knowledge accumulation. In this model knowledge accumulation stems from both the available stock of knowledge as well as from knowledge spill-overs from other countries. This accumulation process is assumed to be driven by the degree to which each country is able to absorpt and apply the knowledge spill-overs coming from its trading partners to its own knowledge stock.

What are the gains for economic policy from exposure to endogenous innovation growth models? Governments have an important role to play when clear external spill-over effects exists (social return on R&D-investments), in achieving the right balance between patent regulation (knowledge protection and disclosure), in competition policy (enhancing competition) and in technology policy (public investments in education and R&D and in improving the interaction between knowledge producing and using agents). On the other hand the focus on economic decision making processes at the micro-level stresses the importance of taxation and the impact of government policies on the individual decision making at both the firm level and at the level of households.

All kinds of policies (economic, fiscal, trade and competition policies) influence the incentives to innovate. A first example of a possible link between policy and technology is suggested by the industrial organisation literature which is rich on the link between market structure and economic performance (innovation, product quality, and so on; see Scherer and Ross, 1990). In particular this literature relates endogenous technology to competition policy. Smulders (1994) illustrates various interrelations between the degree of competition and innovative activity in a theoretical framework. Modelling R&D as a fixed cost for firms he finds that, on the one hand, competition may harm technological progress as firms have insufficient means to innovate. On the other hand, tougher competition may also boost economic growth as innovation may yield a higher social rate of return. An empirical illustration of the link between technology and competition in the Netherlands is provided by Den Hertog and Thurik (1993) who explore the determinants of the decision to either produce R&D inside the firm or to buy external R&D (consultants, public research centres, and so on). Based on a sample of 446 Dutch firms for 1984 they conclude that external experts are consulted especially by small and medium-sized capital intensive firms that operate on highly competitive markets. A similar relationship is found by Van den Berg (1989), who shows that an increase in concentration or industry size leads to a decrease in innovative activity. These findings suggest that there might be an important positive relation between the level of (external) R&D and the new stance of more vigorous competition policy that the

Netherlands follows since 1993. More competition may yield higher levels of external R&D demand and, in addition, induce more innovative activity within firms. In the Dutch figures for R&D this effet can not yet be noticed, which does not imply that in the longer term this effect will not occure. However, this possible positive effect is in sharp contrast to the Schumpeterian insights, which suggest that some sort of monopoly power may be necessary to provide the funds for R&D activities or to appropriate more fully the yields of technology. If so, then a more vigorous competition policy will reduce innovation and may lead to lower growth (see, for example, Davies, 1988). More research on this phenomenon of competition and R&D is therefore necessary. Empirical research in this field still shows ambiguous results, although most studies show that competition stimulates innovation (Symeonidis, 1996). Indeed, in the Netherlands private R&D is concentrated in a small number of large firms and empirical research for the Netherlands only has shown a weak correlation between market concentration and innovative activities (Roelandt *et al.*, 1996). Besides, the innovative capacity of a small and open economy as the Netherlands strongly depends on its capacity to absorb technology by using (and partly importing) high tech intermediate and capital goods. Recent research has shown that embodied technology flows accounts for about half of the total R&D-intensity in the Netherlands (Papaconstantinou *et al.*, 1995). Foreign technology procurement through imports are substantial sources of diffusion-based productivity gains. Encouraging diffusion of technology needs open borders and competitive pressure as well as a patent system that stimulates investment in R&D by creating reasonable royalties on innovation.

A second relationship between economic policy and technology is suggested by the public finance literature that deals with the effects of taxation on technical progress and human capital formation. This strand of the literature highlights the trade-off that policy makers face between, on the one hand, government spending on R&D and education that may stimulate economic growth and, on the other hand, higher taxes that are required to finance these expenditures but which have detrimental effects on economic performance. To illustrate, labour taxes drive a wedge between before and after-tax wages, thereby reducing the incentives to work (see, for example, Van Sinderen, 1990, 1993). However, not only the quantity but also the quality of labour supply is affected by marginal income taxes. For example, Trostel (1993) shows that income taxation discourages investment in human capital, thereby seriously damaging the upgrading of human skills. Capital income taxes are also distortionary as they drive a wedge between before and after-tax returns to capital, thereby discouraging savings and investments (Boskin, 1978, Van Sinderen, 1990). Accordingly, lower taxes on capital may raise

the after-tax rate of return on capital and thus stimulate investment in physical capital and R&D which is an important engine of growth. In addition,
different categories of public expenditures will have different effects on economic growth, so that the relative effectiveness and efficiency of each public
spending category should also be taken into account when a tax increase is
considered. In this respect it is always important to compare the marginal
costs of an extra guilder of taxation with the benefits form an extra guilder of
government expenditure. Especially in a country like the Netherlands with
high overall and marginal tax rates the marginal costs of an extra guilder taxation are high. In such economies a balanced budget cut in taxes might be
more effective as a comparable extra stimulus in R&D. In the next section we
will analyse these questions further. At the same time it should be stressed
that a policy primarily focusing on stimulating R&D in a knowledge based
economy will be doomed to fail when the macroeconomic conditions are un-
stable or leave too little room for enterpreneurship. Preconditions for such
a stable economic environment are a healthy government sector which does
not overtax or overregulate an economy and which implements a competition policy aiming at stimulating the functioning of both the product and the
labour markets. Because of the reasons mentioned some autors and politicians prefer not to concentrate on an explicit technology policy, but advise
to concentrate on an improvement of the overall economic climate. In this
paper we will investigate the economic impact of several policy options,
analysing a direct technology policy (stimulating R&D) as well as indirect
measures aiming at improving the general macroeconomic climate. In this
respect tax reductions are an important policy option.

13.3 New growth theory elements in macroeconomic modelling

13.3.1 MESEM and MESEMET

13.3.1.1 MESEM

MESEM (MacroEconomic Semi Equilibrium Model) is an applied general
equilibrium model for a small open economy that is parameterized for the
Netherlands in 1992 (Van Sinderen, 1993 and Ministry of Economic Affairs,
1995). A main characteristic of MESEM is that it takes into account the
'incentive effect' of taxation through explicit modelling of the tax wedges
on labour and capital. The utility function of households contains both in-

come and leisure as arguments and is of a nested CES type. The production function is also of the CES type containing labour, both public and private capital and technical progress. We assume constant returns to scale. In the model entrepreneurs maximize profits and consumers maximize utility. As a result supply and demand functions for captial and labour are specified. In the MESEM model equations are modelled in such a way that the impact of marginal tax rates on the choice between labour supply and leisure is not only a microeconomic phenomenon, but also yields macroeconomic effects. MESEM contains an aggregated labour market and a capital market which contains supply and demand of both bonds and of equity capital. Capital is mobile, labour is not. The nominal interest rate on government bonds is determined abroad while the rate of return on investments in fixed capital depends on the interaction between supply and demand on the market for equity capital. Given labour and capital costs, entrepreneurs demand labour services and capital services until the marginal product of these factors of production equals the marginal cost. Employees supply labour services while capitalists supply capital services. Supply of the factors of production depends, apart form scale factors, on the net income earned per unit of labour and capital supplied. Taxes drive a wedge between the costs for the entrepreneur and the net income for the suppliers of the factors of production. Labour income is entirely spent on consumption while capital income is partly consumed and partly saved. Accordingly, only the suppliers of capital save (see for an extensive discription of MESEM, Van Sinderen, 1993). In MESEM government investment in infrastructure is included in the production function and is at the same time an imperfect substitute for private investment. Investment in R&D and human capital depends on the change in the marginale tax rate. In this respect MESEM contains elements of the new- growth theory. Still, MESEM asumes decreasing returns to scale in the accumulation of capital. Attractive features of MESEM are further that it offers an analytical framework for separating supply and demand effects of various policy shocks (Van Sinderen, 1992, Van Sinderen and Waasdorp, 1994 and Van Bergeijk, Van Sinderen and Waasdorp, 1993) and that it allows for the empirical analysis of the impact of different assumptions about the speed of price adjustments (Van Bergeijk, Haffner and Waasdorp, 1993 and 1994 and Van Sinderen *et al.*, 1994).

MESEM is not a traditional endogenous growth model where economic agents are represented by dynamic optimising behaviour and are modeled by intertemporal equilibria; the core of the model is a description of a sequence of static equilibria. By assuming perfect market-clearing in the long run, the economy ultimately reaches a new equilibrium in which there is no under-utilisation of inputs. Dynamics enters the model as the economy needs time to adjust after an external shock. Hence, in the short and medium term, markets do not clear instantaneously. Indeed, this adjustment process makes MESEM a powerful tool for economic policy analysis as it allows to analyse both the medium-term and the long-term economic consequences of various policy measures.

13.3.1.2 MESEMET

MESEM is rather limited as technology is concerned. In order to get more insights into the effects of technology on economic development and also in order to analyze different policy options concerning stimulating technological progress, the MESEM- model has been elaborated with endogenous technology. This model is called MESEMET (MESEM with Endogenous Technology). The MESEMET model production function is of the nested-CES type. Such a structure implies that we still assume constant returns to scale. In this respect MESEMET does not follow some important mainstream new growth theories which assume increasing returns to scale. The MESEMET production structure is given in equation (13.1)-(13.5).

$$Y_p = Y_p(L_{eff}, K_{eff}) \qquad (13.1)$$

$$L_{eff} = L_{eff}(HC, L) \qquad (13.2)$$

$$K_{eff} = K_{eff}(KT, K_g) \qquad (13.3)$$

$$KT = KT(TC, K) \qquad (13.4)$$

$$TC = TC(TC_p, TC_g) \qquad (13.5)$$

In which
- Y_p = production capacity
- L_{eff} = effective labour
- K_{eff} = effective capital (including productive public capital)
- KT = capital in the private sector
- TC = technological capital
- HC = human capital
- L = labour
- K_g = public capital
- K = physical capital
- TC_p = private technological capital
- TC_g = public technological capital

In this model structure it is important to note that human capital (HC) as well as both public and private technological capital (TC_p and TC_g) are endogenous specified. Technology in our model is embodied in physical capital. The technological component which is disembodied is included in human capital. The specification for investments in human capital reads as follows:

$$HC = (1 - \delta) \, HC_{-1} + \delta f \, (I_p, RD_p, RD_g, E_g, T_i) \qquad (13.6)$$

In which: I_p = Private investments
RD_p = Private R&D
RD_g = Public R&D
E_g = number of people employed in education
T_i = taxes on labour income

Human capital is defined as a stock with a constant depreciation rate (δ) of 0.12. There is a positive relation between investments in human capital, total private investment (I_p), R&D investments in both the private (RD_p) and the government sector (RD_g) and the total number of people employed in the education sector (E_g). High marginal tax rates on labour income (T_i) will have a negative impact on the investment in human capital. The latter mechanism was already included in MESEM. The government can stimulate human capital formation by investing in public R&D, by increasing expenditure on education and stimulating private R&D or by reducing marginal tax rates.

Technological capital is divided into private and public R&D. In our model, private R&D depends on wage and capital costs, while public R&D is exogenous determined by government. Technological capital accumulates with the investments in R&D. At the same time the stock of technogical capital depreciates at depreciation rates δ_p and δ_g for private and public capital, respectively.

13.3.1.3 Parameterization

The MESEMET model parametrization can be found in more detail in Van Sinderen (1993), Van Hagen (1994) and the Ministry of Economic Affair (1995). As an example we analyse the parameterization of equation (13.6). In order to guess the parameters of the formation of human capital we first calculate the gross government and private expenditure on the different categories of investment included in the specification of investment in human capital. At the same time a guesstimate of the marginal productivity index is made based on the literature. As a result a implied elasticity is calculated (see Table 13.2).The depreciation rate of human capital is set at .12 which implies that of the education of 20 years ago some 10 per cent is still relevant. The coefficient of the influence of the taxrate is 0.2. Equation (13.6)

Determinants of human capital	Gross expenditures (bns of Hfl)	Share (%)	Marginal productivity index[a]	Product	Implied elasticity (rounded)
	(1)	(2)	(3)	(4)=(2)·(3)	(5)
Education	15	16	3	48	0.45
Public R&D	5	5.5	3	17	0.15
Private R&D	6	6.5	1	7	0.05
Private investment	65.5	72	0.5	36	0.35
Totals	91.5	100		108	1

[a] Private R&D is set equal to 1

Source: Ministry of Economic Affairs, 1995.

Table 13.2: Determinations of long-run elasticities with respect to human capital

reads with the parameterization as follows:

$$HC = 0.88HC_{-1} + 0.12\left[.35I_p + .05RD_p + .15RD_g + .45E_g - 0.2T_i\right] \tag{13.7}$$

Equation (13.7) shows that private investments and public education are the most important determinants of the formation of human capital. It is also shown that the positive elasticity of public R&D with respect to human capital is higher than that of private R&D.

13.3.2 Policy instruments

MESEMET incorporates various elements from new growth theory with a special bias towards technological development and human capital formation. In particular, it deals with public investment, human capital, technology capital and various spill-over effects of these activities. Both MESEM and MESEMET can be used to analyse the effects of taxation and different types of government expenditure. However, MESEMET allows for a more explicit examination of the relationship between, on the one hand, fiscal policy and competition policy and, on the other hand, technological progress and economic performance.

We use this model to examine the effectiveness of various policy instruments such as subsidies on private R&D, public investments in R&D and private and public physical capital. Furthermore, the various indirect relationships between the economic and institutional environment, endogenous technological progress and economic performance can be explored.

What can the government do in the MESEMET world? Three direct policy options are open to stimulate the accumulation of knowledge. Firstly, the government can create additional jobs in either public schools or in public R&D institutes. In this way, quality and quantity of, respectively, educational services and academic output can be raised. Scientific knowledge (that is, public R&D) spills over to human capital (and hence labour productivity) as well as to the innovative capacity of enterprises (see (13.7)). Public education is also an important determinant of the stock of human capital. Second, the government may start subsidising a fraction of the R&D activities undertaken by private enterprises. Several indirect mechanisms can be discerned as well. According to the OECD (1987), the 'capabilities' of a country are not sufficient in generating economic growth by themselves. It is also necessary, for example, that the organisation of the tax system allows these 'capabilities' to be transformed into performance through 'incentives.' So a third policy option is a general reduction in the tax on labour that might encourage innovation and growth. Taxation, for example, influences private sector decisions concerning R&D activities and investments in physical capital and human capital. Taxes have a number of effects on the economy. For a start, they reduce disposable income in the private sector. This influence on effective demand plays a major role in the traditional Keynesian type of analysis. These negative demand effects, however, are essentially a short-term phenomenon. In the medium term the supply side effects of taxes may gain importance. This is because the supply of the factors of production depends, among other things, on the net income which can be earned per unit of labour or capital supplied. Taxes drive a wedge between the costs for the entrepreneur and the net income for the supplier of the respective production factor. As a result high taxes act as a disincentive on the propensities to invest, to innovate, to work and to (re)train. Economic policy also impacts the primary income distribution which offers an additional indirect channel through which government influences technological change. Since investments are financed from profits, macroeconomic performance also determines the volumes of physical investment and private R&D. A fourth possibility is to stimulate the functioning of markets and to deregulate the economy in order to increase competition. It should be stressed that all policy measures should not necessarily be seen as exclusive. It might be a very promising policy to create a healthy fiscal climate on the one hand and stimulate R&D directly on the other. The MESEMET framework allows for the investigation of the impact of technology and economic policies in different institutional settings. Essentially, four regimes can be discerned depending on the speed of price adjustment and the importance of supply side elements (Table 13.3). First, in the original MESEMET version (in the lower left part of Table 13.3), both product and labour markets clear in the long run, while

	Market Clearing	Market Inertia
No supply side elements	Short-term neo-classical model (SN)	New Keynesian model (NK)
Supply and demand side elements	MESEMET original model	Empirically relevant context (ER)

Source: Ministry of Economic Affairs, 1995, p. 21

Table 13.3: The four regimes of MESEMET

both supply and demand effects of taxation are taken into account. As said in Section 3.1, MESEMET allows for the possibility to separate the demand effects from the supply effects. Switching off the supply effects therefore results in a more or less demand driven equilibrium model (the upper left part of Table 13.3). Third, the institutional setting of market clearing on the product market can be changed by introducing market inertia. This results in the model version at the right part of Table 13.3. This version of the model seems to be more in line with most empirical evidence, at least for the Netherlands for the short and medium term. The model with market inertia can be combined with either the demand oriented equilibrium model (upper part) or the model including supply-side elements (lower part). Such a regimes approach stands for the economic environment and institutional setting in which the other policies are carried out.

An important implication of this 'regimes-approach' is that we can explore the effects of various policies in different institutional settings and that we no longer consider one particular version of the MESEMET model as 'true.' Thus, we can get a better insight in the relevance of, on the one hand, supply and demand effects of taxation and technology and, on the other hand, inflexibility of markets (or market inertia) on economic and technological processes. For the purpose of this article, assessing the practical policy implications of insights based on endogenous growth models, the new-Keynesian model and the short-run neoclassical model are rather theoretical. These regimes exclude elements that are important for policy makers and open economies; like for instance the effect of tax rates on private R&D activities, the relative international technology position and the influence of productive government expenditures (physical infrastructure, R&D and education) on the production capacity. In this paper we will concentrate on comparing the MESEMET-regime (market clearing) to the empirically relevant context (ER) (with market inertia).

Four policy options are evaluated:

• A rise in public R&D expenditures of 0.1 per cent NI,

- A rise in public expenditures on education of 0.1 per cent NI,
- A subsidy on private R&D by 0.1 per cent NI,
- A reduction in the tax on labour by 0.1 per cent NI.

Production growth (P) and employment (E)	Empirically relevant context (ER)		original MESEMET model	
	P	E	P	E
Public R&D expenditures (0,1% NI)	0.5	0.1	0.7	0.1
Public Expenditure on Education (0,1% NI)	0.3	0.1	0.3	0.1
R&D subsidies (0,1% NI)	0.6	0.0	0.5	0.0
Lowering labour taxes (0,1% NI)	0.1	0.1	0.1	0.1
Source: Ministry of Economic Affairs, 1995.				

Table 13.4: Impact of four policy instruments on production growth and employment for two macroeconomic regimes (policy shocks of 0,1% NI) in the long run

Table 13.4 presents the results for production growth and employment. Investment in public R&D, education as well as R&D subsidies have a positive impact on production growth. In particular, technology policy, either through public R&D activities or by subsidising private R&D, is a very attractive policy option for increasing economic growth in all simulations. Public education, although less effective, may also be favourable for economic growth. Comparing the relevant regimes of market clearing (MESEMET) to market inertia (ER) indicates that the effectiveness of increasing public R&D-expenditures might be improved by combining it with a policy aiming at improving the functioning of markets (competition policy). Private R&D will drop a little when competition is stimulated, but only marginal. For other policy measures there are not much differences between these regimes.

The simulation results should be interpreted with caution. Generally speaking we are more confident about the direction in which the economy is moving than in the exactness of the numerical outcomes of our exercise. They do not present exact estimates of the consequences of policy measures but are aimed at obtaining a better understanding of the trade-offs in economic policy. Some parameters in the model are based on less sound empirical evidence and must be seen as guesstimates because empirical evidence is not yet available. For instance, the importance of external effects of

R&D investment for human capital are very important for the effectiveness of technology policies although it is very hard to determine the numerical values of such spill-overs. Empirical estimations of the private and social return on R&D-investments differ significantly and show ambiguous results (Roelandt *et al*, 1996).

Overall the MESEMET outcomes are quite robust with respect to the assumptions related to flexibility of the goods market. In particular the various policy measures are very sensitive to supply-side elements. Neglecting these supply-side elements may dramatically underestimate the effectiveness of several policy instruments. This might be a problem when interpreting the possible impact of R&D policies in the more traditional analysis.

To obtain insight in the robustness of the MESEMET findings sensitivity analyses were employed.[1] The case that endogenizes technological progress in an empirically relevant context is quite robust. This is illustrated by the fact that the extensive simulations both with respect to the macroeconomic regime and the technology parameters allow us to determine the signs of the reported effects in most cases unambiguously. Only in the case of increased expenditures on public education we cannot determine the sign of the multiplier with respect to production and likewise we find contradictory movements in the physical capital stock when R&D expenditures (both public and private) change. In the other cases, however, we can determine the sign of the multipliers within the framework of the twofold sensitivity analysis of the MESEMET model

The sensitivity analysis provides important insights with respect to the relevant parameters for the effectiveness of technology policies. In particular, the spill-over effects from R&D on human capital seem to be crucial for the economic consequences of public R&D and subsidies on private R&D. Furthermore, the degree of complementarity between (physical and technology) capital and human capital are important for the degree in which public expenditures crowd out private investments in physical and technology capital. Indeed, a higher degree of complementarity raises the effectiveness of technology policies as it does not replace activities but stimulates these investments.

13.4 Evaluation and discussion

Empirical growth economics has relied thus far heavily on regression analysis of international cross-sections drawn from the recent past (Crafts, 1996). This has been useful but has yet revealed ambiguous outcomes on the actual empirical impact of technological progress on economic growth. For

356   *Jarig van Sinderen and Theo J.A. Roelandt*

instance, estimations of the private and social return on R&D-investments vary considerably (between zero and 800 per cent) (Roelandt *et al*, 1996). Undue attention is paid to R&D as opposed to learning, human capital and other types of knowledge as a source of productivity growth and opposed to other determinants of the profitability of innovation. Similarly a more explicit focus on the role of international technology transfer in the growth process seems overdue (Crafts, 1996, p. 45).

Due to both data deficiencies and the fact that no full agreement has yet emerged amongst economists about either the potential relevance of the actual empirical impact of (endogenous) technological progress, modelling endogenous growth suffers from many uncertainties. Future improvement of modelling endogenous economic growth can profit from microeconomic research on the influence of technology on productivity and on value adding activities at the micro level of firms. To improve the policy relevance of endogenous growth theory, this research must be concentrated on estimating the importance of intersectoral spill-overs, technology flows, diffusion and institutional variables. For policy making it is important to gain more unambiguous insights in the complex relation between competition and technology policy by analysing the effectiveness of technology policy at different market regimes (flexible *versus* inflexible market conditions).

Our analysis has illustrated the importance of technology policy for production growth. Investment in public R&D, education as well as R&D subsidies have a positive impact on production growth. In particular, technology policy, either through public R&D activities or by subsidising private R&D, is a very attractive policy option for increasing economic growth in all simulations. Public education, although less effective, may also be favourable for economic growth. Altogether, our results indicate that an explicit technology policy (stimulating public as well as private R&D) contribute positively to production and employment growth. Although the evidence is not very conclusive at the present state of art in modelling the economic impact of technology, an economic policy aiming at improving the functioning of markets might improve the effectiveness of an explicit technology policy directed towards public R&D-institutions.

Acknowledgments

Ministry of Economic Affairs, Economic Policy Directorate, Research Unit. This article draws heavily on: Ministry of Economic Affairs, *Economic Policy, Technology and Growth*, Discussion Paper 9502, The Hague, July 1995.

Notes

1. For details: Ministry of Economic Affairs, 1995:35-48.

References

Aghion, P. and Howitt, P. (1995), "Structural Aspects of the Growth Process," paper presented to Seventh World Congress of the Econometric Society, Tokyo.
Barro, R.J. (1990), "Government Spending in a Simple Model of Economic Growth," *Journal of Political Economy*, 98, part 2: S103-S125.
Ben-David, D. and M.B. Loewy (1996), "Knowledge Dissemination, Capital Accumulation, Trade and Endogenous Growth," *CEPR Discussion Paper* 1335, London.
Berg, F.J.M. van den (1989), "Patenting Activity and Elements of Market Structure in the Dutch Manufacturing Industry," *De Economist*, 137:476-92.
Bergeijk, P.A.G. van, R.C.G. Haffner and P.M. Waasdorp (1993), "Measuring the Speed of the Invisible Hand: The Macroeconomic Costs of Price Rigidity," *Kyklos*, 46:529-44.
Bergeijk, P.A.G. van, R.C.G. Haffner and P.M. Waasdorp (1994), "Market Inertia in the Netherlands," Paper for the Fifth Conference of the International Schumpeter Society, Münster.
Bergeijk, P.A.G. van, J. van Sinderen and P.M. Waasdorp (1993), "Over de dynamiek van belastingverlagingen," *Openbare Uitgaven*, 25:13-9.
Boskin, M.J. (1978), "Taxation, Saving and the Rate of Interest," *Journal of Political Economy*, 86:S3-S27.
Crafts, N.F.R. (1996) "Endogenous Growth: Lessons For and From Economic History," *CEPR Discussion Paper* 1333, London.
Davies, S. (1988), "Technical Change, Productivity and Market Structure," in: S. Davies and B. Lyons (eds.), *The Economics of Industrial Organisation*, Longman, London: 192-241.
Englander, A.S. and Mittelstädt (1988), "Total Factor Productivity: Macroeconomic and Structural Aspects of the Slowdown," *OECD Economic Studies*, spring, (11):7-42.
Fagerberg, J. (1987), "A Technology Gap Approach to Why Growth Rates Differ," *Research Policy*, (August):98-99.
Fagerberg (1994), "Technology and International Differences in Growth Rates," *Journal of Economic Literature*, 32:1147-75.
Grilliches, Z. (1986), "Productivity, R&D, and the Basic Research at the Firm Level in the 1970s," *American Economic Review*: 141-154.
Grossman, G.M., and E. Helpman (1990), "Comparative Advantage and Long-Run Growth," *American Economic Review*: 76:141-54.
Grossman, G.M., and E. Helpman (1991), *Innovation and Growth in the Global Economy*, MIT Press, Cambridge.
Hagen, G.H.A. van (1995), "Kennis, technische vooruitgang en groei," Graduate thesis, Erasmus University Rotterdam and Ministry of Economic Affairs, The Hague.
Hertog, R.G.J. den and A.R. Thurik (1993), "Determinants of Internal and External R&D: Some Dutch Evidence," *De Economist*, 141:279-289.
Long, J.B. de and L. Summers (1991), "Equipment Investment and Economic Growth," *Quarterly Journal of Economics*, 61:445-502.
Lucas, R.E. (1988), "On the Mechanics of Economic Growth," *Journal of Monetary Economics*, 93:395-410.
Ministry of Economic Affairs (1995), "Economic Policy, Technology and Growth," *Discussion Paper* 9502, The Hague, July.
OECD (1987), *Structural Adjustment and Economic Performance*, OECD, Paris.
OECD (1992), *Technology and the Economy: The Key Relationships*. OECD, Paris.
Papaconstantinou, G., N. Sakurai and A. Wyckoff (1995), *Technology Diffusion, Productivity and Competitiveness*, EIMS, EC/OECD.
Rebelo, S., (1991), "Long-Run Policy Analysis and Long-Run Growth," *Journal of Political Economy*, 99:500-521.
Roelandt T.J.A., P.W.L. Gerbrands, H.P. van Dalen and J. van Sinderen (1996), *Onderzoek naar*

technologie en economie: over witte vlekken en zwarte dozen. The Hague, Ministry of Economic Affairs.

Romer, P.M. (1986), "Increasing Returns and Long Term Growth," *Journal of Political Economy*, 94:1002-1037.

Romer, P.M. (1990), "Endogenous Technological Change," *Journal of Political Economy*, 98, part 2: S71-S102.

Sakurai, N., G. Papaconstantinou and E. Ioannidis (1995), *The Impact of R&D and Technology Diffusion on Productivity Growth: Evidence for 10 OECD Countries in the 1970s and 1980s*, OECD, Paris.

Scherer, F.M. and D. Ross (1990), *Industrial Market Structure and Economic Performance*, Houghton Mufflin, Boston.

Schneider,J. and T. Ziesemer (1994), "Whats New and What's Old in New Growth Theory: Endogenous Technology, Microfoundation and Growth Predictions - A Critical Overview," *MERIT*, 2/94-029, Maastricht.

Sinderen, J. van (1990), *Belastingheffing en economische groei* (Taxation and Economic Growth), Wolters Noordhoff, Groningen.

Sinderen, J. van (1992), *Over pre-economen, beleidseconomen en wetenschappers*, Inaugural Lecture, OCFEB, Erasmus University Rotterdam.

Sinderen, J. van (1993), "Taxation and Economic Growth," *Economic Modelling*, 13:285-300.

Sinderen, J. van, and P.M. Waasdorp (1994), "Het aanbodbeleid van drie kabinetten- Lubbers," *ESB*, 79:848-52.

Sinderen, J. van , P.A.G. van Bergeijk, R.C.G. Haffner and P.M. Waasdorp (1994), "De kosten van economische verstarring op macro-niveau," *ESB, 79* :274-279.

Smulders, S. (1994), "Growth, Market Structure and the Environment: Essays on the Theory of Endogenous Economic Growth," *OCFEB Research memorandum* 9503.

Symeonidis, G. (1996), "Innovation, Firm Size and Market Structure: Schumpeterian Hypotheses and Some New Themes," *OECD Working Paper* 161, Paris.

Toen-Gout, M.W. and J. van Sinderen (1995), "The Impact of Investment in Infrastructure on Economic Growth," *OCFEB Research Memorandum* 9503.

Trostel, P.A. (1993), "The Effect of Taxation on Human Capital," *Journal of Political Economy*, 101:327-350.

Verspagen, B. (1994), "Technology and Growth: The Complex Dynamics in the European Area," in: G. Silverberg and L. Soete (eds.), *The Economics of Growth and Technical Change*, MERIT, 2/94-007, Maastricht.

14 Modelling Government Investment and Economic Growth on a Macro Level: A Review

Jan-Egbert Sturm, Gerard H. Kuper and Jakob de Haan

3.1 Introduction

During the 1970s and 1980s many OECD countries have offset increases in debt interest payments and rising social security transfers by winding back public investment. Figure 14.1 shows government investment (excluding residential buildings) as a share of GDP for 18 OECD countries over the period 1970–92. The data relate to consolidated general government and have been taken from the Standardised National Accounts compiled and published by the OECD. As follows from Figure 14.1, public capital spending as a share of GDP declined or remained stable in almost all countries between 1970 and 1992. Spain and Portugal are exceptions. In order to become more competitive within the European Union, these countries undertook extensive programmes of upgrading their stock of public capital. A small rise occurred also in Italy. Another conclusion that can be drawn from figure 14.1 is that the level of government investment spending varies considerably across countries, ranging between 1.3 per cent of GDP for the UK and 5.8 per cent for Switzerland in 1992.

According to Oxley and Martin (1991, p. 161) the decline of government investment reflected 'the political reality that it is easier to cut-back or postpone investment spending than it is to cut current expenditures.' De Haan *et al.* (1996) report evidence that during fiscal contractions government capital spending is indeed reduced more than other categories of government spending. According to some authors this decline in public capital spending has important growth retarding effects. For instance, Aschauer (1989a) has hypothesized that the decrease in productive government services in the

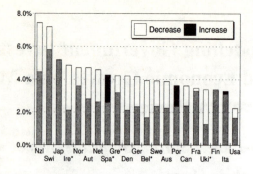

Figure 14.1: Public investment as a share of GDP between 1970 and 1992 (* data available until 1991; ** data available until 1989)

United States may be crucial in explaining the general decline in productivity growth. Taking a production function approach and using annual data for the period 1949 to 1985, Aschauer found a strong positive relationship between productivity and the ratio of the public to the private capital stock. Based on his results, a 1 per cent increase in the public capital stock might raise multifactor productivity by 0.39 per cent. The implications of these results for policymakers seem to be clear: public investment should go up to give a boost to the economy. Because of these well-received policy implications— higher infrastructure spending formed a major part of president Clinton's economic plans—the findings of Aschauer have sparked research into the impact of public sector capital spending on private sector output.

This paper reviews empirical research on the impact of government capital spending on economic growth and presents some new estimation results for illustrative purposes. The pros and cons of five different ways to model the relationship between public investment and economic growth are reviewed. We start with the production function approach (Section 14.2). Instead of adding the public capital stock as an additional input factor in a production function, a cost or profit function in which the public capital stock is included could be estimated. This so-called behavioural approach is discussed in the third section. Section 14.4 reviews Vector Auto Regressions (VARs) which also have been employed to model the relationship between government investment and economic growth. By imposing as few economic restrictions as possible this modelling technique tries to solve some of the problems related to the first two approaches. The previous lines of research are all based on time-series (or panel data). An alternative way to model the growth effects of public capital spending is to include government investment spending in cross-section growth regressions (Section 14.5). The sixth section reviews attempts to estimate the growth effects of public investment spending using structural econometric models. We review here only the impact of public capital on the supply side; demand raising effects of government investment spending in these models are not taken into account. The final section offers some concluding comments.

14.2 The production function approach

The stock of public capital may enter the production function in two ways. First it may influence multifactor productivity (A). Second, it may enter the production function directly, as a third input:

$$Q_t = A(G_t)f(K_t, L_t, G_t) \tag{14.1}$$

where Q_t is real aggregate output of the private sector, L_t is (aggregate hours worked by) the labour force of the private sector and K_t is the aggregate non-residential stock of private fixed capital. It depends on the functional form of the production function whether both effects can be identified. At the end of this section we will elaborate on this problem.

In the first type of analysis generally an aggregated Cobb-Douglas production function in which the public capital stock is added as an additional input factor is used:

$$Q_t = A_t L_t^\alpha K_t^\beta G_t^\gamma \tag{14.2}$$

Dividing both sides of Equation (14.2) by K_t, taking the natural logarithm, and assuming constant returns to scale across all inputs, gives:

$$\ln \frac{Q_t}{K_t} = \ln A_t + \beta \ln \frac{L_t}{K_t} + \gamma \ln \frac{G_t}{K_t} \tag{14.3}$$

Aschauer (1989a) introduces a constant and a trend variable as a proxy for $\ln A_t$. The capacity utilization rate is added to control for the influence of the business cycle. Most authors have used this specification. Only a few studies have used a translog function, which is more general than the Cobb-Douglas function.[1] Another drawback of the estimated production functions is that labour and capital are exogenous; it is implicitly assumed that both factors are paid according to their marginal productivity. Table 14.1 in Appendix A summarizes all production function studies that we are aware of.

As can be seen from both tables the estimate of γ in the studies included there varies between -0.11 and 0.73, and it is found to be insignificant in many cases.

Various authors have used regional or local data in their analysis. The use of such data may circumvent various implicit assumptions that are made when national data are employed, like: the marginal productivity is the same in all states; the rate of technological progress is uniform across states (Aaron, 1990). Unfortunately, many authors (including Munnell and Cook, 1990 and Garcia-Milà and McGuire, 1992) employ estimation techniques such as OLS that ignore state-specific effects—representing differences in underlying productivity stemming from location, climate, and endowments—thereby producing biased and inconsistent estimates.[2] Because more prosperous states are likely to spend more on public capital, there will be a positive correlation between the state-specific effects and public sector capital (Holtz-Eakin, 1992).

Studies using aggregated national data for the U.S.A. apparently find higher production elasticities of public capital than those relying on more disaggregated data. Munnell (1992, pp. 193–194) ascribes this to possible spillover effects: 'because of leakages, one cannot capture all of the payoff to an infrastructure investment by looking at a small geographic area.' However, Holtz-Eakin (1992) and Holtz-Eakin and Schwartz (1994), who take the effects of aggregation explicitly into account, and Holtz-Eakin and Schwartz (1995), who explicitly estimate the spillover effects of motorway infrastructure in the US, find little evidence of spillovers.[3]

A conclusion initially drawn by various authors from the studies under review was that in the United States public capital seems to have a significant positive effect on private sector productivity. Still, several economists questioned the plausibility of the values of the estimated marginal productivity of public capital on the grounds that they are implausibly high (Aaron, 1990; Gramlich, 1994). Aschauer points out, however, that a rate of return between 50 and 60 per cent, while substantial, is in line with estimates of the rate of return on R&D capital. Still, expenditure on R&D is riskier and should yield higher rates of return. A related criticism is that the wide range of reported estimates of public capital's impact on output makes the empirical linkages fragile at best. Although Munnell (1993, p. 32) claims that 'in almost all cases the impact of public capital on private sector output and productivity has been positive and statistically significant,' Table 14.1 shows that she is wrong; even for the case of the U.S.A., there are quite diverging outcomes.

The work of Aschauer (1989a)—and subsequent research by Munnell (1990)—has also been criticized on other grounds.[4] A first issue is *causality*: does the levelling off of public capital reduce the growth of output, or does the reduced growth of output diminish the demand for public capital (Eisner, 1991; Gramlich, 1994)? To this criticism, Aschauer (1990, p. 35) has responded by stating that 'this argument must confront the simple fact that public non-military investment expenditure ... reached a peak in the period between 1965 and 1968, while the usual dating of the onset of the productivity decline is around 1973.'[5]

More recently Finn (1993) and Ai and Cassou (1995) have captured the causality issue by applying a Generalized Method of Moments (GMM) estimator. This estimation technique resembles an instrumental-variables procedure and therefore avoids the possible reverse-causation bias. Finn (1993) reports a significant elasticity of the stock of public highways of 0.16. The elasticity estimates of Ai and Cassou (1995) for the total stock of public capital range between 0.15 and 0.2.[6]

There is also another possible form of simultaneity bias (Gramlich, 1994). Even if the true aggregate supply effect of public capital were zero, a rise in public investment may raise aggregate demand and output in the short run, leading to an inappropriate inference of the productivity effects of public capital.

Second, various authors have taken issue with the *specification* of Aschauer's model. Tatom (1991), for instance, argues that the reported large output elasticities are due to misspecification of the production function. He contends that the rising price of oil during the seventies made some private capital obsolete and therefore negatively affected productivity. Using another specification with energy prices included and capacity utilization entered multiplicatively to both the private and public capital stock, Tatom finds little evidence that the public capital stock raises productivity. However, Duggal *et al.* (1995) criticize Tatom's approach arguing that the relative price of energy is a market cost factor that would be included in the firm's cost function and therefore also in the factor input demand functions. So, the relative price of energy should not be included directly in the production function but the energy input instead.

Another specification issue is the inclusion of the degree of capacity utilization. As already pointed out, Aschauer included this variable to account for business cycle fluctuations. Duggal *et al.* (1995) criticize studies in which this approach is followed—including Hulten and Schwab (1991a) and Sturm and De Haan (1995)—since it is an additive factor in the estimated log equation, which implies that it is a multiplicative factor in the production function. A change in the capacity utilization causes an across the board change in the usage of all three factor inputs, such that the ratio of their marginal products remains the same. This is a very restrictive assumption. However, a specification in which the capacity utilization is not included at all is, of course, the other extreme. Sturm and De Haan (1995) show that this issue is not crucial for their conclusions.

A final specification issue that we would like to address is the specification in the widely cited paper by Ford and Poret (1991). These authors have estimated production functions in which the public capital stock is included as an input factor for eleven OECD countries. However, they combine private capital and labour into one private sector input variable, thereby imposing some restrictions which are not tested for. The same approach is followed by Toen-Gout and Jongeling (1994). Whereas Ford and Poret report very mixed outcomes for their sample of countries, Toen-Gout and Jongeling conclude that 'infrastructure would ... appear to have a significant and positive influence on private output' (Toen-Gout and Jongeling, 1994, p. 13). This result contrasts sharply with the findings of Sturm and De Haan (1995). We have therefore estimated similar specifications of the production function as in Toen-Gout and Jongeling (1994), also taking into account the point raised above with respect to the capacity utilization rate. The results are reported in Appendix B. Taken at face value, these estimates do not support Aschauer's thesis that public capital spending influences productivity.

A third reason to question the results of Aschauer (1989a) and others is the *non-stationarity* of the data which may render a spurious correlation between the public capital stock and output growth (Aaron, 1990; Tatom, 1991). Sturm and De Haan (1995) argue that if Aschauer's (1989a) model is estimated in first differences—which is necessary as the variables used are neither stationary nor cointegrated—the model produces only ambiguous results. However, first differencing also has its problems. As pointed out by Munnell (1992) it is assumed that output growth in one year is only correlated with input growth in that same year; it eliminates the ability to estimate the underlying long-term relationship between production and factor inputs.[7] Duggal et al. (1995, p. 6) argue: 'The fact that first-differenced equations generate a priori implausible labour and capital output elasticities is enough to question the capability of first-differenced equations to capture the long run relationships.' Indeed, researchers should examine not simply the extent to which variables are non-stationary, but also whether they grow together over time and converge to their long-run relationship, *i.e.* whether they are cointegrated. Various authors have followed this approach, with rather mixed results. While Sturm and De Haan (1995) conclude, for instance, that public capital and private sector output in the U.S.A. and the Netherlands are not cointegrated, Bajo-Rubio and Sosvilla-Rivero (1993) and Otto and Voss (1996) find that both variables are cointegrated in Spain and in Australia, respectively.

In a recent paper Garcia-Milà *et al.* (1996) systematically test for various specification problems (including non-stationarity of the data) for the case of a state-level production function with public capital as an input. Although likely to be less of a problem with panel data sets, it is still possible that these kinds of estimates are contaminated by non-stationarity of the variables. Indeed, in the preferred specification, which is in first differences with fixed states effects, Garcia-Milà *et al.* report no significant effects of public capital, thereby refuting earlier evidence, including their own.

A final problem according to Duggal *et al.* (1995) is that in all studies in the production function approach public capital is treated as a factor input, like private capital and labour. This violates standard marginal productivity theory in that it assumes a market determined per unit cost of infrastructure that is known to individual firms and can be used in calculating total costs. Aaron (1990) argues that absence of a market price, coupled with government pricing inefficiencies makes it impossible to assume that infrastructure as a factor input will be remunerated based upon its marginal product.[8] The basic problem here is, however, that in a Cobb-Douglas function (estimated in log levels) it does not make any difference whether public capital is treated as a third production factor or as influencing output through the factor representing technology. Both ways of modelling the influence of public capital yield similar equations to be estimated. In other words, in these kinds of empirical models the direct and indirect impact of public capital cannot be disentangled.

A somewhat different approach, which also employs a production function and which is therefore finally taken up in this section, is 'growth accounting' or 'sources of growth analysis.' This method can be explained as follows. Assume a production function like Equation (14.1) without the inclusion of public capital. If it is also assumed that each input is paid the value of its marginal product than the output elasticities are equivalent to income shares. In that case, the well-known 'multifactor productivity' (MFP) can be estimated as a residual, since all other terms are directly observable. Apart from the rate of change of technical efficiency, all sorts of other things like errors in measurement and omitted variables are included in the MFP estimate. Public capital can affect MFP in two ways: indirectly, by enhancing productivity of all or some of the private inputs, or as a direct factor of production, as in Aschauer (1989a). Both factors lead to biases in the estimate of MFP. By comparing MFP-growth with public capital growth one might find indications for the importance of these biases. Hulten and Schwab (1991b) conclude that public capital is not a key determinant of multifactor productivity growth. Productivity grew slightly faster in the older, declining regions of the U.S.A. during 1951–86, whereas in contrast, the public capital stock grew substantially faster in the Sun Belt. They also do not find any significant link between MFP and public capital in their pooled regression for 16 regions over the period 1970–86. Similar results are reported in Hulten and Schwab (1993). However, Eberts (1990), who analysed MFP-growth in 36 metropolitan areas between 1965 and 1977, reports somewhat different results. He finds that variation across these areas in public capital stock growth has a positive and statistically significant effect on MFP-growth over the period 1965–73; this significant relationship disappears during the second period (1973–77). Eberts also concludes that the public capital stock affects output, but only when private inputs are not included in the regression. When they are included, the size of the coefficient of public capital falls and becomes insignificant. According to the author, the primary channel through which public capital influences output growth is via private inputs.

14.3 The behavioural approach

Some of the drawbacks of the production function approach can be eliminated by describing optimizing behaviour of economic agents (that is, firms) either by maximizing profits or minimizing costs. Given the cost function we can under certain regularity conditions derive a unique production function by applying duality theory (Diewert, 1974). For instance, one can specify a cost function for the private sector (C) in which firms are envisaged as attempting to produce a given level of output (Q) at minimum private cost. Because the input prices (p^i) are exogenously determined, the instruments of the firm are the quantities of the private inputs (q^i). Alternatively, firms are assumed to maximize their profits (P) given the output (p^Q) and input prices.

$$
\begin{aligned}
C\left(p_t^i, q_t^i, A_t, G_t\right) &= \min_{q_t^i} \sum_i p_t^i q_t^i \\
&\quad \text{s.t. } Q_t = f\left(q_t^i, A_t, G_t\right) \\
P\left(p_t^Q, p_t^i, q_t^i, A_t, G_t\right) &= \max_{Q_t, q_t^i} p_t^Q Q_t - \sum_i p_t^i q_t^i \\
&\quad \text{s.t. } Q_t = f\left(q_t^i, A_t, G_t\right)
\end{aligned}
\tag{14.4}
$$

When private firms optimize, they take into account the environment in which they operate. One of these environmental variables is the state of technical knowledge (A).[9] Another environmental variable affecting production relationships to the firm is the amount of public infrastructure capital (G) available. The public capital stock enters the production function and thus also the cost or profit function as an unpaid fixed input.[10]

The dual function that satisfies one of the two optimization problems defined in equation (14.4) is normally approximated by a second-order Taylor approximation, like the transcendental logarithmic (translog) or the generalized Leontief function.[11] First-order conditions can be derived which result into input share (translog) or input demand equations (generalized Leontief).[12] To increase efficiency, these first-order conditions and the cost or profit function are normally estimated together using a system estimator.[13] From these estimates, several elasticity measures can be revealed which fully describe the underlying production function. Besides estimating several elasticities, the behavioural approach also makes it possible to estimate the shadow value of public capital—as a proxy for the unknown market price—and therefore the long-run desired level of public capital, that is, one can assess whether the amount of public capital is insufficient or excessive.[14]

Two differences are conspicuous when comparing the behavioural approach with the production function approach. First, the use of a flexible functional form hardly enforces any restrictions on the production structure. For example, *a priori* restrictions placed upon substitutability of production factors—as encountered in the production function approach—do not apply. Apart from the direct effect that is focused upon when production functions are estimated, public capital might also have indirect effects. Firms might adjust their demand for private inputs, if public capital is a substitute or a complement to these other production factors. It seems very plausible that, for instance, a larger stock of infrastructure raises the quantity of private capital used and therefore indirectly increases production. By using a flexible functional form, the influence of public capital via private inputs can be determined.

Second, as pointed out in the previous section, production function estimates may suffer from a simultaneous equations bias. Specifically, the right-hand variables in the various equations estimated by Aschauer (1989a) and Munnell (1990) include measures of labour input and utilization, and strong arguments have been made that such variables should be treated as endogenous. This problem does not arise in the behavioural approach, because costs or profits, and therefore input shares or demands, are directly represented. In the behavioural approach inputs are no longer exogenous to the level of output. However, input prices and—in case of the cost function approach, the level of output—are the exogenous variables. Therefore, the problem of possible endogeneity of variables remains, but does not concern the same variables as before.

Table 14.2 in Appendix A summarizes all studies we found in the literature which apply cost and/or profit functions. A flexible function not only consists of many parameters which need to be estimated,[15] but also of many second-order terms which are cross-products of the inputs. These second-order variables can create multicollinearity problems. Therefore, the data set not only has to be relatively large, but must also contain a lot of variability so that multicollinearity can be dealt with. Most studies therefore use panel data which combine a time dimension with either a regional dimension or a sectoral dimension.

In most studies summarized in Table 14.2 it is concluded that public capital reduces private sector costs or increases private sector profits. However, the estimated effects are generally significantly smaller than those reported by Aschauer (1989a). Only Deno (1988) and Lynde (1992) come up with a larger effect of public capital on the private sector. The remaining studies roughly estimate less than half the impact Aschauer (1989a) reported. In a variable cost function framework, the cost saving effects of public capital only arise if the substitution effect of some private inputs outweigh the complementary effects of other private inputs. Several studies, however, find that intermediate inputs (and sometimes also labour) are substitutes for public capital.

Most authors clearly reject the hypothesis of constant returns to scale to all inputs. Exceptions are Conrad and Seitz (1992, 1994), Lynde (1992) and Lynde and Richmond (1992). Conrad and Seitz (1992, 1994) even reject the hypothesis of homogeneity of the cost function,[16] which is assumed in most studies. However, imposing homogeneity results in accepting the hypothesis of constant returns to scale to all inputs. It can easily be seen that the behavioural approach boils down to allocating economic profits to increasing returns to scale, imperfect competition, and public capital. Therefore, imposing constant returns to scale and perfect competition on the model implies that all profits are automatically allotted to the public capital stock. However, as already noted, Lynde and Richmond (1992, 1993a, 1993b) only estimate share equations, and because these share equations only contain some parameters of the cost function, the structure of the underlying production function cannot be resolved without imposing several restrictions as constant returns to scale and perfect competition. This makes clear that estimating all parameters of the dual function is crucial.

Many authors adjust the stock of public capital by an index, such as the capacity utilization rate, to reflect their usage by the private sector (Conrad and Seitz, 1994; Deno, 1988; Nadiri and Mamuneas, 1994b; Seitz, 1994; and Shah, 1992). The impression exists that this is mainly done to artificially increase the variability of the data in order to cope with multicollinearity problems. On a theoretical basis two reasons have been advocated for adjusting the stock of public capital. First, public capital is a collective input which a firm must share with the rest of the economy. However, since most types of public capital are subject to congestion, the amount of public capital that one firm may employ will be less than the total amount supplied. Moreover, the extent to which a capacity utilization index measures congestion is dubious. Second, firms might have some control on the usage of the public capital stock in existence. For instance, a firm may have no influence on the level of highways provided by the government, but it can vary its usage of existing highways by choosing routes. Therefore there are significant swings in the intensity with which public capital is used. Other authors explicitly '... refrain from all of these possible adjustment procedures because of their ad hoc character and because 'proper' adjustment makes virtually all results possible' (Seitz, 1993, p. 230).

Despite the fact that time series properties are a main issue in the production function approach, they are hardly addressed in studies using the behavioural approach. Nevertheless, also in this line of research the results of many standard inference procedures are invalidated by non-stationary series. To make econometrically justifiable estimations one has to filter the time series to make them stationary or apply some kind of cointegration technique. The latter approach is followed by Lynde and Richmond (1993a, 1993b). They apply an error correction model (ECM) to capture the non-stationarity of the data.

Using an ECM approach also introduces dynamics in this framework. The standard behavioural approach assumes that all endogenous variables adjust to their equilibrium level within one period. Of course, it is hard to imagine that for instance the private capital stock fulfils this prerequisite. Not surprisingly, Sturm and Kuper (1996) report severe autocorrelation using the standard behavioural approach. Furthermore, they show that this can be overcome by adopting an ECM representation within a translog cost function. However, during the empirical implementation, they came up with other difficulties; several first-order conditions were no longer satisfied. Probably the increased flexibility was too much to be asked for given the available data.

In estimating the optimal stock of public capital, the assumption on the public good character of infrastructure is crucial. In the case of pure public goods, one could define total marginal benefits of public capital as the sum of the shadow values over all private sector firms, plus the summation of corresponding marginal benefits over all final consumers, yielding what might be called the social or total marginal benefit of public capital. Alternatively, if there is no congestion in the consumption of public goods, the total marginal benefit could be the largest benefit accruing to any one or set of consumers and producers, rather than the addition over all consumers and producers. The simplest rule to determine the optimal provision of public capital is to calculate the amount of infrastructure for which social marginal benefits just equal marginal costs. The difficulty in the empirical implementation of this rule lies in approximating the marginal costs of public capital. Therefore, only a few studies have estimated the optimal amount of public capital and compared it with the actual stock. Berndt and Hansson (1991) report an excess in public capital in the US that has declined over time. In contrast, Shah (1992) concludes that there was under-investment in public capital in Mexico. These studies use some measure for the cost of borrowing, such as the government bond yield, to approximate the marginal costs of public capital. To get around the problem Conrad and Seitz (1994) interpret the case in which the social marginal benefit of public capital is greater than the price of private capital as a shortage of public capital, whereas the reverse indicates over-investment in public capital. These authors find that during 1961–79 the social marginal benefit of public capital in Germany is larger than the user cost of private capital, whereas in the 1980–88 period the opposite is true.

Summarizing, the most appealing feature of the behavioural approach induces also the biggest problem; the flexibility of the functional form requires a tremendous amount of information to be included in the database. Furthermore, several problems raised in the production function approach still remain. In most of the studies using the behavioural approach energy is not included as an additional input, and time series properties are not taken into account.

14.4   The VAR approach

Another line of research to examine the effect of public capital spending on the economy, which we will label the VAR approach, is primarily data-oriented. By imposing as little economic theory as possible, this approach tries to solve some problems raised by the production and behavioural studies. In a VAR model a limited number of variables is distinguished that are explained by their own lags and lags of the other variables, meaning that all variables are treated as jointly determined. If necessary, deterministic variables, such as a constant or a trend, are included.[17] In this approach usually Granger-causality tests are carried out. Some variable is said to 'Granger-cause' another variable, if the time series prediction of the latter from its own past improves when lags of the former are added to the equation. Therefore, Granger-causality is a statistical concept of antecedence, or predictability. Granger-causality tests address the question of whether one variable helps to explain the subsequent time path of another. This interpretation of causality is, of course, intuitively attractive. It has therefore become widely accepted, although some of its implications are still under debate (Granger, 1980).

Besides conducting Granger causality tests, a VAR model can be analyzed by observing the reactions over time of different shocks on the estimated system (impulse-response analysis). Rewriting the VAR into its Vector Moving Average (VMA) representation allows us to trace out the time path of various shocks on the variables contained in the VAR system. However, there are many equivalent VMA representations for one VAR model. Because of this identification problem some economic structure has to be imposed, which reduces an important advantage of the VAR analysis. Furthermore, the model needs to be stable in order to make the conversion. A sufficient condition that makes the model stable is that the variables used are stationary or cointegrated. Problems concerning non-stationarity and cointegration of the data may be solved within this framework by applying the Johansen (1988, 1991) cointegration technique (Gonzalo, 1994). This method consists of a VAR model in which an error correction mechanism is included.

The production function studies derive single-equation models from first principles, which are then estimated while conclusions are based on the estimated elasticities. The multi-equation regressions in the behavioural approach are also deduced from first principles. In order to derive these first principles, some economic structure has to be assumed. First of all, this implies that the causal relationships are determined by theory. However, causality is often an issue when discussing the results of the production function approach. An advantage of VARs is that no *a priori* causality directions are imposed or other identifying conditions derived from economic theory are needed. For instance, the causality might run from output to public capital, which is the opposite of what is usually assumed.

An additional advantage, compared to the production function approach, is that public capital might indirectly influence output by its effect on a third variable, for instance private investment. Some authors, like Aschauer (1989b) and Erenburg (1993), report evidence for a complementary relationship between public capital and private capital, which suggests the existence of these indirect effects.

Because the VAR approach does not completely reveal the underlying production process, it is somewhat harder to get elasticity estimates. The only way to get specific elasticity estimates is via the impulse-response functions; they result in estimates of the long-run effects of different shocks.

Despite these advantages of the VAR approach, so far it has hardly been applied to the problem at hand. Besides the preliminary results presented in Sturm and De Haan (1997), we have traced only four studies in which some variant of the VAR approach is conducted to test the effects of public capital spending on the private sector: Clarida (1993); McMillin and Smyth (1994); Sturm *et al.* (1995); and Otto and Voss (1996). We will discuss these papers in chronological order. Table 14.3 provides a summary.

Clarida (1993) compares multifactor productivity (MFP) measures and public capital stocks for the United States (US), France, Germany, and the United Kingdom (UK). Clarida rejects stationarity in levels for both series in all four countries. The Johansen method leads to the conclusion that MFP and public capital cointegrate, that is, there exists a long-run equilibrium relationship between these two variables. However, the causation is unclear. The Granger causality tests produce evidence that public capital influences productivity as well as that productivity determines public capital.

McMillin and Smyth (1994) claim to stick to the production function framework as closely as possible in order to be able to compare their results with the production function literature. The variables that enter their VAR model are the private sector output per unit of private capital (Y/K), the hours of work per unit of private capital (H/K), government capital per unit of private capital (G/K), the relative price of energy (P^E/P^Y), and the inflation rate (π). However, not all of these variables can be described as typical for the single-equation estimates of the aggregate production function. For instance, inflation is included to incorporate another line of research, but how it can be seen as an additional production factor is unclear, and including it is definitely not common for this line of literature. As already discussed in the section concerning the production function approach the inclusion of the relative price of energy is also debatable. Furthermore, both private and public capital are adjusted by the capacity utilization rate which may also be questioned (see Section 14.3). The data relate to the US for the 1952–90 period. McMillin and Smyth do not test whether the variables are stationary or not. They decided to estimate the VAR model with both levels and first-differences of the variables. As already mentioned, this may lead to inconsistent estimates or spurious regressions. To account for apparent non-stationarity, a trend term was added to each equation in the model estimated in levels. McMillin and Smyth analyze both VAR models by observing the reactions over time of different shocks on the estimated system (impulse-response analysis) and by computing the share of the forecast error variance for each variable that can be attributed to its own innovations and to shocks to the other variables in the system (variance decompositions).

Using impulse-response analysis and variance decompositions, McMillin and Smyth reach the same conclusion in both models, that is, in levels and first-differences: Public capital per unit of private capital has no significant role in explaining output per unit of private capital. However, McMillin and Smyth (1994) apply the Choleski decomposition of the covariance matrix in order convert their VAR into a VMA representation. This approach implies an ordering of the variables from the most pervasive—a shock to this variable affects all the other variables in the current period—to least pervasive—a shock does not affect any other variables in the current period. Unfortunately, there are many ways to order the variables, and the choice of one particular ordering might not be innocuous. The ordering employed is P^E/P^Y, π, G/K, H/K, Y/K. Placing Y/K last is consistent with the production function studies, because these studies treat this variable as the endogenous variable. The importance of the ordering depends on the correlation between the residual terms. If the estimated correlations are almost zero, the ordering is immaterial. However, if a correlation coefficient is almost unity then a single shock in the system contemporarily affects two variables. McMillin and Smyth (1994) report large absolute correlations between several residual terms. To what extent the results change when other orderings are imposed is not discussed.

The third paper in which the VAR approach is employed, covers the Netherlands for the 1853–1913 period. In contrast to the above-mentioned studies, Sturm *et al.* (1995) do not use capital stocks but utilize investment series in their analysis. ADF tests show that the series for machinery investment, infrastructure investment, and output are trend-stationary over the period considered. Therefore cointegration techniques like the Johansen method are not necessary. The tri-variate VAR model that these authors estimate includes a trend in each equation. Granger-causality tests indicate large effects of infrastructure on output. No evidence is found for indirect effects of infrastructure via machinery investments. These findings are confirmed by impulse-response analysis and variance decompositions. They also apply an ordering of the variables that stick to the production function approach. However, in contrast to McMillin and Smyth (1994), Sturm *et al.* (1995) check whether changing the ordering alters their conclusions. Because there is hardly any correlation between the residual terms in their VAR model, the ordering does not influence the outcomes in a significant way.

Finally, Otto and Voss (1996) employ VAR techniques to Australian quarterly data covering the 1959:III–92:II period. Augmented Dickey-Fuller tests indicate that all four variables—output, working hours, private and public capital—are non-stationary. However, these authors decided to estimate a VAR model using levels of all variables. Variance decomposition and impulse-response functions are used to arrive at some conclusions about the short-run dynamics. They find that public capital is not affected by—and has hardly any effect on—both labour and output. The impulse-response functions show that a positive shock to public capital tends to have a positive lagged effect on private capital. Variance decompositions, however, indicate that private capital is largely exogenous, which argues against the conclusion that public investment can influence private investment. The other way around, public capital responds positively to private capital innovations. This is confirmed by the variance decompositions. As the correlations between the residuals are very close to zero, alternative orderings do not produce qualitatively different results.

To summarize these four papers, only the third paper of Sturm et al. (1995) finds clear evidence for the thesis that public capital spending influences output or productivity. Furthermore, not all four papers are on solid econometric ground or come up with unambiguous results. McMillin and Smyth (1994) and Otto and Voss (1996) do not take account of possible cointegrating relationships and present results that are possibly inconsistent or based on spurious regressions. Clarida (1993) reports that MFP and public capital influence each other mutually.

14.5 Public capital in cross-section models

In the previous sections we discussed the impact of government capital on economic growth in models using mainly a time-series framework. Although it is generally accepted that public capital and economic growth are related somehow, the literature is not unambiguous with respect to the size of the effects. Here we review the empirical literature on the relationship between public capital and growth based primarily on the theory of (endogenous) economic growth.

Since the mid-1980s the study of economic growth and its policy implications vigorously re-entered the research agenda (see Romer, 1986 and Baumol, 1986). A diverse body of literature appeared trying to explain, both theoretically and empirically, why differences in income over time and across countries did not disappear as the neoclassical models of growth—developed by Solow (1956) and Swan (1956)—of the 1950s and 1960s predicted. The idea that emerged from this literature is that economic growth is endogenous. That is, economic growth is influenced by decisions made by economic agents, and is not merely the outcome of an exogenous process. Endogenous growth assigns a central role to capital formation, where capital is not just confined to physical capital, but includes human capital, infrastructure and knowledge capital.

The econometric work on growth is dominated by cross-country regressions (Barro, 1991; Mankiw et al. 1992). In these studies the model of growth collapses to a single growth equation by log-linearizing the model around the steady state. Empirically, growth of real per capita GDP is estimated by a catch-up variable, human capital, investment, and population factors like fertility. Public investment might be a relevant factor in these kinds of models. Indeed, some studies employ government investment as an explanatory variable. Table 14.4 provides a summary.[18] The equations estimated in various studies can be summarized as follows:

$$\Delta \ln \left(\frac{Y}{L}\right)_{0,T} = \alpha + \beta \left(\frac{Y}{L}\right)_0 + \gamma \left(\frac{I^G}{Y}\right)_{0,T} + \delta(\text{conditional variables})$$

$$(14.5)$$

where $(Y/L)_{0,T}$ is the average per capita GDP over a period $[0,T]$, $(Y/L)_0$ is the initial level of real per capita GDP, and $(I^G/Y)_{0,T}$ is the average rate of public investment (% GDP) over a period $[0,T]$. Obviously, the regression should include public capital and not public investment. In empirical work, however, public investment is used because data on public capital are difficult to come by or are not available at all. Furthermore, again due to (lack of) availability of data, the period over which investment rates are averaged may deviate from the period used to average the growth rates of per capita GDP. The set of conditional variables include averages of primary and/or secondary enrolment (as a proxy of human capital), measures of political instability (assassinations, revolts and coups, and war casualties), measures of economic freedom, and the ratio of government consumption to GDP. Parameter β in the equation above measures technological catch up (if negative). Parameter γ measures the effect of public investment on growth and is not the same as the marginal productivity of public capital. The data used are often based on the dataset compiled by Summers and Heston (1988, 1991) for a large sample of countries extended with investment data and information on other variables gathered by Barro and Wolf (1989) and Barro and Lee (1994).

The model of Barro (1989) builds on Romer (1989), Lucas (1988), Rebelo (1991). Public services enter the production function: some infrastructure activities of government are inputs to private production and also raise the marginal product of private capital. Preliminary, but suggestive, findings are that public investment tends to be positively correlated with growth and private investment. In Barro (1991) growth is inversely related to the share of government consumption in GDP, but insignificantly related to the share of public investment. Once the total investment ratio is included, there is no separate effect on growth from the breakdown of total investment between private and public components.

Easterly and Rebelo (1993) run pooled regressions (using decade averages for the 1960s, 1970s and 1980s) of per capita growth on (sectoral) public investment and conditioning variables (decade averages of primary and secondary enrolment, measures of political instability, and the ratio of government consumption to GDP). The share of public investment in transport and communication (infrastructure) is robustly correlated with growth in a Barro-type cross-section regression (including a set of conditioning variables like initial level of income, enrolment rates, measures of political instability and the ratio of government consumption to GDP). This type of activity is likely to enhance the productivity of the private sector. Using instrumental variables the size of the effect of investment in transport and communication on growth as reported by Easterly and Rebelo becomes disturbingly high. This issue requires more research.

Until now we have discussed cross-country models. In section two we have already referred to one study in which the growth approach has been applied using regional data (Holtz-Eakin, 1992). Table 14.5 summarizes these studies. In the remainder of this section the studies of Mas et al. (1994b, 1995a, 1995b) and Crihfield and Panggabean (1995) will be discussed, as they have not been addressed in previous sections.

Mas et al. (1994b, 1995a, 1995b) estimate growth equations directly. They conclude that for the Spanish regions the initial stock of public capital relative to gross value added affects output per capita positively. However, their result only applies to the period until 1967. Crihfield and Panggabean (1995) use quite a different approach to arrive at conclusions which are in line with those of other authors: public infrastructure surely plays a role, but its contribution may be less than that of other forms of investment. Crihfield and Panggabean (1995) adopt a two-stage estimation technique, since exogeneity of labour and capital, often implicitly assumed, is rejected in most of the cases they consider. In the first stage they estimate reduced-form equations for population growth and investment equations, including public capital data (state or local investment in various types of public capital like education, streets and highways and sewerage and sanitation) as explanatory variables. The predicted values from these estimations enter the GDP per capita equation in the second stage. The implied elasticities of public capital are negative for both local and state public capital and ambiguous for most types of public capital, except for streets and highways.

Various authors have pointed at problems associated with cross-country regressions. There are biases due to omitted variables, reverse causation (Levine and Renelt, 1992, Levine and Zervos, 1993) and sample selection (see De Long, 1988). Measurement may also be problematic (Dowrick, 1992). Furthermore, the interpretation is often tempted by wishful thinking (Solow, 1994), perhaps because the interpretation becomes blurred due to the number of variables that are included in the regressions (Pack, 1994). Cross-section regressions, especially in a cross-section of heterogeneous countries, are often not very robust (Levine and Renelt, 1992). Finally, growth regressions are single-equation models. Economic theory may indicate, and economic data may not reject (see Crihfield and Panggabean, 1995), that there is more than one endogenous variable in the system. This calls for a more structural approach which also reduces the problem of multicollinearity.

14.6 Public capital in structural models

The literature reviewed so far aimed at identifying and estimating the effect of (an increase in) government investment (public capital) on output (growth). Except for the VAR approach most models are single-equations models which are sensitive to problems of causation, multicollinearity and so on. In this section we will take a look at structural models for the Netherlands designed to illustrate the role of government investment. This means that we do not include macroeconometric models for the Netherlands, like the models of the CPB Netherlands Bureau of Economic Policy Analysis, the CCSO Centre for Economic Research and the Central Bank in the Netherlands.

Of the various papers belonging to this approach, only Westerhout and Van Sinderen (1994) try to estimate the marginal productivity of public capital. The other papers *assume* a positive relationship between government investment and the performance of the economy simply by referring to Aschauer (1989a). These authors more or less neglect the criticism associated with this kind of modelling. Strictly speaking, this type of modelling is reduced-form modelling. The gain is that there is some more endogeneity. Multicollinearity between government investment and private investment is avoided by endogenizing private investment. However, the specification and the estimates are easily tainted by wishful thinking.

Using a small linearized macroeconomic model for the Netherlands, Westerhout and Van Sinderen (1994) assess the indirect effect of both government policies and external factors on economic growth through their effect on private investment. The model, which covers the past thirty years (1958–89), consists of four reduced-form equations. The rate of output growth depends on the private gross investment rate, whereas the private gross investment rate is assumed to be positively related to the rate of growth of public investment with the causality running from public to private investment. From the estimation results the long-run coefficient on the rate of growth of public investment in the equation for the private gross investment rate appears to be 0.23, while the coefficient for the gross private investment rate in the output growth equation is 0.48. This implies that in the long run the elasticity of output with respect to public investment equals 0.11 (0.23 × 0.48). In the short run it is only 0.02. If the ratio of public investment over output equals about 0.03, then the marginal product of public investment in the long run equals 3.7 which seems to be highly unrealistic. This might be caused by the fact that Westerhout and Van Sinderen (1994) use public investment data rather than data on public capital. Or perhaps the gross investment rate in the output growth equation picks up other influences as well.

In the following papers public capital is included as an argument in the production function in a general equilibrium framework with optimizing agents. In all cases the production function is linearized and is written down in terms of rates of growth. Most papers only report the output elasticity of public capital. The marginal product of capital can be derived as the ratio between the output elasticity of public capital and the average productivity of public capital. Only De Mooij *et al.* (1996) explicitly report the marginal productivity of public capital. Van de Klundert (1993) and De Mooij *et al.* (1996) assume constant returns to scale (CRS) to all private inputs and increasing returns to all inputs. Van Hagen *et al.* (1995) assumes CRS to all inputs private and public. This has consequences for the way public inputs are financed. Except for Toen-Gout and Van Sinderen (1995) all the papers focus their attention on the long-run effects of public capital which seems the right thing to do since '... the benefits reaped from public investment are far ahead in the future...' (Van de Klundert, 1993, p. 273).

Toen-Gout and Van Sinderen (1995) concentrate on the effects of infrastructure investment in the Netherlands on economic growth and employment in the short and medium run using a general equilibrium model. They assume a positive relationship between government investment (especially infrastructure) and the performance of the economy and a negative impact due to financing these outlays. The production function includes public investment instead of public capital. Public investment is assumed to have the same elasticity as private investment (0.34). Public investment is supposed to have a direct (positive) effect on growth by raising private productivity. There is a negative effect between private and public capital due to substitution. They estimate the overall elasticity to be about 0.04. Even when the method of financing is taken into consideration they still find a positive effect between public capital and growth. Based on simulation exercises they favour a reduction of income transfers to pay for public investment.

Unlike Toen-Gout and Van Sinderen (1995), Van de Klundert (1993) focuses on long-run effects of crowding out of private and public capital accumulation. Van de Klundert uses a two-country general equilibrium model (with optimizing finitely lived agents) with imperfect commodity substitution and imperfect capital mobility. Consumers face an intertemporal optimization problem. Furthermore they have to decide between consumption of home and foreign goods. Firms produce under perfect foresight and maximize the net present value of the cash flow subject to the production function. Public capital is included in the production function (Aschauer, 1989a) with CRS to private factors. The output elasticity of public capital is set equal to 0.2.

Van Hagen, Haffner and Waasdorp (1995) focus on long-run effects for the Dutch economy of an increase in public expenditures on education financed by either a cut in other expenditure categories or by taxes. They use a general equilibrium model including endogenous accumulation of human capital. They assume that government expenditures (be it infrastructure investment or investment in R&D and education) have a positive impact on productive capacity. Van Hagen *et al.* are inspired by Mankiw, Romer and Weil (1992) by assuming CRS to all inputs, private and public. The output elasticity of infrastructure capital is set equal to 0.05. The parameters are fixed rather arbitrarily. In their model labour is paid its marginal product so the model implies that the rents for public capital are paid for by private capital.

Van Hagen *et al.* (1995) conclude that the long-run effects of a policy change aimed at a faster accumulation of skills on production and employment are potentially large, especially when financed by reducing income transfers or by reducing the number of civil servants. The authors indicate that the results are sensitive to the degree of substitutability between skilled and unskilled labour, the importance of human capital in production, and the effect of education on human capital.

Finally, De Mooij, Van Sinderen and Toen-Gout (1996) focus on welfare effects of public expenditures using a general equilibrium model. The linearized production function implies that the marginal productivity of public capital equals 0.12. The output elasticity of public capital equals 0.04. This means that the rate of public capital to output is only 0.3 which does not seem to be very realistic. Here we have CRS to private inputs just as in the model by Van de Klundert (1993). The production function includes labour-augmenting technical progress which is assumed to be negatively related to labour taxes.

De Mooij *et al.* (1996) conclude that cutting back public investment may seriously harm economic performance. Furthermore, they show that investment in infrastructure has a positive effect on economic welfare and economic growth, especially when financed with lump sum taxes or labour taxes. Capital taxation results in negative effects. Again substitutability between private and public capital is important.

14.7 Conclusions

In this final section of our review, we will first outline our main conclusions with respect to the literature surveyed here. Next we add some general observations which are relevant for most studies surveyed. Finally, we will briefly discuss policy issues.

The academic debate on public infrastructure was stimulated by Aschauer. As Gramlich (1994, p. 1176–7) put it: 'He wrote a series of papers ... that put these movements [the US productivity slowdown and the neglect of public capital] together econometrically ... His work hit the magic button ... and Aschauer's papers were followed by an unusual amount of attention, from politicians and economists.' This review takes stock of this literature. This implies that many of the issues dealt with are of an econometric nature. It also implies that some other research which has been used to assess whether there is a shortage of public capital (like engineering needs assessments, political voting outcomes and cost-benefit analyses) have not been discussed in the paper (see Gramlich (1994) for a discussion of these other methods).

We have reviewed the pros and cons of five different ways to model the relationship between public investment and economic growth. We started with the production function approach in which the public capital stock is added as an additional input factor in a production function, which is then estimated at a national or regional level. Initially, this approach yielded results 'that were just too good to be true' (Aaron, 1990, p. 62). Indeed, the results of Aschauer (1989a) and Munnell (1990) have been criticized on various grounds. The most serious objections are related to the assumed causality between public capital and output, the specification and restrictiveness of the estimated model and the time-series characteristics of the data. With respect to this last issue, the proper way to proceed is to analyze whether the data are stationary or cointegrated. Applying this procedure for the case of the Netherlands, it is concluded that public capital and private sector inputs are only cointegrated if the variables are in first differences. This specification eliminates the ability to estimate the underlying long-term relationship between production and factor inputs. A serious shortcoming of much of this literature is that most authors employ a Cobb-Douglas production function, thereby simply following the lead of Aschauer. This implies, however, that various restrictions are introduced.

An alternative for the production function approach is to estimate a cost or profit function in which the public capital stock is included. Some of the drawbacks of the production function approach can be eliminated by using this so-called behavioural approach. However, the flexibility of the functional form requires the database used to contain a tremendous amount of information. Furthermore, most problems raised in production function estimates still remain. In most of the studies using the behavioural approach energy is not included as an additional input, and time series properties are not taken into account. The issue of causality is also problematic. Most studies following the behavioural approach conclude that public capital reduces private sector costs or increases private sector profits. However, the estimated effects are generally significantly smaller than those reported by Aschauer (1989a).

A third way to examine the relationship between government investment and economic growth is the so-called VAR approach. By imposing as little economic restrictions as possible this way of modelling tries to solve some of the problems raised by the production and behavioural approach. An advantage of VAR models is, for instance, that no a priori causality directions are imposed or other identifying conditions derived from economic theory are needed. Indirect effects of public capital are also taken into account. As the VAR approach does not completely reveal the underlying production process, only impulse-response functions yield estimates of the long-run effects of different shocks. So far, there are only a few studies in which VARs have been used to analyze the problem at hand; therefore it is too early to reach definite conclusions. Still, it seems a promising approach.

The first three approaches are all based on time-series (or panel data). A fourth way to model the growth effects of public capital spending is to include government investment spending in cross-section growth regressions. Problems associated with these cross-section regressions include biases due to omitted variables and reverse causation. The two-step approach adopted by Crihfield and Panggabean (1995) seems a promising way to deal with problems related to the single-equation nature of most cross-section regression models. Conclusions based on cross-section regressions, especially in a cross-section of heterogeneous countries, are often not very robust and this is also true for the outcomes with respect to the growth-raising effects of public investment.

Finally some attempts to estimate the growth effects of public investment spending using structural econometric models are discussed. The conclusions from this approach to model the impact of public capital on economic growth are twofold. First, in most cases it is simply assumed that there exists a causal and positive relationship between public capital on the one hand, and economic growth on the other. However, this assumption may not be warranted as the previous sections have shown. Second, even if we assume that such a positive relationship exists, the conclusions of studies in this approach should be interpreted with great care. Not only the size of the effect of public capital on output varies considerably among the various studies, but some authors also point at the importance of the degree of substitutability in the system in order to assess the importance of public capital for the performance of the economy. Here, we enter another discussion—which actually takes us back to Cambridge-Cambridge controversy in the 1960s—since economists do not agree on the degree of substitution in an economy.

Summarizing, we come up with only very modest conclusions. First, public capital probably enhances economic growth, a conclusion that most economists intuitively would ascribe to. Second, we are less certain about the magnitude of the effect and this is a disappointing outcome, given the enormous amount of research in this field.

An issue that is not always dealt with carefully is that the concept of the stock of public capital includes rather diverging ingredients, like highways and streets, gas, water and electricity facilities, water supply, bridges, water transportation systems, *etc*. Most authors employ data in their analyses which are generally chosen on the ground of their availability, without analyzing whether their conclusions are sensitive not only to the concept of the public capital stock (narrow versus broad definitions), but also to the way the capital stock has been constructed. For instance, most data on the capital stock are constructed using the perpetual inventory method, in which assumptions about the expected life of the assets are crucial. Few authors experiment with different definitions of the stock of public capital, which indeed, sometimes lead to diverging outcomes (Sturm and De Haan, 1995; Garcia-Milà *et al.*, 1996). Although some authors, including Aschauer (1989a), differentiate between the total stock of non-military public capital and the stock of infrastructure, one may wonder whether this suffices. It is likely that regions and industries react differently to various types of public capital. Indeed, Pinnoi (1994) finds strong evidence in support of this view.

Another issue to which hardly any attention has been paid is the fact that what really matters from a theoretical perspective is the amount of services provided by the public capital stock. In all empirical research it is implicitly assumed that these can be proxied by the stock of public capital or the level of government investment spending, which may not be true. For instance, the amount of services provided is also determined by the efficiency with which services are provided from the stock of public capital. Indeed, according to Munnell (1993) there is substantial room for improving the efficiency.

A further general observation that we would like to make concerns the implicit assumptions about the time it takes for public capital to affect GDP growth. It may well be that there is a substantial lag before the existence of say a new road leads to new set ups of businesses. A simple Cobb-Douglas production function as often applied in the literature will probably not reflect this effect. Similarly, it does not allow for network effects, whereby the quality of the connections facilitated by infrastructure investments may be more important than the level of the public capital stock (Garcia-Milà *et al.*, 1996). It may also make quite a difference whether the investment concerns infrastructure which previously did not exist at all, or simply more public capital (compare: a new two-lane road versus a two-lane road turned into a four-lane road). Indeed, the evidence of Sturm *et al.* (1995) suggests that the former may be more important than the latter.

The simple fact that public investment relative to GDP has declined in most OECD countries in itself is no evidence that public capital is currently undersupplied. However, according to Aschauer (1993) there is an underprovision of public capital in the US. Indeed, infrastructure issues have moved to the forefront of the policy agenda in many OECD countries.[19] The enthusiasm among policy-makers for increased infrastructure spending has been matched, if not surpassed by scepticism on the part of many economists (Munnell, 1993). Our review of the huge amount of empirical research only adds to this scepticism. Decisions on public capital spending should not be motivated by alleged growth enhancing effects of public investment. Other approaches not discussed in the present paper, like cost-benefit analyses, are probably better suited to justify capital expenditures (Munnell, 1993; Gramlich, 1994).

Appendix A   Overview of empirical studies

This appendix provides a systematic overview of all empirical studies that we are aware of, following the various approaches that have been used to model the relationship between public capital spending and economic growth. The first table—Table 14.1, which continues on three pages—summarizes all production function studies.

Table 14.2—which continues over eight pages—summarizes empirical studies using the so-called behavioural approach. VAR studies are listed in Table 14.3, while cross-section growth models are summarized in Table 14.4 (models using national data) and Table 14.5 (models based on regional data).

Table 14.1: Estimation of the production elasticity of public capital based on production functions

Study	Aggregation level	Specification	Data	Output elasticity of public capital
Ratner (1983)	National	Cobb-Douglas; log level	Time series, 1949–73	0.06
Aschauer (1989a)	National	Cobb-Douglas; log level	Time series, 1949–85	0.39
Ram & Ramsey (1989)	National	Cobb-Douglas; log level	Time series, 1949–85	0.24
Munnell (1990)	National	Cobb-Douglas; log level	Time series, 1949–87	0.31–0.39
Aaron (1990)	National	Cobb-Douglas; log level & delta log	Time series, 1952–85	not robust
Ford & Poret (1991)	National	Cobb-Douglas; delta log	Time series, 1957–89	0.39–0.54
Tatom (1991)	National	Cobb-Douglas; delta log	Time series, 1949–89	insignificant
Hulten & Schwab (1991a)	National	Cobb-Douglas; log level & delta log	Time series, 1949–85	not robust: 0.21 & insignificant
Finn (1993)	National	Cobb-Douglas; delta log	Time series, 1950–89	0.16
Eisner (1994)	National	Cobb-Douglas; log level	Time series, 1961–91	0.27
Sturm & De Haan (1995)	National	Cobb-Douglas; log level & delta log	Time series, 1949–85	0.41 & insignificant,
Ai & Cassou (1995)	National	Cobb-Douglas; delta log	Time series, 1947–89	0.15–0.20
Costa *et al.* (1987)	48 states	Translog; level	Cross-section, 1972	0.19–0.26
Merriman (1990)	48 states	Translog; level	Cross-section, 1972	0.20
Munnell & Cook (1990)	48 states	Cobb-Douglas; log level	Pooled cross-section, 1970–86	0.15
Aschauer (1990)	50 states	Cobb-Douglas; log level	Cross-section, averaged 1965–83	0.055–0.11

(continued)

Study	Aggregation level	Specification	Data	Output elasticity of public capital
Eisner (1991)	48 states	Cobb-Douglas & translog; log level	Pooled cross-section, 1970–86; Pooled time series, 1970–86	0.17; insignificant
Garcia-Milà & McGuire (1992)	48 states	Cobb-Douglas; log level	Panel data, 1969–82	0.04–0.05
Holtz-Eakin (1992)	48 states & 9 regions	Cobb-Douglas; log level	Panel data, 1969–86	insignificant
Munnell (1993)	48 states	Cobb-Douglas; log level	Pooled cross-section, 1970–86 (90)	0.14–0.17
Pinnoi (1994)	48 states	Translog; level	Panel data, 1970–86	−0.11–0.08
Evans & Karras (1994a)	48 states	Cobb-Douglas & translog; log level & delta log	Panel data, 1970–86	insignificant
Baltagi & Pinnoi (1995)	48 states	Cobb-Douglas; log level	Panel data, 1970–86	insignificant
Garcia-Milà et al. (1996)	48 states	Cobb-Douglas; delta log	Panel data, 1970–83	insignificant
Eberts (1986)	38 metropolitan areas	Translog; level	Panel data, 1958–78	0.03–0.04
Mera (1973)	9 Japanese regions	Cobb-Douglas; level	Panel data, 1954–63	0.12–0.50
Aschauer (1989c)	G-7	Cobb-Douglas; delta log	Panel data, 1966–85	0.34–0.73
Merriman (1990)	9 Japanese regions	Translog; level	Panel data, 1954–63	0.43–0.58
Ford & Poret (1991)	11 OECD countries	Cobb-Douglas; delta log	Time series, 1960–89	only significant in Belgium, Canada, Germany, Sweden
Berndt & Hansson (1991)	Sweden	Cobb-Douglas; log level	Time series, 1960–88	mixed & implausible results

(continued)

Study	Aggregation level	Specification	Data	Output elasticity of public capital
Bajo-Rubio & Sosvilla-Rivero (1993)	Spain	Cobb-Douglas; log level	Time series, 1964–88	0.19, cointegrated
Mas *et al.* (1993)	17 Spanish regions (manufacturing)	Cobb-Douglas; log level	Panel data, 1980–89	0.21
Evans & Karras (1994b)	7 OECD countries	Cobb-Douglas; delta log	Panel data, 1963–88	estimates are fragile & generally insignificant
Mas *et al.* (1994a)	17 Spanish regions	Cobb-Douglas; log level	Panel data, 1980–89	0.24
Otto & Voss (1994)	Australia	Cobb-Douglas; log level	Time series, 1966–90	0.38–0.45 (poor results at sectoral level)
Toen-Gout & Jongeling (1994)	Netherlands	Cobb-Douglas; delta log	Time series, unknown	0.37
Sturm & De Haan (1995)	Netherlands	Cobb-Douglas; log level & delta log	Time series, 1960–90	estimates are fragile, no cointegration
Dalamagas (1995)	Greece	Translog; level	Time series, 1950–92	0.53 (if budget deficit is included, otherwise negative)
Mas *et al.* (1996)	17 Spanish regions	Cobb-Douglas; log level	Panel data, 1980–89	0.08
Otto & Voss (1996)	Australia	Cobb-Douglas; log level	Time series, 1959:III–92:II	0.17, cointegrated
Wylie (1996)	Canada	Cobb-Douglas, translog; log level	Time series, 1946–91	0.11–0.52

Table 14.2: Studies using the behavioral approach

Study	Berndt & Hansson (1991)	Berndt & Hansson (1991)	Conrad & Seitz (1992)
Function	Cost: generalized Leontief	Cost: generalized Leontief	Cost: translog
Data	U.S.A., time series, 1960–88 (private business)	U.S.A., time series, 1960–88 (manufacturing)	Germany, panel, 4 sectors, 1961–88
Public Capital	Core	Total	Core
Other Inputs	K, L	K, L, E, M	K, L, M
Subst. to G			M
Compl. to G			K, L
Constant Returns	Rejected	Rejected	If homogeneity imposed then CRS accepted
Elasticities	MFP: (0.06;0.17)		Cost: $(-0.07; -0.22)$
Conclusions	Increase in G reduces costs; Excess G declined.	Increase in G reduces costs; Excess G declined.	Largest impact G in trade & transport sector; No productivity slowdown when taking account of G
Comments	K is a quasi-fixed input; Government bond yield used as discount rate for G; Problems implementing a dynamic model. Only cost function estimated.		Estimating sector by sector leads to multicollinearity problems. Use mark-up to capture imperfect competition.

(continued)

Study	Conrad & Seitz (1994)	Dalamagas (1995)	Dalamagas (1995)
Function	Cost: translog	Cost: translog	Cost: translog
Data	Germany, panel, 3 sectors, 1961–88	Greece, time series, 1950–92 (manufacturing)	Greece, time series, 1950–92 (manufacturing)
Public Capital	Core	Total public investment	Infrastructure; Education; Health
Other Inputs	K, L, M	K, L, E	K, L, Q
Subst. to G	L	K, L	Edu: K, L
Compl. to G	K		Infra: K, L
Constant Returns	If homogeneity imposed then CRS accepted	Imposed	
Elasticities		Cost: -2.35	Infra: Output: 1.06 Profit: 1.06
Conclusions	Increase in G reduces costs; Slowdown growth G partially responsible for productivity slowdown	Government investment is important in determining the production cost in manufacturing.	Spending on infrastructure leads to higher profits. Negative relationship between education and output.
Comments	Use $CU \cdot G$ to measure flow of services and remove multicollinearity problems.	All variables converted into indices. 3 models estimated: standard; using instrumental variables; augment system by reaction functions.	All variables converted into indices.

(continued)

Study	Deno (1988)	Keeler & Ying (1988)	Kitterer & Schlag (1995)
Function	Profit: translog	Cost: translog	Cost: translog
Data	U.S.A., panel, 36 SMSA's, 1970–78 (manufacturing)	U.S.A., panel, 9 regions, 1950–73 (road freight transport sector)	Germany, time series, 1961–88 (private business)
Public Capital	Motorways; Water; Sewers; Total	Motorways	Total; Core
Other Inputs	K, L, Q	K, L, E, M, Transport	K, L
Subst. to G			L
Compl. to G	K, L		K
Constant Returns		Rejected, IRS	
Elasticities	Output: 0.69		Cost: -0.01
Conclusions	G more effective in declining regions. G plays important role in output supply and input demand.	G improved productivity; Benefits for trucking sector cover at least 33 per cent of total capital costs of G.	The effect G on costs is significant but decreases over time.
Comments	G multiplied by percentage of population employed in manufacturing in order to capture congestion. Only share equations estimated.	All variables converted into indices.	ECM used to capture non-stationarity of data.

(continued)

Study	Lynde (1992)	Lynde & Richmond (1992)	Lynde & Richmond (1993a)
Function	Profit: Cobb-Douglas	Cost: translog	Cost: translog
Data	U.S.A., time series, 1958–88 (non-financial corporate sector)	U.S.A., time series, 1958–89 (non-financial corporate sector)	U.K., time series, 1966:I–90:II (manufacturing)
Public Capital	Total; Federal; State \& local	Total; Federal; State & local	Total excl. dwellings
Other Inputs	K, L	K, L	K, L, M, Y
Subst. to G		L	
Compl. to G		K	
Constant Returns	Accepted	Accepted	Rejected
Elasticities	Profit: 1.2		Output: 0.20
Conclusions	Significant share of profits attributable to G (especially State & local); Profit elasticity of K is higher.	G has a positive marginal product.	Higher G in the 1980s could have increased labour productivity growth by 0.5 per cent per year.
Comments	Perfect competition imposed; Correction for heteroscedasticity and serial correlation carried out. Only profit function estimated.	Perfect competition imposed; Statistical analysis indicates problems in the interpretation of the estimates. Only share equations estimated.	ECM of Phillips & Hansen (1990) used to capture non-stationarity of data. Only share equations estimated.

(continued)

Study	Lynde & Richmond (1993b)	Morrison & Schwartz (1992)	Morrison & Schwartz (1996)
Function	Profit: translog	Cost: generalized Leontief	Cost: generalized Leontief
Data	U.S.A., time series, 1958–89 (non-financial corporate sector)	U.S.A., panel, 48 states, 1970–87 (manufacturing)	U.S.A., panel, 6 New England states, 1970–78 (manufacturing)
Public Capital	Total excl. dwellings	Motorways & water & sewers	Motorways & water & sewers
Other Inputs	K, L, M, Y	$K, Lprod, Ln-prod, E$	$K, Lprod, Ln-prod, E$
Subst. to G			short run: $K; Y$-fixed: $Lprod, Ln-prod$
Compl. to G			E; long run: K; profit max: $Lprod, Ln-prod$
Constant Returns	Rejected	Rejected	
Elasticities	Output: 0.20	Cost: (0.07; -0.17)	
Conclusions	40 per cent of productivity slowdown is explained by G/L.	G has been below social optimum; G growth must more than keep up with output growth to have a positive productivity impact.	K is more valuable for society than G. Public investment is warranted if public policy is ineffective at increasing private investment.
Comments	Perfect competition imposed. ECM of Phillips & Hansen (1990) used to capture non-stationarity of data. Only share equations estimated.	K is a quasi-fixed input; Problems concerning heteroscedasticity.	K is a quasi-fixed input; Minor auto-correlation problems. Add short-run profit-maximization equation to the system. Calculate user cost of public capital similar to that of private capital.

(continued)

Study	Nadiri & Mamuneas (1994a)	Nadiri & Mamuneas (1994b)	Seitz (1993)
Function	Cost: generalized Cobb-Douglas	Cost: generalized Cobb-Douglas	Cost: generalized Leontief
Data	U.S.A., panel, 12 sectors, 1956–86 (manufacturing)	U.S.A., panel, 12 sectors, 1956–86 (manufacturing)	Germany, panel, 31 sectors, 1970–89 (manufacturing)
Public Capital	Total; R&D	Total; R&D	Roads (monetary); Motorways (physical)
Other Inputs	K, L, M	K, L, M	K, L
Subst. to G		M	L
Compl. to G		K, L	K
Constant Returns	Rejected, IRS		
Elasticities	Cost: $(-0.11; -0.21)$	Cost: $(-0.10; -0.21)$	
Conclusions	R&D and G are not major contributors to MFP; Contribution of G to MFP is twice as large as that of R&D.	R&D and K both have higher rates of return than G;	Shadow values of G are significant for 22 sectors. Complementary relationship between K and G is negligible.
Comments	Use $CU \cdot G$.	Results do not change if $CU \cdot G$ is used.	Results using monetary measure for roads are comparable to the results using the physical measure for motorways.

Study	Seitz (1994)	Seitz & Licht (1995)	Shah (1992)
Function	Cost: generalized Leontief	Cost: translog	Cost: translog
Data	Germany, panel, 31 sectors, 1970–89 (manufacturing)	Germany, panel, 11 states, 1971–88 (manufacturing)	Mexico, panel, 26 sectors, 1970–87 (manufacturing)
Public Capital	Total; Core	Total	Core
Other Inputs	K, L	K^M, K^B, L	K, L, M
Subst. to G	L	L	M
Compl. to G	K	K^M, K^B	K, L
Constant Returns	Rejected	Rejected	Rejected, IRS
Elasticities		Cost: $(-0.10; -0.36)$	Output: 0.05
Conclusions	G has a stabilizing but decreasing impact on private input demand, due to low G formation.	Distinction between K^M and K^M is crucial, former more affected by G than latter	K has larger impact than G; Underinvestment in K and G
Comments	Use $CU \cdot G$ to get variation over industries.		K is a quasi-fixed input; G is like a private or highly congested public good.

(continued)

394 *Jan-Egbert Sturm et al.*

Study	Sturm & Kuper (1996)	Sturm (1997)
Function	Cost: translog	Cost: symm. gen. McFadden
Data	Netherlands, panel, 5 sectors, 1953–87 (manufacturing)	Netherlands, exposed versus sheltered sector, 1953–93
Public Capital	Total; Other; Buildings; Infrastructure	Infrastructure
Other Inputs	K, L, E	K, L
Subst. to G		
Compl. to G	L, E	
Constant Returns		imposed
Elasticities	Output: (0.20;0.25) insign.	Cost: $(-0.2; -0.3)$
Conclusions	Multicollinearity and autocorrelation problems produce weird coefficients (not induced by public capital.).	Increase in G mainly reduces costs of sheltered sector.
Comments	All variables converted into indices. Autocorrelation problems exist with static version. Dynamic version results in violation of first order conditions.	ECM used; relationship with private inputs unclear.

G: public capital stock; K: private capital stock; K^M: private stock of machinery; K^B: private stock of buildings; L: private labour; E: energy; M: intermediate goods; Y: value added; Q: output; $Lprod$: production labour; $Ln-prod$: non-production labour; CU: capacity utilization rate.

Table 14.3: A summary of VAR studies

Study	Data	Model	Variables[a]	Conclusions
Clarida (1993)	U.S.A., France, Germany, U.K.	VECM	MFP, G	MFP and public capital are cointegrated, but direction of causality is unclear
McMillin & Smyth (1994)	U.S.A., 1952–90	VAR, levels and first differences	H/K, P^E/P^Y, G/K, inflation	No significant effect of public capital
Sturm et al. (1995)	Netherlands, 1853–13	VAR	Y, K, G, L	Infrastructure Granger-causes output
Otto & Voss (1996)	Australia, 1959:III–92:II	VAR	Y, K, G, H	No relationship between public capital and labour or output. Private capital affects public capital positively

[a]MFP: multifactor productivity; K: private capital stock; L: private labour; H: number of working hours; Y: private sector GDP; G: public capital stock; H/K: hours of work per unit of capital; P^E/P^Y: relative price of energy.

Table 14.4: A summary of cross country growth regressions

Study	Countries	Sample	Government capital concept	Conclusions
Barro (1989)	72 non-OPEC countries	1960–85	public investment	significant effect
Barro (1991)	76 countries	1960–85	public investment	no effect
Easterly & Rebelo (1993)	about 100 countries	1970–88	public investment or transport and communication spending	first variable not significant, second variable is significant

Table 14.5: A summary of cross states growth regressions

Study	Countries	Sample	Public capital concept	Marginal productivity of public capital
Crihfield & Panggabean (1995)	282 U.S.A. metropolitan areas	1960–77	local public capital, state public capital, types of public capital	negative, negative, ambiguous
Holtz-Eakin & Schwartz (1994)	48 U.S. states	1971–86	public capital infrastructure	fragile, insignificant
Mas *et al.* (1994b)	17 Spanish regions	1955–91, 1955–67, 1967–79, 1979–91	public capital	0.005, 0.0097, insignificant, insignificant
Mas *et al.* (1995a)	17 Spanish regions	1955–91, 1955–61, 1961–67, 1967–73, 1973–79, 1979–85, 1985–91	public capital	0.005, 0.0061, 0.017, insignificant, insignificant, insignificant, insignificant resp.
Mas *et al.* (1995b)	50 Spanish regions	1955–91, 1955–67, 1967–79, 1979–91	public capital	0.0043, 0.0094, 0.009, insignificant resp.

Appendix B   Estimation results using the production function approach

This appendix contains our estimation results of the production function approach for the Netherlands over the period 1953–91, following two different specifications. The first specification originates from Aschauer (1989a), whereas the second specification is the one used by Toen-Gout and Jongeling (1994) which is a slightly modified form of the specification initially used by Ford and Poret (1991):[20]

$$\ln \frac{Y_t}{K_t} = c_0 + c_1 t + \beta \ln \frac{L_t}{K_t} + (\alpha + \beta + \gamma - 1) \ln K_t +$$

$$\gamma \ln \frac{G_t}{K_t} \ldots [+ c_2 \ln CU_t] \ldots + \varepsilon_t, \tag{14.6}$$

$$\ln Y_t = c_0 + c_1 t + (\alpha + \beta)[s_L \ln L_t + s_K \ln K_t] +$$

$$\gamma \ln G_t + c_2 \ln CU_t + \varepsilon_t, \tag{14.7}$$

where Y_t is the real aggregate output of the private sector, K_t is the real aggregate non-residential stock of private capital, L_t is the labour force of the private sector, G_t is the real total stock of public capital, and CU_t is the capacity utilization rate. Following previous research, a trend variable (t) is also included in both specifications. In the specification of Toen-Gout and Jongeling the bundle of private-sector inputs is not freely estimated, but computed by weighting private-sector capital and employment by sample-average factor shares. The factor share of private-sector capital (s_K) is calculated as 1 minus the factor share of labour (s_L). We expect the coefficient ($\alpha + \beta$) to lie somewhere around one, which would indicate constant returns to scale to the private inputs.

Except for both capital stocks, all data are extracted from databases kept by the CPB Netherlands Bureau of Economic Policy Analysis. Both unpublished and preliminary estimates of the capital stocks have been kindly provided by the Central Bureau of Statistics.

As has been explained in Section 14.2, the inclusion of the capacity utilization rate is an issue under debate. Therefore, as indicated by the square brackets in Equation (14.6), we have estimated two versions of Aschauer's model. Furthermore, economists disagree about the appropriate definition of the public capital stock. For that reason, we decided to also use the stock of public infrastructure (G_t^i) in Equation (14.7).

Before estimating the above-defined equations we examined the order of integration of the series. The augmented Dickey-Fuller tests show that only the utilization rate is stationary ($I(0)$). Both endogenous variables, $\ln(Y_t/K_t)$ and $\ln Y_t$, are first-difference stationary ($I(1)$). The remaining variables, $\ln(L_t/K_t)$, $\ln(G_t/K_t)$, $\ln K_t$, $[s_L \ln L_t + s_K \ln K_t]$, $\ln G_t$ and $\ln G_t^i$ need to be differenced twice in order to become stationary ($I(2)$). Because in each equation more than one variable is integrated of the order two, it is possible that these variables cointegrate into a $I(1)$ variable. These $I(1)$ variables might also cointegrate, which lead to stationary residuals. Only in this hypothetical case does estimation of the above equations not lead to spurious regressions.

To test whether the above prevails, we first estimated the equations in levels. The results are shown in Table 14.6. The results do not differ much whether we include the capacity utilization rate in Aschauer's model or not. In case the capacity utilization rate is included (column 14.6(a)) the estimates imply an output elasticity of private capital (α) of 0.69, an output elasticity of labour (β) of 0.77, and an output elasticity of public capital (γ) of 0.51. Constant returns to scale to all inputs are highly rejected.

In the Toen-Gout and Jongeling model the effect of public capital even becomes larger. However, it is not clear how to interpret the negative coefficient of the aggregated private input variable.

Table 14.6: Estimates of aggregate production functions in levels for the Netherlands, 1953-1991

| | Specification[a] | | | |
| | Equation (14.6) | | Equation (14.7) | |
	(a)	(b)	(a)	(b)
Dependent variable:	$\ln(Y_t/K_t)$	$\ln(Y_t/K_t)$	$\ln Y_t$	$\ln Y_t$
t	0.002	0.001	0.021	0.009
	(0.46)	(0.25)	(3.10)	(1.64)
$\ln(L_t/K_t)$	0.772	0.885		
	(8.03)	(10.61)		
$\ln K_t$	0.969	0.911		
	(9.87)	(9.23)		
$\ln(G_t/K_t)$	0.505	0.630		
	(2.79)	(3.51)		
$\ln G_t$			1.496	
			(12.42)	
$\ln(G_t^i/K_t)$				1.178
				(13.30)
$[s_L \ln L_t + s_K \ln K_t]$			-1.197	-0.507
			(-4.30)	(-2.37)
$\ln CU_t$	0.248		0.790	0.731
	(2.09)		(4.73)	(4.62)
adjusted R^2	0.962	0.958	0.997	0.997
Durbin-Watson	1.38	1.25	0.82	0.88
Series cointegrated?	No	No	No	No

[a]In all models a constant and a trend (t) are included; t-statistics in parentheses; the Engle-Granger cointegration test is conducted. Y: private sector value added; K: private capital stock; L: private labour; G: public capital stock; G^i infrastructure capital; CU: capacity utilization rate. Columns entitled 'Equation (14.6)' and 'Equation (14.7)' refer to the formulae above.

As a first look at the Durbin-Watson statistics already indicates, the residuals of the estimated equations might not be stationary, that is, the variables are not cointegrated. To explicitly test this, we conducted the Engle-Granger cointegration test to all equations and conclude that in neither case the variables are cointegrated. Therefore estimation in levels has led to spurious correlation, and interpretation of the outcomes is not very useful. To cope with this problem, we estimated the models in first-differences. However, as noted in Section 14.2, using a specification in first differences eliminates the ability to interpret the outcomes as a long-term relationship, which makes the interpretation of the outcomes as a production function very difficult beforehand.

Nevertheless, the results are shown in Table 14.7. As the bottom row reveals, now the series are cointegrated. Aschauer's model including the capacity utilization rate implies an output elasticity of private capital (α) of -0.73, an output elasticity of labour (β) of 0.84, and an output elasticity of public capital (γ) of 1.13. The incredibly large effect of public capital on output and the insignificant estimate of $(\alpha + \beta + \gamma - 1)$—which implies constant returns to scale to all three inputs—cause the output elasticity of private capital to become negative. We also tested whether constant returns to scale to both private inputs is rejected by the data. As this was not the case, we re-estimated the model and restricted $\alpha + \beta$ to equal 1, that is, the parameters in front of $\ln K_t$ and $\ln(G_t/K_t)$ are assumed equal (not shown). As a result the output elasticity of public capital decreases to 0.85, whereas the output elasticity of labour (β) rises to 1.03 which still implies a negative output elasticity of private capital (α). Economic interpretation of these equations remains in our opinion impossible.

Table 14.7: *Estimates of aggregate production functions in first-differences for the Netherlands, 1954-1991*

	Specification[a]				
	Equation (14.6)		Equation (14.7)		TJ
	(a)	(b)	(a)	(b)	
Dependent variable:	$\Delta \ln Y_t/K_t$	$\Delta \ln Y_t/K_t$	$\Delta \ln Y_t$	$\Delta \ln Y_t$	$\Delta \ln Y_t$
Δt	0.015	0.014	0.028	0.020	1.22
	(1.10)	(1.06)	(1.80)	(1.29)	(3.16)
$\Delta \ln L_t/K_t$	0.842	0.955			
	(3.74)	(4.83)			
$\Delta \ln K_t$	0.234	0.348			
	(0.53)	(0.82)			
$\Delta \ln G_t/K_t$	1.125	1.112			
	(4.24)	(4.19)			
$\Delta \ln G_t$			1.216		
			(3.93)		
$\Delta \ln G_t^i/K_t$				0.980	0.48
				(3.83)	(5.92)
$\Delta [s_L \ln L_t + s_K \ln K_t]$			-1.031	-0.539	0.37
			(-1.81)	(-1.09)	(2.48)
$\Delta \ln CU_t$	0.137		0.374	0.369	0.82
	(1.05)		(2.78)	(2.72)	(7.62)
adjusted R^2	0.547	0.546	0.373	0.363	0.84
Durbin-Watson	2.42	2.25	2.26	2.27	1.98
Series cointegrated?	Yes	Yes	Yes	Yes	NA[b]

[a] In all models a constant (Δt) is included; t-statistics in parentheses; the Engle-Granger cointegration test is conducted. Y: private sector value added; K: private capital stock; L: private labour; G: public capital stock; G^i infrastructure capital; CU: capacity utilization rate. Columns entitled 'Equation (14.6)' and 'Equation (14.7)' refer to the formulae above. Column 'TJ' is taken directly from Toen-Gout and Jongeling (1994). [b] Not available.

The results of the specification of Toen-Gout and Jongeling roughly stay the same. The only difference is that this time the negative effect of the aggregated private input measure becomes insignificant. However, even a coefficient of zero bares no economic meaning at all. Toen-Gout and Jongeling (1994) presented rather different results. As shown in the last column of Table 14.7 their aggregated private input variable turned out to be significantly positive, indicating increasing returns to scale to the private inputs. Their estimated effect of public infrastructure on output is approximately half the effect we found. These differences may stem from other definitions of the various variables.

In our opinion it is very difficult to interpret the estimation results in light of the underlying production function framework. Several elasticities become implausible large or have the theoretically incorrect sign. Drawing policy conclusions from these estimates—as Toen-Gout and Jongeling do—would in our view be irresponsible.

Notes

1. For instance, the Cobb-Douglas production function restricts—by definition—the sub-

stitution elasticities of the production factors to be equal to one.

2. Hulten and Schwab (1993) estimate the level of multifactor productivity (*MFP*) in each region in four versions of their model: constrained (assuming the same level and growth of *MFP* across regions), including only regional intercepts, including only regional time effects and including both regional intercepts and time effects. The results of the various specifications are quite different.

3. Holtz-Eakin (1992) and Holtz-Eakin and Schwartz (1994) incorporate infrastructure (streets and motorways, sanitation and sewage, and electric, gas, and water utilities) as a component of aggregate production. Holtz-Eakin (1992) estimates the model without fixed effects and including both state and time effects. The estimated values for γ imply that the macro-economic impact of public sector capital on private productivity has been small. Holtz-Eakin and Schwartz (1994) first estimate the model without fixed effects to find negative values for γ. Controlling for state-specific differences in productivity changes parameters dramatically, but the effects are still not very significant. Both Holtz-Eakin (1992) and Holtz-Eakin and Schwartz (1994) conclude that there is little support for claims of a dramatic productivity decline due to decreased infrastructure spending. Holtz-Eakin and Schwartz (1995) examine the degree to which motorways provide productivity benefits beyond the confines of each state's border. They find no evidence of quantitatively important productivity spillovers.

4. Earlier, Ratner (1983) had already reached similar conclusions. Costa et al. (1987) also found a significant impact of the public capital stock on output.

5. Aschauer (1989a, 1990) also employed instrumental variables to estimate his production functions. This did not affect his basic conclusions.

6. Formally, as several Euler equations have to be specified, the GMM procedure implies that a more behavioural approach is used. However, as essentially a production function is estimated, we opted to include these two papers in this section.

7. The GMM procedure adopted by Finn (1993) and Ai and Cassou (1995) also follows the first-differencing suggestion. However, these authors suggest that, because the GMM procedure tightly identifies the parameters, their estimates are not sensitive to the differencing. Holtz-Eakin and Schwartz (1995) on the other hand use so-called long differences; the data for each year is entered as differences from the initial year. In this way cumulative growth rates are used. All three studies report labour and capital elasticities which are in accordance with theoretical predictions.

8. Duggal *et al.* (1995) develop an alternative approach in which infrastructure is incorporated as part of the technological constraint, rather than as a factor input. Infrastructure reduces production costs, thereby raising the productivity of factor inputs, and it expands markets, thereby lowering production costs through economies of scale and also allowing the implementation of technological innovations that do not become cost effective until an output threshold is reached. However, estimating their theoretical model for the U.S.A. over the period 1960–89 these authors find estimates of the marginal product of infrastructure which even exceeds those of Aschauer and Munnell. In the authors' view these results are not implausible when one stops thinking of infrastructure as a factor input that siphons off its factor share of income.

9. Most studies approximate the state of technical knowledge by a time trend. The studies by Nadiri and Mamuneas (1994a, 1994b) are exceptions in which also public Research and Development (R&D) capital is taken into account.

10. There is, of course, a price to be paid for the services of public capital through the tax system. However, by assuming that firms do not have direct control over how much capital the government supplies to them, all studies treat these services as 'unpaid' factors of production.

11. Exceptions are Lynde (1992) who employs a standard Cobb-Douglas function and thereby eliminates most advantages of the behavioural approach, as well as Nadiri and Mamuneas (1994a, 1994b) who use a generalized Cobb-Douglas function which can also be seen as a restricted translog function.

12. Diewert and Wales (1987) have shown that the translog and the generalized Leontief functions frequently fail to satisfy the appropriate curvature conditions. This is confirmed by Sturm and Kuper (1996) who have used a translog specification. Sturm (1997) reports results using a modified symmetric generalized McFadden (SGM) function on which the appropriate curvature conditions can easily be imposed without destroying the flexibility properties. Preliminary results using Dutch data suggest that this modified SGM specification outperforms the translog specification.

13. Exceptions are Berndt and Hansson (1991) and Lynde (1992) who only estimate the dual function, and Lynde and Richmond (1992, 1993a, 1993b) who only estimate share equations.

14. The shadow value of public capital measures the impact on the private firm's costs of there being an exogenous increase in the amount of available public capital.

15. If there are n production factors in the cost functions, then $2n + n(n-1)/2$ coefficients need to be estimated besides the constant and possible dummy-variables.

16. Homogeneity of the cost functions means that doubling the input prices implies a doubling of the cost.

17. In case n variables are included with each p lags, $n^2 \times p$ coefficients need to be estimated, besides the deterministic variables. Using for instance Akaike's (1969, 1970) Final Prediction Error (FPE) criterion one can select the appropriate lag specification for each explanatory variable in each equation and save a substantial number of degrees of freedom.

18. Since we are interested in government investment, we restrict ourselves to models including some measure of government investment. Consequently we do not include studies that include total investment (Dowrick, 1992), nor do we review articles on government consumption expenditures. Furthermore, as we concentrate on the OECD countries, studies of, *e.g.* Devarajan et al. (1996), Hulten (1996), Khan and Reinhart (1990) and Ram (1996)—which focus solely on developing countries—are not included in Table 14.4.

19. Whether this will actually lead to higher spending levels remains to be seen. Indeed there is evidence that public capital spending is a relatively easy target during periods of fiscal contraction (De Haan *et al.*, 1996). As many countries still have to redress their public finances, the lip-service of many politicians in favour of more public investment should not be taken for granted.

20. Ford and Poret (1991) subtract the term $[s_L \ln L_t + s_K \ln K_t]$ from both sides of the equation and therefore take multifactor productivity as their endogenous variable.

References

Aaron, H.J. (1990), "Discussion," in: A.H. Munnell, (ed.), *Is There a Shortfall in Public Capital Investment?*, Federal Reserve Bank of Boston, Boston.

Ai, C. and S.P. Cassou (1995), "A Normative Analysis of Public Capital," *Applied Economics*, 27: 1201–1209.

Akaike, H. (1969), "Fitting Autoregressive Models for Prediction," *Annals of the Institute of Statistical Mathematics*, 21: 243–247.

Akaike, H. (1970), "Statistical Predictor Identification," *Annals of the Institute of Statistical Mathematics*, 22: 203–217.

Aschauer, D.A. (1989a), "Is Public Expenditure Productive?," *Journal of Monetary Economics*,

     23: 177–200.
Aschauer, D.A. (1989b), "Does Public Capital Crowd Out Private Capital?," *Journal of Monetary Economics*, 24: 171–188.
Aschauer, D.A. (1989c), "Public Investment and Productivity Growth in the Group of Seven," *Federal Reserve Bank of Chicago, Economic Perspectives*, 13: 17–25.
Aschauer, D.A. (1990), "Why is Infrastructure Important?," in: A.H. Munnell (ed.), *Is There a Shortfall in Public Capital Investment?*, Federal Reserve Bank of Boston, Boston.
Aschauer, D.A. (1993), "Genuine Economic Returns to Infrastructure Investment," *Policy Studies Journal*, 21: 380–390.
Bajo-Rubio, O. and S. Sosvilla-Rivero (1993), "Does Public Capital Affect Private Sector Performance? An Analysis of the Spanish Case, 1964–88," *Economic Modelling*, 10: 179–184.
Baltagi, B.H. and N. Pinnoi (1995), "Public Capital Stock and State Productivity Growth: Further Evidence from an Error Components Model," *Empirical Economics*, 20: 351–359.
Barro, R.J. (1989), "A Cross-Country Study of Growth, Saving, and Government," *NBER Working Paper* 2855.
Barro, R.J. (1991), "Economic Growth in a Cross Section of Countries," *Quarterly Journal of Economics*, 106: 407–443.
Barro, R.J. and J.-W. Lee (1994), "Sources of Economic Growth," *Carnegie-Rochester Conference Series on Public Policy*.
Barro, R.J. and S. Wolf (1989), "Data Appendix for Economic Growth in a Cross-Section of Countries," unpublished manuscript, Harvard University, Cambridge.
Baumol, W.J. (1986), "Productivity Growth, Convergence and Welfare: What the Long-Run Data Show," *American Economic Review*, 76: 1072–1085.
Berndt, E.R. and B. Hansson (1991), "Measuring the Contribution of Public Infrastructure Capital in Sweden," *NBER Working Paper* 3842.
Boarnet, M.G. (1997), "Infrastructure and the Productivity of Public Capital: The Case of Streets and Highways," *National Tax Journal*, 50: 39–58.
Clarida, R.H. (1993), "International Capital Mobility, Public Investment and Economic Growth," *NBER Working Paper* 4506.
Conrad, K. and H. Seitz (1992), "The Public Capital Hypothesis: The Case of Germany," *Recherches Economiques de Louvain*, 58: 309–327.
Conrad, K. and H. Seitz (1994), "The Economic Benefits of Public Infrastructure," *Applied Economics*, 26: 303–311.
Costa, J. da Silva, R.W. Ellson and R.C. Martin (1987), "Public Capital, Regional Output, and Developments: Some Empirical Evidence," *Journal of Regional Science*, 27: 419–437.
Crihfield, J.B. and M.P.H. Panggabean (1995), "Is Public Infrastructure Productive? A Metropolitan Perspective Using New Capital Stock Estimates," *Regional Science and Urban Economics*, 25: 607–630.
Dalamagas, B. (1995), "A Reconsideration of the Public Sector's Contribution to Growth," *Empirical Economics*, 20: 385–414.
De Long, J.B. (1988), "Productivity Growth, Convergence and Welfare: Comment," *American Economic Review*, 78: 1138–1154.
Deno, K.T. (1988), "The Effect of Public Capital on U.S. Manufacturing Activity: 1970 to 1978," *Southern Economic Journal*, 55: 400–411.
Devarajan, S., V. Swaroop and H. Zou (1996), "The Composition of Public Expenditure and Economic Growth," *Journal of Monetary Economics*, 37: 313–344.
Diewert, W.E. (1974), "Applications of Duality Theory," in: M.D. Intriligator and D.A. Kendrick (eds.), *Frontiers of Quantitative Economics*, Volume II, North-Holland, Amsterdam:106–120.
Diewert, W.A. and T. Wales (1987), "Flexible Functional Forms and Global Curvature Conditions," *Econometrica*, 55: 43–68.
Dowrick, S. (1992), "Technological Catch Up and Diverging Incomes: Patterns of Economic Growth 1960–88," *Economic Journal*, 102: 600–610.
Duggal, V.G., C. Saltzman and L.R. Klein (1995), "Infrastructure and Productivity: A Nonlinear Approach," paper presented at the 7th World Congress of the Econometric Society, Tokyo, Japan.
Easterly, W. and S. Rebelo (1993), "Fiscal Policy and Economic Growth," *Journal of Monetary*

Economics, 32: 417–458.

Eberts, R.W. (1986), "Estimating the Contribution of Urban Public Infrastructure to Regional Growth," *Federal Reserve Bank of Cleveland Working Paper* No. 8610.

Eberts, R.W. (1990), "Cross-Sectional Analysis of Public Infrastructure and Regional Productivity Growth," *Federal Reserve Bank of Cleveland Working Paper* No. 9004.

Eisner, R. (1991), "Infrastructure and Regional Economic Performance: Comment," *New England Economic Review*, Sep/Oct: 47–58.

Eisner, R. (1994), "Real Government Saving and the Future," *Journal of Economic Behavior and Organization*, 23: 111–126.

Erenburg, S.J. (1993), "The Real Effects of Public Investment on Private Investment," *Applied Economics*, 25: 831–837.

Evans, P. and G. Karras (1994a), "Are Government Activities Productive? Evidence from a Panel of U.S. States," *Review of Economics and Statistics*, 76: 1–11.

Evans, P. and G. Karras (1994b), "Is Government Capital Productive? Evidence from a Panel of Seven Countries," *Journal of Macroeconomics*, 16: 271–279.

Finn, M. (1993), "Is all Government Capital Productive?," *Federal Reserve Bank of Richmond, Economic Quarterly*, 79: 53–80.

Ford, R. and P. Poret (1991), "Infrastructure and Private-Sector Productivity," *OECD Economic Studies*, No. 17: 63–89.

Garcia-Milà, T. and T.J. McGuire (1992), "The Contribution of Publicly Provided Inputs to States' Economies," *Regional Science and Urban Economics*, 22: 229–241.

Garcia-Milà, T., T.J. McGuire and R.H. Porter (1996), "The Effects of Public Capital in State-Level Production Functions Reconsidered," *Review of Economics and Statistics*, 78: 177–180.

Gonzalo, J. (1994), "Five Alternative Methods of Estimating Long Run Equilibrium Relationships," *Journal of Econometrics*, 60: 203–233.

Gramlich, E.M. (1994), "Infrastructure Investment: A Review Essay," *Journal of Economic Literature*, 32: 1176–1196.

Granger, C.W.J. (1980), "Testing for Causality: A Personal Viewpoint," *Journal of Economic Dynamics and Control*, 2: 1176–1196.

Haan, J. de, J.E. Sturm and B.J. Sikken (1996), "Government Capital Formation: Explaining the Decline," *Weltwirtschaftliches Archiv*, 132: 55–74.

Hagen, G.H.A. van, R.C.G Haffner, and P.M. Waasdorp (1995), "How Strong is the Case for Public Investments in Human Capital? Assessments with an AGE Model," Paper for the 51st Congress of the International Institute of Public Finance, Lisbon, Portugal.

Holtz-Eakin, D. (1992), "Public-Sector Capital and the Productivity Puzzle," *NBER Working Paper* 4122, published in *Review of Economics and Statistics*, 76: 12–21 (1994).

Holtz-Eakin, D. (1994), "Public-sector Capital and the Productivity Puzzle," *Review of Economics and Statistics*, 76: 12–21.

Holtz-Eakin, D. and A.E. Schwartz (1994), "Infrastructure in a Structural Model of Economic Growth," *NBER Working Paper* 4824; published in *Regional Science and Urban Economics*, 25: 131–151 (1995).

Holtz-Eakin, D. and A.E. Schwartz (1995), "Spatial Productivity Spillovers from Public Infrastructure Evidence from State Highways," *NBER Working Paper* 5004; published in *International Tax and Public Finance*, 2: 459–468 (1995).

Hulten, C.R. (1996), "Infrastructure Capital and Economic Growth: How Well You Use It May Be More Important than How Much You Have," *NBER Working Paper* 5847.

Hulten, C.R. and R.M. Schwab (1991a), "Is There Too Little Public Capital?," Conference Paper, American Enterprise Institute.

Hulten, C.R. and R.M. Schwab (1991b), "Public Capital Formation and the Growth of Regional Manufacturing Industries," *National Tax Journal*, 44: 121–134.

Hulten, C.R. and R.M. Schwab (1993), "Endogenous Growth, Public Capital, and the Convergence of Regional Manufacturing Industries," *NBER Working Paper* 4538.

Johansen, S. (1988), "Statistical Analysis of Cointegrating Vectors," *Journal of Economic Dynamics and Control*, 12: 231–254.

Johansen, S. (1991), "Estimation and Hypothesis Testing of Cointegration Vectors in Gaussian Vector Autoregressive Models," *Econometrica*, 59: 1551–1580.

Keeler, T.E. and J.S. Ying (1988), "Measuring the Benefits of a Large Public Investment. The

404 *Jan-Egbert Sturm et al.*

Case of the U.S. Federal-Aid Highway System," *Journal of Public Economics*, 36: 69–85.

Khan, M.S. and C.M. Reinhart (1990), "Private Investment and Economic Growth in Developing Countries," *World Development*, 18: 19–27.

Kitterer, W. and C.-H. Schlag (1995), "Sind öffentliche Investitionen produktiv? Eine empirische Analyse für die Bundesrepublik Deutschland," *Finanzarchiv*, 52: 460–477.

Klundert, T. van de (1993), "Crowding Out of Private and Public Capital Accumulation in an International Context," *Economic Modelling*, 10: 273–284.

Levine, R. and D. Renelt (1992), "A Sensitivity Analysis of Cross-Country Growth Regressions," *American Economic Review*, 82: 942–963.

Levine, R. and S.-J. Zervos (1993), "Looking at the Facts: What We Know About Policy and Growth from Cross-Country Analysis," *Working Paper* 1115, World Bank, Washington, D.C.

Lucas, R.E. (1988), "On the Mechanics of Development Planning," *Journal of Monetary Economics*, 22: 3–42.

Lynde, C. (1992), "Private Profit and Public Capital," *Journal of Macroeconomics*, 14: 125–142.

Lynde, C. and J. Richmond (1992), "The Role of Public Capital in Production," *Review of Economics and Statistics*, 74: 37–45.

Lynde, C. and J. Richmond (1993a), "Public Capital and Total Factor Productivity," *International Economic Review*, 34: 401–414.

Lynde, C. and J. Richmond (1993b), "Public Capital and Long-Run Costs in U.K. Manufacturing," *Economic Journal*, 103: 880–893.

Mankiw, N.G., D. Romer, and D.N. Weil (1992), "A Contribution to the Empirics of Economic Growth," *Quarterly Journal of Economics*, 107: 407–437.

Mas, M., J. Maudos, F. Pérez and E. Uriel (1993), "Competitividad, Productividad Industrial y Dotaciones de Capital Público," *Papeles de Economía Española*, 57: 144–160.

Mas, M., J. Maudos, F. Pérez and E. Uriel (1994a), "Capital Público y Crecimiento Regional Español," *Moneda y Crédito*, 198: 163–193.

Mas, M., J. Maudos, F. Pérez and E. Uriel (1994b), "Disparidades Regionales y Convergencia en las Comunidades Autónomas," *Revista de Economía Aplicada*, 2: 129–148.

Mas, M., J. Maudos, F. Pérez and E. Uriel (1995a), "Public Capital and Convergence in the Spanish Regions," *Entrepreneurship and Regional Development*, 7: 309–327.

Mas, M., J. Maudos, F. Pérez and E. Uriel (1995b), Growth and Convergence in the Spanish Provinces, in: Armstrong, H.W. and R.W. Vickerman (eds.), *Convergence and Divergence Among European Regions*, European Research in Regional Science, Pion Limited, chapter 13, 66–88.

Mas, M., J. Maudos, F. Pérez and E. Uriel (1996), "Infrastructures and Productivity in the Spanish Regions", *Regional Studies*, 30: 641–650.

McMillin, W.D. and D.J. Smyth (1994), "A Multivariate Time Series Analysis of the United States Aggregate Production Function," *Empirical Economics*, 19: 659–673.

Mera, K. (1973), "Regional Production Functions and Social Overhead Capital: An Analysis of the Japanese Case," *Regional and Urban Economics*, 3: 157–186.

Merriman, D. (1990), "Public Capital and Regional Output. Another Look at Some Japanese and American Data," *Regional Science and Urban Economics*, 20: 437–458.

Mooij, R.A. de, J. van Sinderen and M. Toen-Gout (1996), "Welfare Effects of Different Public Expenditures and Taxes in the Netherlands," *Research Memorandum* No. 9602, Research Centre for Economic Policy, Erasmus University Rotterdam.

Morrison, C.J. and A.E. Schwartz (1992), "State Infrastructure and Productive Performance," *NBER Working Paper* 3981.

Morrison, C.J. and A.E. Schwartz (1996), "Public Infrastructure, Private Input Demand, and Economic Performance in New England Manufacturing," *Journal of Business and Economic Statistics*, 14: 91–101.

Munnell, A.H. (1990), "Why Has Productivity Growth Declined? Productivity and Public Investment," *New England Economic Review*, Jan/Feb: 2–22.

Munnell, A.H. (1992), "Policy Watch. Infrastructure Investment and Economic Growth," *Journal of Economic Perspectives*, 6: 189–198.

Munnell, A.H. (1993), "An Assessment of Trends in and Economic Impacts of Infrastructure Investment," in: *Infrastructure Policies for the 1990s*, OECD, Paris.

Munnell, A.H. and L.M. Cook (1990), "How Does Public Infrastructure Affect Regional Economic Performance," in: A.H. Munnell (ed.), *Is There a Shortfall in Public Capital Investment?*, Federal Reserve Bank of Boston, Boston.

Nadiri, M.I. and T.P. Mamuneas (1994a), "The Effects of Public Infrastructure and R&D Capital on the Cost Structure and Performance of U.S. Manufacturing Industry," *Review of Economics and Statistics*, 76: 22–37.

Nadiri, M.I. and T.P. Mamuneas (1994b), "Infrastructure and Public R&D Investments, and the Growth of Factor Productivity in US Manufacturing Industries," *NBER Working Paper* 4845.

Otto, G. and G.M. Voss (1994), "Public Capital and Private Sector Productivity," *Economic Record*, 70: 121–132.

Otto, G. and G.M. Voss (1996), "Public Capital and Private Production in Australia," *Southern Economic Journal*, 62: 723–738.

Oxley, H. and J.P. Martin (1991), "Controlling Government Spending and Deficits: Trends in the 1980s and Prospects for the 1990s," *OECD Economic Studies* No. 17, 145–189.

Pack, H. (1994), "Endogenous Growth Theory: Intellectual Appeal and Empirical Shortcomings," *Journal of Economic Perspectives*, 8: 55–72.

Phillips, P.C.B. and B.E. Hansen (1990), "Statistical Inference in Instrument Variables Regression with I(1) Processes," *Review of Economic Studies*, 57: 99–125.

Pinnoi, N. (1994), "Public Infrastructure and Private Production. Measuring Relative Contributions," *Journal of Economic Behavior and Organization*, 23: 127–148.

Ram, R. (1996), "Productivity of Public and Private Investment in Developing Countries: A Broad International Perspective," *World Development*, 24: 1373–1378.

Ram, R. and D.D. Ramsey (1989), "Government Capital and Private Output in the United States. Additional Evidence," *Economics Letters*, 30: 223–226.

Ratner, J.B. (1983), "Government Capital and the Production Function for U.S. Private Output," *Economics Letters*, 13: 213–217.

Rebelo, S. (1991), "Long Run Policy Analysis and Long Run Growth," *Journal of Political Economy*, 99: 500–521.

Romer, P.M. (1986), "Increasing Returns and Long-Run Growth," *Journal of Political Economy*, 94: 1002–1037.

Romer, P.M. (1989), "Human Capital and Growth: Theory and Evidence," *NBER Working Paper* 3173.

Seitz, H. (1993), "A Dual Economic Analysis of the Benefits of the Public Road Network," *Annals of Regional Science*, 27: 223–239.

Seitz, H. (1994), "Public Capital and the Demand for Private Inputs," *Journal of Public Economics*, 54: 287–307.

Seitz, H. and G. Licht (1995), "The Impact of Public Infrastructure Capital on Regional Manufacturing Production Cost," *Regional Studies*, 29: 231–240.

Shah, A. (1992), "Dynamics of Public Infrastructure, Industrial Productivity and Profitability," *Review of Economics and Statistics*, 74: 28–36.

Solow, R.M. (1956), "A Contribution to the Theory of Economic Growth," *Quarterly Journal of Economics*, 70: 65–94.

Solow, R.M. (1994), "Perspectives on Growth Theory," *Journal of Economic Perspectives*, 8: 45–54.

Sturm, J.E. (1997), "The Impact of Public Infrastructure Capital on the Private Sector of the Netherlands: An Application of the Symmetric Generalized McFadden Cost Function," *CPB Research Memorandum* 133, CPB Netherlands Bureau for Economic Policy Analysis, The Hague.

Sturm, J.E. and J. de Haan (1995), "Is Public Expenditure Really Productive? New Evidence for the US and the Netherlands," *Economic Modelling*, 12: 60–72.

Sturm, J.E. and J. de Haan (1997), "Public Capital Spending in the Netherlands: Developments and Explanations," *Applied Economics Letters* [forthcoming].

Sturm, J.E., J.P.A.M. Jacobs and P. Groote (1995), "Productivity Impacts of Infrastructure Investment in the Netherlands 1853–1913," *SOM Research Report* No. 95D30, Groningen.

Sturm, J.E. and G.H. Kuper (1996), "The Dual Approach to the Public Capital Hypothesis: The Case of The Netherlands," *CCSO Series* No. 26, Groningen.

Summers, R. and A. Heston (1988), "A New Set of International Comparisons of Real Product and Price Levels: Estimates for 130 Countries, 1950–1985," *Review of Income and Wealth*,

34: 1–25.

Summers, R. and A. Heston (1991), "The Penn World Table (Mark 5): An Expanded Set on International Comparisons, 1950–1988," *Quarterly Journal of Economics*, 106: 327–368.

Swan, T. (1956), "Economic Growth and Capital Accumulation," *Economic Record*, 32: 334–361.

Tatom, J.A. (1991), "Public Capital and Private Sector Performance," *Federal Reserve Bank of St.Louis Review*, 73: 3–15.

Toen-Gout, M.W. and M.M. Jongeling (1994), "Investment in Infrastructure and Economic Growth," *OCFEB Research Memorandum* No. 9404.

Toen-Gout, M.W. and J. van Sinderen (1995), "The Impact of Investment in Infrastructure on Economic Growth," Research Memorandum No. 9503, *Research Centre for Economic Policy*, Erasmus University Rotterdam.

Westerhout, E.W.M.T. and J. van Sinderen (1994), "The Influence of Tax and Expenditure Policies on Economic Growth in the Netherlands: An Empirical Analysis," *De Economist*, 142: 43–61.

Wylie, P.J. (1996), "Infrastructure and Canadian Economic Growth 1946–1991," *Canadian Journal of Economics*, 29: S350–S355.

Index